THERMAL INSULATION HANDBOOK FOR THE OIL, GAS, AND PETROCHEMICAL INDUSTRIES

THERMAL INSULATION HANDBOOK FOR THE OIL, GAS, AND PETROCHEMICAL INDUSTRIES

ALIREZA BAHADORI, PhD

School of Environment, Science & Engineering,
Southern Cross University, Lismore, NSW, Australia

Amsterdam • Boston • Heidelberg • London
New York • Oxford • Paris • San Diego
San Francisco • Singapore • Sydney • Tokyo

Gulf Professional Publishing is an imprint of Elsevier

Gulf Professional Publishing is an imprint of Elsevier
225 Wyman Street, Waltham, MA 02451, USA
The Boulevard, Langford Lane, Kidlington, Oxford, OX5 1GB, UK

Library of Congress Cataloging-in-Publication Data
Bahadori, Alireza.
  Thermal insulation handbook for the oil, gas, and petrochemical industries / Alireza
Bahadori.
     pages cm
  Includes bibliographical references and index.
  ISBN 978-0-12-800010-6
1. Petroleum pipelines–Insulation–Handbooks, manuals, etc. 2. Petroleum industry and
trade–Equipment and supplies–Protection–Handbooks, manuals, etc. I. Title.
  TN879.57.B34 2014
  621.402'4--dc23
                           2013046422

British Library Cataloguing in Publication Data
A catalogue record for this book is available from the British Library

ISBN: 978-0-12-800010-6

For information on all Gulf Professional Publishing
publications visit our web site at store.elsevier.com

Printed and bound in USA
14 15 16 17 18   10 9 8 7 6 5 4 3 2 1

CONTENTS

Dedicated to my wife (Megan), the loving memory of my parents, grandparents and to all who contributed so much to my work over the years

PREFACE

In times of high energy prices and great strains on the environment, efficient insulation systems are of the utmost importance. Fast worldwide population growth is leading to quickly increasing energy demand. To maintain the standard of living in industrialized countries and to improve the situation in developing countries, the issues surrounding energy consumption cannot be avoided and energy can be used much more efficiently and with a higher share of renewable sources.

The most important part of the energy strategy of any country is energy saving. Because of the limited energy sources and environmental pollution arising from the use of fuels, energy saving has become compulsory.

Oil, gas, and petrochemical processing plants contain intricate and costly piping configurations. Piping systems are also employed in many other situations including water supply, fire protection, and district cooling-heating applications. In the petroleum production industry, the hot oil/gas composition flows up at the wellhead and is transported through manifolds, various critical instruments, spools, and flowlines before the riser brings the oil to the surface. Insulation is necessary to avoid the formation of hydrate plugs and wax buildup (paraffin). The formation of wax and hydrates will start when the oil/gas composition is depressurized and exposed to the low seawater temperature at the seabed.

Thermal insulation materials are applied in order to prevent the formation of hydrate/wax during a shutdown scenario. During shutdown the extra insulation gives sufficient time for:

- "Short time" inspection of the pipe/equipment
- Solving production problems, etc.
- Methanol or glycol injection (probably long-term maintenance services).

Thermal insulation materials like other natural or man-made materials vary with the nature of material and the influencing temperature range. The use of thermal insulation should be so arranged that the optimum effectiveness is derived from all components of the complete system, bearing in mind always any economic limitations that may be imposed by the purchaser.

This engineering book provides the minimum requirements for thermal insulation of pipework, vessels, tanks, and equipment in temperatures up to $800^\circ C$ but excludes structural insulation of buildings, fireproofing structures, refractory lining of plants, and external underground mains. On the basis of the temperature the thermal insulating is divided into two different systems as follows:

- Hot system, which is applicable in the temperature range of +5°C to 800°C.
- Cold system, which is applicable in the temperature range of -165°C to +5°C.

The book is intended for use in the oil, gas, and petrochemical and similar industries mainly for refineries, chemical, petrochemical, and natural gas plants.

The book gives minimum requirements for insulation systems, including insulation materials of sufficient quality and thicknesses, weatherproofing, and finishing.

The required life of the insulation systems should be considered because this affects the annual cost and hence the economic thickness. If the plant has only a short life, an inexpensive insulation system may be considered; if the plant has a longer life maybe a more expensive thermal insulation type should be considered as it may actually be more economical.

In order to guarantee the liquefied natural gas (LNG) cold temperature of −165°C, high-quality insulation installation in accordance with strict specifications is essential. Cryogenic insulation restricts the inflow of atmospheric heat into the pipe or process equipment, keeping the liquid cold and allowing it to retain its form.

Also, consideration should be given individually to systems for pipework, to flat surfaces (e.g., air ducts, gas flues, walls of drying ovens, walls of large boilers, etc.), and to vessels, tanks, and large curved surfaces. The components of the system should be appropriate to the specific requirements of use. Generally, materials to be exposed to outdoor weather will differ from those used indoors, if only in the type of finish employed. Different types of system are likely to be required for the various ranges of application.

In addition, there is discussion on the design basis of thermal insulation including selection of a system, corrosion under thermal insulation, and general application of insulation. Also, the characteristics and selection of insulation and accessory materials will be presented.

Thermal insulation is a critical element in the design and operation of flowlines in deep waters due to a combination of low temperatures and high pressure; as a result of these conditions, there are stringent requirements for optimal insulation.

This technical book contains all matters relating to heat and cold insulation and provides engineering information for technically demanding solutions. The scope of the book includes:
- Dimensioning of insulation systems in accordance with ecological and economic considerations
- Development of high-temperature and cryogenic insulation concepts
- Qualifying materials and systems

The required thickness of insulation for any specific application depends upon the characteristics of the insulating material and the purpose of the equipment. When the sole object is to achieve the minimum total cost, the appropriate thickness is known as the economic thickness. If the economic thickness is required to be calculated the following additional information will be necessary:

Cost of heat to be used for calculation purposes, e.g., US $ per useful megajoule

Evaluation period (working hours)

Whether or not the cost of finish is to be included in the calculation

When the base of insulation is conservation of energy, the economically accepted insulation thicknesses should be used.

For any specific set of installation requirements, the properties of a material determine its suitability. If there were only one, or a limited number of sets of installation requirements, selection of a material would be simple, and the need for all the various types of insulations and weather barriers reduced. However, this is not the case. Each installation must be considered, and its requirements evaluated to allow the selection of the best-suited material (or materials) for the individual installation under consideration.

Not only do the installation requirements change with the individual case, but also the relative importance of the requirements also vary. So all the above technical issues are discussed in the book.

Moreover, this book covers the minimum requirements for composition, sizes, dimensions, physical properties, inspection, packaging, and marking of various types of insulation for use on different surfaces.

Alireza Bahadori
School of Environment, Science & Engineering,
Southern Cross University, Lismore, NSW, Australia
August 20th, 2013

ABOUT THE AUTHOR

Alireza Bahadori, PhD, is a research staff member in the School of Environment, Science & Engineering at Southern Cross University, Lismore, NSW, Australia. He received his PhD from Curtin University, Western Australia. For the better part of 20 years, Dr. Bahadori has held various process engineering positions and been involved in many large-scale projects at NIOC, Petroleum Development Oman (PDO), and Clough AMEC PTY LTD.

He is the author of over 200 articles and 8 books which have been accepted and/or published by many prestigious publishers. Dr. Bahadori is the recipient of the highly competitive and prestigious Australian Government's Endeavour International Postgraduate Research Award as part of his research in the oil and gas area. He also received a Top-Up Award from the State Government of Western Australia through the Western Australia Energy Research Alliance (WA:ERA) in 2009. Dr. Bahadori serves as a member of the editorial board for a number of journals such as the *Journal of Sustainable Energy Engineering.*

CHAPTER ONE

Design and Application of Thermal Insulation

Contents

Alireza Bahadori, Thermal Insulation Handbook for the Oil, Gas, and Petrochemical Industries
© 2014 Elsevier Inc.
http://dx.doi.org/10.1016/B978-0-12-800010-6.00001-0

1.1. PURPOSE OF THERMAL INSULATION

Various thermal insulation systems taking advantage of different types of thermal insulation materials on both an organic (such as expanded plastics, wood, wool, cork, straw, technical hemp) and inorganic (such as foamed glass, glass, and mineral fibers) basis are being designed and tested, and new methods for analyzing the properties of both insulation materials and insulation systems are being devised. The particular products differ in their shape, flammability, composition, and structure, which in relation to designers' requirements define the possibilities of their application in engineering practice.

Researchers in thermal science are attempting to minimize capital and operation costs as well as heat loss. In previous works multiple objective functions were applied by researchers for the design analysis of a piping system to minimize the heat loss and the amount of insulation used.

In these types of complicated methods, a common approach is to sum all objective functions with appropriate weighting factors, and minimize the resulting composite function. However, the analytical solution should only be attempted if a very precise value of thickness is needed because it takes into account specific details and often is not a requirement from a practical viewpoint as many types of insulation are available only in certain specific sizes.

The required thickness of insulation for any specific application depends on the characteristics of insulating material as well as the purpose of the equipment. If a process is critical, the most important single consideration may be reliability. If conservation of heat or power is the deciding factor, the savings per year as compared to the installed cost is the most important factor.

In contrast, when insulation is to be used for a temporary function such as holding the heat in while a lining is being heat cured, then the lowest possible installed cost would be decisive. Thus, because of conflicting requirements, there can be no multipurpose insulation. Nor is there a "perfect" insulation for each set of requirements.

A low thermal conductivity is desirable to achieve a maximum resistance to heat transfer. Therefore, for any given heat loss, a material of low thermal conductivity will be thinner than an alternative material of high conductivity. This is of particular advantage for process pipes because thinner layers of insulation reduce the surface area emitting heat and also reduce the outer surface that requires protection. The main purpose of insulation is to limit the transfer of energy between the inside and outside of a system.

A thermal insulator is a poor conductor of heat and has a low thermal conductivity. Insulation is used in buildings and in manufacturing processes to prevent heat loss or heat gain. Although its primary purpose is an economic one, it also provides more accurate control of process temperatures and protection of personnel. It prevents condensation on cold surfaces and the resulting corrosion. Such materials are porous, containing large number of dormant air cells. Figure 1.1 shows a sample application of thermal insulation in industry.

Figure 1.1 Sample thermal insulation applications. *(Source: Trelleborg).*

Thermal insulation may be applied for one or a combination of the following purposes:

- Saving energy by reducing the rate of heat transfer
- Maintenance of process temperature
- Prevention of freezing, condensation, vaporization, or formation of undesirable compounds such as hydrates
- Protection of personnel from injury through contact with equipment
- Prevention of condensation on surface of equipment conveying fluids at low temperature
- Avoidance of increase in equipment temperature from outside fire
- To conserve refrigeration
- Offers better process control by maintaining process temperature
- Prevention of corrosion by keeping the exposed surface of a refrigerated system above the dew point
- Absorption of vibration.

1.2. DESIGN CONSIDERATIONS

Below is a discussion of parameters that must be considered when designing insulation materials.

1.2.1. Location

- Whether indoors, outdoors but protected, outdoors exposed to weather, enclosed in ducts or trenches below ground level, underground and/or underwater
- Difficult or unusual site conditions that will influence the selection or application of insulating materials, or both, e.g., in regard to transport, scaffolding, weather protection or excessive humidity
- Type of material to be insulated, with details of special or unusual materials.

1.2.2. Dimensions of Surfaces

If these are adequately detailed on drawings (preferably colored to indicate areas to be insulated), the provision of copies of the drawings may be sufficient; otherwise detailed information will be required, e.g.,

- Surface dimensions of flat or large curved areas
- External diameters of pipes
- Lengths of each size of pipe

- Number and type of pipe fittings, e.g., flanged joints, valves, tees, expansion bends.

1.2.3. Temperature Conditions

- Normal working temperature for each portion of the plant to be insulated
- Maximum temperature for each hot surface, if higher than the normal working temperature
- Ambient temperature: where a specified temperature is required on the outer surface of the insulation.

A specified temperature on the surface of the insulation may be required for the following:

- To protect personnel, e.g., local insulation of hot pipes
- To provide comfortable conditions at certain locations, e.g., at control panels and in operating galleries
- To provide a means of indicating the effectiveness of the insulation. This should generally be avoided, as the surface temperature will depend upon the diameter over the insulation, on the ambient conditions, and on the nature of the outer surface.
- Pipes of small diameter will show relatively low surface temperatures, but high rates of heat loss; air flow over the outer surface will tend to reduce the surface temperature but will increase the rate of heat loss; a polished metal surface will show appreciable increase in surface temperature compared with a nonmetallic finish, although the rate of heat loss will be reduced.
- It may be necessary also to give the conditions of ambient air for which the surface temperature is to be calculated, e.g., the velocity of air passing over the surface.
- Normally, the theoretical heat loss will be based on the manufacturer's declared value of thermal conductivity and, unless otherwise stated, it will refer to conditions of ambient still air at 20°C.
- When making use of theoretical figures, therefore, allowance has to be made for the effect of the ambient conditions at the site and for supplementary heat loss that will occur through supports, hangers, valve control wheels, and other fittings. As it is extremely difficult to measure with accuracy either the surface temperature or the heat loss from the surface of the insulation under site conditions, some reservation should be made when interpreting site measurements.

- Any requirement to prevent condensation on the warm face of an insulated pipe or vessel containing cold media.
- Details of any pipework sections that are to be trace heated, of the trace-heating method, and of the arrangements of insulation required
- Details of any sections to be left uninsulated to facilitate testing, e.g., welded and flanged joints
- Confirmation, with details, that heat will be available in insulated pipework for drying out any plastic insulating material or finishing composition.

1.2.4. Preparation of Surfaces

Requirements for the preparation of surfaces, including special requirements, e.g., for the removal of works–applied protective paint or lacquer, or for the application at the site of paint or other protective coating to the surface to be insulated, should be clearly stated.

1.2.5. Type of Insulation Required

There are two types of insulation required:
- Main insulation for each portion of a plant, e.g., preformed, plastic composition, flexible, loosefill, insulating concrete
- Insulation for bends, flanges, valves, hangers, and other fittings.

1.2.6. Type of Finish Required

The finish required could include, for example, hard–setting composition or self-setting cement, weatherproofing compound, or sheet metal.

1.2.7. Special Service Requirements

This could include, for example, resistance to compression, resistance to fire, and resistance to abnormal vibration. If there is any special hazard from contact with chemicals or oils in the plant, attention should be drawn to this.

1.2.8. Basis on Which the Thickness of Insulation Is Determined

- Specified temperature on outer surface of insulation
- Specified heat loss per unit dimension, linear or superficial
- If the economic thickness must be considered, the following additional information will be necessary:
    - Cost of heat to be used for calculation purposes, e.g., dollar per useful megajoule

- Evaluation period (working hours)
- Whether or not the cost of the finish is to be included in the calculation
- Specified temperature conditions for the surfaces to be insulated. Insulation to provide specified conditions at the boundary surfaces of the containment system may be required for reasons such as:
  - To avoid differential thermal expansion between the insulated surface and adjacent structures
  - To prevent condensation of moisture on the internal surfaces of the containment system, e.g., in waste gas flues
  - To prevent the condensation of moisture on the external surface of insulated plant containing cold media
  - To ensure that the walls of the containment system are not subjected to excessive temperatures
- Specified conditions of fluid at point of delivery
- Special thickness requirements.

1.2.9. Information to Be Supplied by the Manufacturers of Insulation

- The manufacturer's declared value of thermal conductivity appropriate to the temperature of use, plus the corresponding bulk density. The manufacturer's declared value should include any necessary commercial tolerances. When the thermal conductivity is liable to change on aging, the aged value should be stated.
- Limitations of use, physical and chemical
- The overall thickness, with details of the thickness of the individual layers
- Information regarding the surface preparation
- The appropriate section of the code (to be specified) with which the following are in accord:
  - Insulating material
  - Reinforcement (if any)
  - Fixing devices and finishes.

1.2.10. Types and Application

- **Temperature ranges**

  Insulation can be classified into three groups according to the temperature ranges for which they are used.

- **Low-temperature insulation (up to 90°C)**

  This range covers insulating materials for refrigerators, cold and hot water systems, storage tanks, etc. The commonly used materials are cork, wood, 85% magnesia, mineral fibers, polyurethane and expanded polystyrene, etc.

- **Medium-temperature insulation (90–325°C)**

  Insulators in this range are used in low-temperature heating and steam raising equipment, steam lines, flue ducts, etc. The types of materials used in this temperature range include 85% magnesia, asbestos, calcium silicate, mineral fibers, etc.

- **High-temperature insulation (325°C and above)**

  Typical uses of such materials are superheated steam systems, oven dryers, furnaces, etc. The most extensively used materials in this range are asbestos, calcium silicate, mineral fibers, mica- and vermiculite-based insulation, fireclay- or silica-based insulation, and ceramic fibers.

- **Insulation material**

  Insulation materials can also be classified into organic and inorganic types. Organic insulations are based on hydrocarbon polymers, which can be expanded to obtain high void structures. Examples include Thermocol (expanded polystyrene) and polyurethane foam (PUF).

  Inorganic insulation is based on siliceous/aluminous/calcium materials in fibrous, granular, or powder forms. Examples include mineral wool, calcium silicate, etc.

  Properties of common insulating materials follow:

  - **Calcium silicate**

    This is used in industrial process plant piping where high service temperature and compressive strength are needed (Figure 1.2). Temperature range varies from 40 to 950°C.

  - **Glass mineral wool**

    This is available in flexible forms, rigid slabs, and preformed pipe work sections. It is good for thermal and acoustic insulation for heating and chilling system pipelines. The temperature range of application is −10 to 500°C (Figure 1.3).

  - **Thermocol**

    This is mainly used as cold insulation for piping and cold storage construction.

  - **Expanded nitrile rubber**

    This is a flexible material that forms a closed cell integral vapor barrier. Originally it was developed for condensation control in

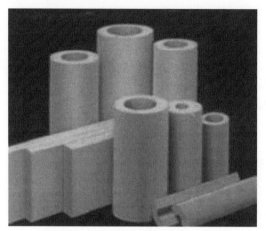

Figure 1.2 Calcium silicate insulators.

refrigeration pipe work and chilled water lines; nowadays it is also used for ducting insulation for air conditioning.

- **Rock mineral wool**

    This is available in a range of forms from lightweight rolled products to heavy rigid slabs including preformed pipe sections. In addition to good thermal insulation properties, it can also provide acoustic insulation and is fire retardant.

Figure 1.3 Glass mineral wool.

Table 1.1 Thermal Conductivity of Hot Insulation

Mean Temperature, °C	Calcium Silicate	Resin Bonded Mineral Wool	Ceramic Fiber Blankets
100	–	0.04	–
200	0.07	0.06	0.06
300	0.08	0.08	0.07
400	0.08	0.11	0.09
700	–	–	0.17
1000	–	–	0.26
Specific heat kJ/(kg°C)	0.96 @40°C	0.921 @20°C	1.07 @980°C
Service temperature, °C	950	700	1425
Density, kg/m^3	260	48 to 144	64 to 128

- **Use of molded insulation**

  Lagging materials can be obtained in bulk, in the form of molded sections (semi-cylindrical for pipes, slabs for vessels, flanges, valves, etc.). The main advantage of the molded sections is the ease of application and replacement when undertaking repairs for damaged lagging.

  The *thermal conductivity* of a material is the heat loss per unit area per unit insulation thickness per unit temperature difference. The unit of measurement is W-m^2/m°C or W-m/°C. The thermal conductivity of materials increases with temperature. So thermal conductivity is always specified at the mean temperature (mean of hot and cold face temperatures) of the insulation material.

  The thermal conductivities of typical hot and cold insulation materials are given in Tables 1.1 and Table 1.2.

1.3. SELECTION OF THERMAL INSULATING SYSTEMS

1.3.1. Optimum Effectiveness

The use of thermal insulation shall be arranged so that the optimum effectiveness is derived from all components of the complete system, bearing in mind always any economic limitations that may be imposed by the purchaser. Where optimum effectiveness cannot be obtained, as a result of economic or other limitations, it is the duty of the insulation experts to advise the relevant company to this effect, giving their reasons for their opinion. The required life of the insulation systems should be considered because this affects the annual cost and hence the economic thickness. If the

Table 1.2 Thermal Conductivity of Cold Insulation

Materials	Thermal Conductivity, W/m°C
Mineral/glass fiber blanket	0.039
Board or Slab	
Cellular glass	0.058
Cork board	0.043
Glass fiber	0.036
Expanded polystyrene (smooth) – Thermocol	0.029
Expanded polystyrene (cut cell) – Thermocol	0.036
Expanded polyurethane	0.017
Phenotherm (Trade name)	0.018
Loose Fill	
Paper or wood pulp	0.039
Sawdust or shavings	0.065
Mineral wool (rock, glass, slag)	0.039
Wood fiber (soft)	0.043

plant has only a short life, an inexpensive insulation system may be considered; if the plant has a longer life, durability and other considerations may come into play.

When the technical requirements of the application have been met, the total cost (as distinct from the initial cost) during the life of the insolation should be the prime consideration.

1.3.2. Extent of System

A thermal insulating system shall be understood to include the following:
- The required attachments to the surface to be insulated
- Means of securing the insulation system to those attachments, where appropriate, or to the surface directly
- The types and thicknesses of the insulating materials to be used
- Means of protecting the insulating material, other than the outer finish, e.g., vapor barrier (could also be extended to include initial protection of the surface to be insulated, e.g., from corrosive attack)
- Reinforcing materials
- Finishing materials.

Consideration shall be given individually to systems for pipework, to flat surfaces (e.g., air ducts, gas flues, walls of drying ovens, walls of large boilers, etc.), and to vessels, tanks, and large curved surfaces.

The components of the system shall be appropriate to the specific requirements of use. Generally, materials to be exposed to outdoor weather will differ from those used indoors, if only in the type of finish employed. Different types of system are likely to be required for the following ranges of application:

- Refrigeration (-100°C up to +5°C)
- Chilled and cold water supplies, industrial use
- Central heating, air conditioning, and domestic hot and cold water supplies (10°C up to 200°C)
- Process pipework and equipment (from ambient temperature up to 650°C)
- Dual temperature systems, e.g., refrigeration with occasional use at higher temperatures
- Heat transfer fluid systems.

1.3.3. Factors for Consideration

The main controlling factors that shall be considered at an early design stage are outlined in the following. Attention should be paid to details at the design stage to ensure the effectiveness of the complete system.

- **Temperature**

   Thermal insulation on a plant operating at temperatures below the dew point of the surrounding air has to be kept dry, both before and after application. This means that some form of vapor barrier is essential. For elevated temperatures the insulating material has to be adequately resistant to the highest temperature involved under eventual service conditions.

- **Mechanical stability**

   The system, including the insulating material, the method of fixing, and the finishing material, has to be capable of giving effective service for its design life. This is of special importance for certain plants in which access for repair work may be difficult.

- **Resistance to degradation**

   This requirement can have wide implications, from resistance to vermin and fungoid attack, to freedom from fire hazard. It will also include resistance to the required environmental conditions, e.g., adequate weather resistance for outdoor service as well as resistance to accidental spillage of oils or other chemicals. The insulating material itself should not tend to dissociate or disintegrate.

- **Thermal effectiveness**

  A low thermal conductivity figure for one test sample may not be a reliable guide to the thermal effectiveness of a complete insulation system during prolonged periods of service. There shall be adequate quality control at all stages of the manufacture of materials and at all stages of their application at the site.

- **Type and dimensions of the plant to be insulated**

  The size and configuration of the plant will have an important bearing on the suitability of a particular insulating material, as well as on the reinforcement, means of securing, and type of finishing material to be used. Particular care will be required with large flat areas, especially when elevated temperatures may cause extensive thermal movements.

- **Compatibility of the components of the system**

  Care shall be exercised to avoid the use of solvent-based adhesive or polymeric finishing compounds where these may attack the main insulating material, e.g., polystyrene. Also it is important to avoid corrosion problems, e.g., those caused by electrochemical action between dissimilar metals under possible humid conditions.

- **Total weight of the system**

  Although substantially thick insulating material may be desirable for energy saving, consideration shall be given to the resultant need to provide additional support, especially if a heavy metal finish is to be used for mechanical protection. On the other hand, a fibrous or exfoliated loosefill insulating material of low bulk density, with a lightweight finish, may tend to disintegrate or to settle under service conditions where there is substantial vibration, caused, for example, by road vehicles.

- **Potential hazard to health**

  It is now realized that certain types of fibrous insulating materials, notably asbestos, and certain finely divided aggregates such as crystalline silica, can be a hazard to health when inhaled into the lungs. While only asbestos-free insulation should now be used, appropriate care should be exercised to ensure that the selected insulation system is a safe one, bearing in mind the probable need for subsequent removal of the insulation.

- **Corrosion hazard**

  As no insulating materials, with the possible exception of multiple-foil stainless steel, are substantially free from soluble chlorides, adequate protective measures shall be adopted as an essential part of the insulation system when the surface to be insulated is sensitive to chloride attack, e.g., austenitic steels. In such cases these precautions may include covering the surface with aluminum foil, or treatment with suitable protective compound.

- **Fire hazard**

   It is possible to provide some form of protection for insulation that would otherwise be flammable by careful choice of a finishing material or outer sheet cover, but this may only obscure a dangerous condition within the system. Conversely, some forms of covering materials may themselves introduce a fire hazard that would not otherwise have existed, e.g., certain types of weatherproofing materials.

- **Space for insulating system**

   Sufficient space should be provided during plant design to enable the chosen insulating system to be correctly installed and maintained.

1.4. DESIGN CONSIDERATIONS

Thermal insulation for plant and equipment is too often considered at so late a stage in the design and erection sequence that the insulation contractor has to submit a scheme under pressure of time. The ultimate cost of insulation should be an important factor, but it is by no means the only one and the designer will base his considerations on having the right material, in the right amount, and in the right place.

Purchase, delivery, and application at the right time are not the main concern of this section, but guidance is given for assessing the influence that design considerations will have on the choice of the right material in the right amount. It is unlikely that any problem will arise for which all the factors discussed here will need to be considered, but every factor is dealt with in sufficient detail to ensure that it will be of practical value when required. The following is technical information to help select the appropriate insulating system for most requirements.

1.4.1. Economic Thickness

Where the sole object of applying insulation to a portion of plant is to achieve the minimum total cost during a specific period, the appropriate thickness is known as the economic thickness. To some extent the relevant calculations are unsatisfactory as they relate only to money values rather than to the conservation of energy and they require assumptions that are mainly arbitrary.

The principle is to find at what thickness further expenditure on insulation would not be justified by the additional financial savings on heat to be anticipated during the period (the "evaluation period").

An increase in the amount of insulation applied will raise the initial installed cost, but it will reduce the rate of heat loss through the insulation, thus reducing the total cost during the evaluation period.

1.4.2. Applications for Which Economic Thicknesses Are Not Appropriate

Technical requirements should take precedence over economic considerations when the insulating material is required for the following purposes:

- To maintain fluid inside a plant system within specific temperature limits, e.g., refrigeration plant, or plant containing fluid with a low "freezing point"
- To ensure that a fluid in a pipe has specified physical properties at the point of delivery
- To avoid danger to personnel, e.g., from plant items at low temperature, or plant items carrying low-grade heat such as drains and waste gases
- To control thermal movement of plant items, particularly those exposed to high temperatures
- To limit the temperature of portions of hot plant in order to avoid damage from excessive temperature
- To prevent condensation of moisture on the external surface of the insulation of "cold" plant, and to maintain the internal temperature of a system above a specified minimum in order to avoid corrosive attack, e.g., from condensation of acidic products resulting from the combustion of heavy fuel oil
- To improve ambient comfort conditions.

1.4.3. Effect of Air Spaces

It should be noted that an air space is much less effective for thermal insulation purpose than a space of similar dimensions filled with one of the convectional insulating materials. This fact is of particular significance at elevated temperatures.

The provision of air spaces for thermal insulation purposes is not recommended. But air spaces may be incorporated between insulation and finish for economic or drainage requirements.

An air cavity, set normal to the direction of flow of heat, can act as an insulating material, but its efficiency may vary in accordance with the influence of many factors, of which the following are the most important:

- Width of the cavity between the boundary surfaces
- Extent of ventilation of the cavity
- Nature of the boundary surfaces
- Mean temperature of the boundary surfaces
- Effect of barrier sheets that may be interposed in the path of the flow of heat
- Orientation of the cavity.

1.4.4. Thermal Efficiency

Thermal efficiency as applied to insulation does not represent a self-evident idea; it is not a question of energy being partly used constructively, and partly rejected or wasted. Nevertheless the term has had some popular use for a long time for expressing the effectiveness of some particular piece of insulation in preventing heat loss from some particular surface in given design conditions.

Dimensions are not automatically defined, and this effectiveness of insulation, or thermal efficiency, may be written simply as the ratio or expressed as a percentage. This ratio has been used in specifications to instruct the designer or supplier what the insulation requirement is.

$$\text{Thermal efficiency} = \frac{Q_{without\ insulation} - Q_{with\ insulation}}{Q_{without\ insulation}} \times 100 \qquad (1.1)$$

It is unusual to define anything other than the supposed temperature of the hot surface and possibly the wind speed, and certainly not the many other factors that also influence the heat loss from a surface.

1.4.5. Specified Conditions at the Point of Delivery

When it is necessary that a fluid emerges from a pipeline or duct system under specified physical conditions, the selection of insulating material and the thickness applied require special consideration that shall take into account the rate of mass flow and certain physical properties of the fluid to be conveyed through the system; economic considerations may well become of secondary importance.

1.4.6. Protection of Boundary Surfaces

Although the more conventional applications of insulating materials are concerned with restricting loss of heat to the surroundings or gain of heat from those surroundings, dependent on whether the ambient temperature is less or greater than that of the surface to be insulated, it may be necessary to apply insulating materials to an area of plant for other reasons, the principal ones being:

- To control the extent of thermal movement of the portion of plant
- To protect the boundary material from excessively high or low temperatures
- To regulate the temperature of an area of plant to prevent deterioration
- To protect personnel.

1.4.7. Clearance Between Insulation and the Surrounding Surfaces

One of the chief problems when insulating complex plant is to provide for adequate access, and it is of prime importance that this shall be taken into account when designing the plant and pipe layout.

A minimum clearance of 50 mm beyond the full extent of thermal movement shall be allowed between insulated plant and structural or other insulated surfaces, except where the shielded depth is greater than 300 mm. In such cases, or where pipe banks against walls or ceilings are involved, the designer shall envisage the sequence of fixing of the insulating material and its finish and shall make all necessary provisions.

1.4.8. Provision for Differential Thermal Movement

- **Hot surfaces**

    Due to the difference in expansion coefficients of metals and insulating materials it is necessary to make allowance for the differential movements between the hot surface, the insulant, and the finish. As a guide it is recommended that such allowances or expansion joints be inserted at the intervals shown in Table 1.3.

    Above 250°C preformed sections or flexible insulation shall be applied in two layers and that joints in individual layers shall be staggered. At the junction between preformed insulating materials and fixed steel work, the joint area shall be packed with mineral fiber to accommodate differential thermal movement. With a metal finish it is customary to fit sliding joints. When a plaster-type finish is applied, an expansion joint

Table 1.3 Intervals for Allowances or Expansion Joints for the Differential Movements Between the Hot Surface, the Insulant, and the Finish

Temperature (°C)	Spacing Intervals (m)
Up to 200	5
200 to 300	4
300 to 400	3
400 to 500	2

shall be provided by cutting at a circumferential joint in single layer or in the outer layer of a double-layer system.

These joints shall be covered with mineral-fiber cloth, secured in place, or as a muff in the same way as at a flange. As a rough approximation, radial expansion of pipes and vessels covered with plaster-type finishes is acceptable without longitudinal expansion joints if the value of $D \times t$ is less than 250 for pipes or vessels up to 1.5 m in diameter over insulation or 500 if over 1.5 m (D is diameter in meters and t is temperature in °C).

• **Sliding and bellows expansion joints**

The insulation shall not interfere with the operation of the expansion joints. For this reason, the bellows or joint are usually fitted with a metal cage, fastened at one end only, on which the insulation can be secured. Care shall be taken to see that mild-steel tie rods, etc., are not enclosed within the insulation where they may attain too high a temperature or cannot be adjusted.

• **Cold surfaces**

Insulating materials for use at subambient temperatures may have coefficients of thermal movement that not only vary with different materials, but also differ appreciably from the corresponding movements of the pipe or item of equipment to which they are fitted. In some cases the insulating material will be sufficiently compressible to accommodate this differential movement but, with long straight lengths at extreme temperatures or with non-compressible insulation, contraction joints will be required.

These may take the form of a 10 mm gap in the insulation that is packed with a flexible insulation. Contraction joints shall be provided immediately below insulation support rings on vertical pipes and at suitable intervals on horizontal pipes. Joint intervals shall be chosen with due regard to the pipe and insulation materials and the operating temperature. The joints shall be covered with a preformed insulation vapor seal that shall accommodate movement of this joint.

- **Plastic pipe**

  In the unlikely event that plastic pipe requires thermal insulation, attention is drawn to the fact that the thermal expansion coefficients are generally much higher than those of metals.

1.4.9. Provision for Preventing Settlement and Cracking

On vertical surfaces provision shall be made for insulation supports to take the dead weight of the insulation. These shall project halfway through the insulation thickness or in the case of multilayer work to a line halfway through the thickness of the outer layer.

These supports may take the form of flat bars, angles, or studs as appropriate. Preferably, provision should be made for tying back any reinforcement for the finishing cement. Clay-based compositions, bitumen emulsions, and other similar finishes shall only be applied to "hot" plant when heat is available for drying out the whole insulation from within.

1.4.10. Insulation of Pipes in Ducts and Subways

Pipes in internal ducts and trenches are normally insulated after hydraulic testing and, in the case of pipes for heating services, before heating is commenced; therefore, plastic insulation is not normally used.

Insulation is recommended for steam, condensate, heating, and hot-water supply mains, in order to conserve heat and to avoid high ambient temperatures. It is not usually necessary to insulate cold water pipes, fire mains, etc., against frost when they are enclosed in a duct, unless insulation is required to prevent surface condensation. To prevent condensation of moisture, chilled–water and refrigeration pipes will generally require insulation and adequate vapor seal.

For hot-water heating, and for steam, hot-water, and cold-water services, effective insulation can be provided by using preformed or flexible insulating materials.

For internal ducts where the pipes are hidden from sight, it may not be necessary to apply further finish; where a neater appearance is required it may be advisable to finish with textile fabric or with a thin sheet material. Painting of the finished surface may be necessary.

(Note: Where pipes are to run underground external to buildings, consideration shall be given not only to the insulation required but also to the protection necessary to prevent the penetration of ground water to the insulating material and the pipe).

Advice should be sought on the extent of water penetration so that consideration can be given to the protection methods to be used for protecting the insulation.

1.4.11. Fire Hazards

Not all the thermal insulating materials in common use are nonflammable. Some of them, often used for refrigeration systems, are entirely of organic composition and thus may constitute a fire hazard, or they may emit smoke and toxic fumes. Designers of thermal insulation systems shall therefore consider the process conditions and the plant arrangement before deciding whether or not the proposed thermal insulating material might contribute to the spread of fire, however initiated, and they shall vary their choice of material accordingly.

1.4.12. Insulation Against Freezing

There is no known insulation that will prevent freezing of liquid in pipes and vessels under all conditions. If the outside temperature remains low enough for long enough, and the movement of liquid through the pipe or vessel is slow, then no insulation, however thick, will prevent internal freezing.

Insulation will retard the onset of freezing and if the intervals during which the liquid is static are short enough freezing may be avoided. If more heat is supplied from the liquid passing through the system than is lost through the insulation together with the associated losses through supports and hangers, freezing can be avoided.

When fluids are static in pipes or vessels and the ambient temperate is below the freezing point of the contained fluid, the only sure way to prevent freezing is to supply heat to the system, e.g., by means of tracer pipes or electric heating elements, which should be fitted before the particular item of plant is insulated. The amount of heat thus supplied per unit period of time shall be sufficient to replace that lost from the system during the same period. Each problem of insulation against frost conditions shall receive individual consideration, and due regard shall be paid to the climatic conditions of the area concerned.

1.4.13. Protection Against Surface Condensation

Condensation takes place on piping and equipment held at temperatures below the dew point of the ambient air. Although the application of the insulation can prevent condensation at the surface, it will not necessarily prevent the moisture being drawn through the insulation itself, and frequently the dew point will be reached at some distance inside the layer of insulation.

It is therefore imperative that a vapor barrier be applied on the warm side on the insulation layer. If an insulating material is applied to a cold surface in humid conditions without a vapor barrier, the insulation can become saturated, its heat-insulating properties impaired, and also, probably, its mechanical strength. If the cold surface is at a temperature lower than the freezing point, the moisture will freeze and tend to rupture and break away the insulation.

The object of the vapor barrier is to prevent ingress of moisture. It may be desirable that the thickness of insulation be chosen so that the outside surface temperature of the insulation remains above the dew point. This requirement tends to increase the thickness above the range normally considered for hot insulation work. An adequate space shall be provided for the extra thickness, and due regard should therefore be given to the spacing of pipes and equipment to allow room for insulation.

Certain cold-insulation materials have in themselves a high resistance to the passage of water vapor; even so they will require a vapor seal, and all joints have to be adequately sealed. To avoid condensation on plant situated out of doors, the finish has to be vaporproof as well as weatherproof and it shall be remembered that watertight coatings are not necessarily vapor tight.

1.4.14. Vapor Barriers

Due to lack of precise knowledge and because of variations in practical application, any quantity recommendations are to be taken as approximate guidance only. Care needs to be taken to ensure that the vapor barrier does not add seriously to the fire hazard of the complete system.

The ability of air to hold water in the vapor phase is reduced as the temperature decreases, the vapor condensing when the temperature falls to the dew point; the saturation vapor pressure of water in the atmosphere increases with increase in temperature.

Where the surface temperature of insulation is higher than the plant on which it is used, and some part of the insulation is at a temperature below the dew point of the ambient air, there is a water vapor pressure differential across the insulation. This differential will tend to force the vapor towards the cold surface of the plant where it will condense and, if the temperature is below freezing point, the condensed water will turn into ice.

As a rough guide, the thermal conductivity of water is about three times that of a typical good quality dry insulating material, and that of wet ice may be up to twenty times that of water. This means that internal condensation and ice formation will appreciably reduce the effectiveness of the thermal insulating material. Additionally, the increase in volume of the moisture on freezing can disrupt the physical structure of the material.

Insulating materials that consist substantially of closed cells possess an inherent resistance to the passage of water vapor, but open–cell insulants and loosefill porous materials are readily permeable to water vapor.

Even with materials that have good resistance to the transmission of water vapor, differential movement of plant and insulation can cause joints in the latter to open, thus allowing moisture to penetrate towards the underlying surface. Joint-sealing compounds also may fail to exclude water vapor completely, in which case the contained water or ice may form strongly conducting paths from the surface of the plant to the ambient air.

As a general rule, all insulation on plant working at subambient temperatures shall have a "vapor barrier" layer over the outer (warm face) surface. This barrier shall be resistant to the passage of water vapor and it shall be applied to the dry insulation immediately after the latter has been fitted and before the plant is cooled.

The effectiveness of the vapor barrier is expressed in terms of the rate at which water vapor is transmitted through it under defined conditions, i.e., by a "permeability" figure or a "permeance" figure. Permeability relates to the rate of transmission through unit thickness (normally 1 m) of the material whereas permeance relates to the total rate of transmission through the actual thickness of a particular material, as applied. SI units are:

Permeability: Grams per second per meganewton vapor pressure difference, for one meter thickness: gm/(s MN).

Permeance: Grams per second per meganewton vapor pressure difference: g/(s MN). It shall be noted that whereas "permeability" is a characteristic of a given material, "permeance" relates only to a particular layer of known thickness after application. Typical values for some vapor barrier layers are given in Table 1.4. For conversion factors refer to Table 1.5.

Essentially, the vapor pressure of the ambient air will depend on the actual moisture content, which may be related to the combined effects of the dry

Table 1.4 Typical Values for Water-Vapor Permeance

Material	Approximate Thickness, mm	Permeance (Maximum) (For Conversion Factors See Table 1.5), g/(s MN)
Water-Based Emulsions		
Polymeric mastic emulsion	1.6	0.029
Bitumen emulsion, mineral	3.2	0.046
Bitumen emulsion, fibrated	3.2	0.005
Rubber latex/bitumen emulsion	3.2	0.003
Solvent-Based "Cut-Backs" and Resins		
Asphalt mastic	3.2	0.0006
Polymeric mastic	3.2	0.0025
Polymeric mastic	0.5	0.017
Bitumen mastic	3.2	0.0008
Epoxide resin	0.5	0.003
Elastomer-based mastic	1.6	0.0035
Miscellaneous		
Aluminum foil/kraft paper laminate (0.009 mm foil)	–	0.0024
Aluminum foil/kraft paper laminate (0.018 mm foil)	–	0.0008
Bitumen-impregnated tape	2	0.0057
PVC tape (self-adhesive)	0.18	0.068
PVC tape (self-adhesive)	25	0.057
Aluminum foil (self-adhesive), plain or reinforced	0.018	0.0008
Polyethylene sheet	0.5	0.004

bulb temperature and the percentage relative humidity at the particular barometric pressure. Thus, for example, air at $20°C$ (dry bulb) and 52.6% relative humidity (rh) will exert approximately the same pressure (1230 N/m^2) as air at $10°C$ and 100% rh (i.e., saturated); the dew point in each case will be $10°C$ and the actual moisture content will be approximately 0.00767 kg/kg dry air. These figures are for a barometric pressure of 1013 m bar.

If now we assume the insulated cold metal surface to be at a temperature of $-17°C$, the vapor pressure at that surface would be $(1230 - 125) = 1105$ so that, for ambient air as above, the vapor pressure difference would be $(1230 - 125) = 1105 \text{ N/m}^2$, approximately.

Table 1.5 Water-Vapor Permeance for Some Vapor Barriers and the Conversion Factors

	g/(S . MN)	g/(cm² s mbar)	g/(m² . 24 h mm Hg)	1b (ft² h atm)	gr/(ft² h mbar)	gr/(ft² h inHg) = 1 perm	Temperate g/ (m² 24 h)	Tropical g/ (m² 24 h)
g/(S . MN)	1	1×10^{-8}	1.152×10	7.471×10^{-2}	5.161×10^{-2}	1.749×10	2.052×10^{2}	5.149×10^{2}
g/(m² s mbar)	1×10^{8}	1	1.152×10^{9}	7.471×10^{6}	5.161×10^{7}	1.749×10^{9}	2.052×10^{10}	5.149×10^{10}
g/(m² . 24 h mm Hg)	8.681×10^{-2}	8.681×10^{-10}	1	6.486×10^{-3}	4.481×10^{-2}	1.517	1.782×10	4.472×10
1b (ft² h atm)(this was the term used by the building industry)	1.339×10	1.339×10^{-7}	1.542×10^{2}	1	6.909	2.339×10^{2}	2.747×10^{3}	6.896×10^{3}
gr/(ft² h mbar) (the symbol "gr" refers to grains)	1.937	1.937×10^{-8}	2.233×10	1.447×10^{-1}	1	3.388×10	3.975×10^{2}	9.980×10^{2}
gr/(ft² h in Hg)) = 1 perm	5.719×10^{-2}	5.719×10^{-10}	6.590×10^{-1}	4.275×10^{-3}	2.951×10^{-2}	1	1.174×10	2.948×10
Temperate	4.874×10^{-3}	4.874×10^{11}	$5.6:3\times10^{-2}$	3.641×10^{-4}	2.515×10^{-3}	8.514×10^{-2}	1	No conversion from temperate to tropical is shown for the reasons given in Clause 46 of BS 2972:1989.
Tropical g/(m² 24h)	1.942×10^{-3}	1.942×10^{-11}	2.236×10^{2}	1.450×10^{4}	1.002×10^{-3}	3.392×10^{-2}	No conversion from temperate to tropical is shown for the reasons given in Clause 46 of BS 2972:1989.	1

Even a good-quality vapor barrier is not guaranteed to exclude water vapor entirely, although it will diminish the rate of water-vapor penetration proportional to its permeance value. In essence, only the time factor will be relevant; ultimately the insulation will become wet if the plant is maintained at a low temperature for long enough.

It is necessary to distinguish between the behavior of insulation and vapor barriers over porous structures, e.g., nonmetallic walls, and the corresponding behavior over nonporous surfaces, e.g., metallic pipework and vessels. It is important to remember that when moisture penetrates into a porous structure it is probable that it will re-evaporate through the colder surface, especially if air movement can be arranged.

On the other hand, with insulated metal surfaces condensed moisture tends to remain at the cold surfaces unless it is possible to make provision for it to be collected and drained away; more usually it accumulates at the metallic surface and saturates the insulating material, with consequent danger of increase in thermal conductivity and the possibility of corrosive or electrolytic attack on the metal surface.

If the condensation is transient or periodic, the water may evaporate later, in which case it may be satisfactory either to apply a "breather coat" to the outer surface, or even to dispense with the vapor barrier completely. If, however, the condensation is likely to be of long duration, it will be essential to apply an efficient vapor barrier.

As an example, based on a permeance of 0.001 g/(s MN) for the vapor barrier and assuming a vapor pressure difference of 1105 N/m^2, one could expect a rate of accumulation of water within the insulation in total of 0.095 N/m^2 per 24 h, or approximately 34.7 g/m^2 per year.

With insulation of inherently low permeance, it may be convenient to apply a vapor barrier of higher permeance than would apply otherwise if, by doing so, superior mechanical resistance and easier application is achieved; incorrect methods of application can spoil otherwise good-quality vapor barrier materials. For very low temperatures, the use of multiple vapor barriers may be justified.

For refrigeration work it is preferable that the vapor-barrier layer shall not be exposed to mechanical damage if it is susceptible to easy perforation. Frequently it is possible to use a tough outer finish, e.g., sheet metal or vinylacrylic copolymer, as a protective layer over a more vulnerable barrier material.

This outer finish conveniently could be suitable for the final application of a high gloss white paint to reduce solar absorption. Alternatively, polished metal sheet may be applied as protection over a vulnerable vapor-barrier coat or film. Care shall be taken to ensure that the vapor barrier does not add seriously to the fire hazard of the complete system.

The compatibility of the vapor-barrier material with the chosen insulation shall be established, e.g., solvent-based materials shall not be used directly over polystyrene.

Materials suitable for use as vapor barriers are as follows:

Wet-applied vapor barriers comprise cut-back bitumens, bitumen emulsions with or without elastomer latex, vinyl emulsions, and solvent-based polymers. Frequently these are reinforced by means of cotton scrim cloth or open-mesh glass fabric.

Glass fabric or hessian sheet or tape, impregnated with lanolin or petroleum jelly, can be used, especially where removable insulation and finish are required. Elastomer sheets provide for contraction and other movements while maintaining good resistance to the transmission of water vapor; joints shall be sealed with an appropriate adhesive and the overlaps shall be at least 40 mm.

Polyvinyl chloride, polyethylene, polyisobutylene, or other plastic tapes or sheets are of special value for wrapping bends on insulated pipes, or where a colored descriptive finish is required.

Epoxy and polyester resins give good resistance against mechanical damage, together with protection against the weather and against chemical spillage. Sheet metal can give good protection, provided that the joints are overlapped and sealed, with additional securing devices to maintain the system in vapor tight condition.

Metal foils, if used alone, shall be sufficiently thick to exclude penetration by "pin holes," or they shall be laminated to building paper, building sheet, or plastic film. The joints should have an overlap of at least 40 mm and they shall be sealed by a suitable waterproof adhesive or mastic.

It is important to note that under certain conditions trapped moisture in a construction, e.g., in a cavity, may be allowed to vent or disperse by the use of a so-called "breather material."

This is a material that will permit the passage of water vapor but not of liquid water. There are two essential types: the first an aqueous emulsion coating and the second a breather paper. Cut-back dispersions, elastomer sheets, and metal sheets do not permit "breathing."

Frequently, circumstances will arise when a flame-resisting finish is required over insulation for "cold work." The choice shall be made from sheet metal, a fire-resistant mastic, and a normal type of vapor barrier supercoated with a self-setting mineral-fibered cement, reinforced as may be necessary.

Under certain conditions, an outer finish of sheet metal may cause a hazard by preventing easy access for fire suppression if ignition should occur in the main layer of insulating material.

Before the vapor is applied, the insulant shall be smooth and regular with all joints tightly butted together so that there are no gaps. Adhesives and joint sealants shall be fully cured. If the vapor barrier is a wet-applied coating or mastic, all corners on the insulant should be radiused.

Where possible, supports for pipes and vessels shall be external to the insulation and the vapor barrier. Where this is not possible the vapor barrier shall be returned to the support and effectively sealed. Where insulation is required to be removable, e.g., at flanged joints, manholes, etc., the main insulation shall be cut short of the fitting and the vapor barrier shall be sealed directly to the pipework or shell of the vessel. The removable portion of the insulation shall then be fitted as a separate item, with its vapor barrier overlapping and being sealed to the main vapor barrier.

Where the operating temperature of the plant may cycle above and below that of the ambient air, it is more difficult to design an effective vapor barrier. If the period of operation below ambient temperature is short, or if the difference in temperature is small, it may be advisable to omit the vapor barrier and to use a porous coating that will permit evaporation of the condensed water when the temperature is raised; but this type of arrangement will result in low efficiency when the plant is working under cold conditions.

For operating conditions where the temperatures of cycling involve wide fluctuations, it may be advantageous to use multilayer insulation with a vapor barrier over each layer. Each vapor barrier shall be returned over the edges of the insulating material to the metal surface of the pipe or vessel, and sealed to each other at regular intervals so that the individual layers of insulation are enclosed in separate compartments. With this type of arrangement the correct selection of material for the various vapor barriers may be critical on account of possible restriction by limiting temperatures.

As a general rule, a vapor barrier will be required over the insulation on all plant working at temperatures below the relevant "dew point" of the ambient air. Pipework and equipment essentially provides a nonpermeable base surface that, if excessive accumulation of water is to be avoided, theoretically will require the lowest possible permeance for an insulation system. However, constraints such as flammability, durability, and chemical resistance will often preclude the selection of components with the lowest water–vapor permeance.

Field experience often shows that a suitable selection of mastics and coating compounds, in conjunction with insulating materials of adequately low permeance rating, can give good practical service, provided that they are associated with good joint-sealing technique and, especially for very low temperatures, with the use of mildlayer sealants or vapor barriers.

The lower portions of the shaded zone will be appropriate where membrane, sheet, or laminated vapor-barrier materials are installed, or where the insulating material is of relatively high permeability. With systems of this type it is necessary to ensure that there is adequate sealing and protection against mechanical damage under service conditions, with ample allowance of safety factors.

Mechanical design must incorporate insulation requirements for preventing water ingress such as:

- A continuous welded water-retaining collar shall be installed around all protruding parts of vessels and columns. These collars shall be included in the manufacturer's drawings.
- Edges of tank roofs shall be provided with rainwater shields.
- Hand railings on insulated tank roofs shall be installed at the side of the roof edge instead of on top of it.
- The number of supports and fixings penetrating the insulation shall be reduced to an absolute minimum.
- Fixings to walls and roofs shall be of round shape (pipe) instead of angle steel and shall be installed with a slight outward slope.
- Angle steel used as vacuum/compression rings of columns shall be installed with the vertical part downwards.
- Valves and flanges shall be located in the horizontal part of pipelines instead of the vertical.
- Nameplates shall not interfere with insulation and shall be relocated.
- Installation of gratings on top of columns, vessels, tanks, and over piping, where applicable, to avoid damage to insulation.

Care shall be taken to avoid contact between dissimilar materials which might cause galvanic corrosion. Studs, pins, clips, or other attachments shall not be field welded to the postweld heat-treated vessels (including tanks) or piping without written authorization by the employer.

Expansion and contraction joints shall be provided below vessel insulation support rings and in piping as required. Contraction joints on cold piping shall be at flanged joints if possible. Summaries of the insulation requirements for equipment on hot service and cold service are given in Tables 1.6 and 1.7, respectively.

Table 1.6 Summary of Insulation Requirements for Equipment on Hot Service

Type of Equipment	Form of Insulation	Finish	Insulation Supports
Storage tank shells only	Slab, mattress, or foamed in situ plastics	Corrugated or plain metal sheet	Welding studs, pins, or self support cage
Spheres	Slab, mattress, or foamed in situ plastics	Plain metal sheets or hard setting compound	Welding studs, pins, or self support cage
Vessels, vertical	Slab or mattress	Corrugated metal sheet	Support rings 12" max.
Vessel, horizontal	Slab or mattress	Plain or reeded metal sheet	Vertical pitch, studs, lugs, cage studs, lugs, cage if required
Vessel, sheltered	Slab or mattress	Hard setting compound or sheet metal	Vertical pitch, studs, lugs, cage studs, lugs, cage if required
Manhole covers	Insulation slab or blanket completely enclosed in a metal skin	–	Lugs if required
Exchanger ends	Insulation slab or blanket completely enclosed in a metal skin	–	Lugs if required
Piping, vertical	Rigid sections, lags, or mattress	Plain or reeded metal sheet	Rings or lugs, 12" max. vertical spacing
Piping, horizontal	Rigid sections, lags, or mattress	Plain or reeded metal sheet	Nil for rigid sections, rings or lugs for mattress
Piping, sheltered	Rigid sections, lags, or mattress	Hard setting compound or sheet metal	Nil for rigid sections, rings or lugs for mattress
Bends	Rigid sections and lags	Metal sheet or hard setting compound	Rings and lugs
Steam valves and flanges, external	Box packed with insulation, strip wrap, blanket	Fabricated removable box, molded box	Self fixing
Steam valves and flanges, sheltered	Box packed with insulation, strip wrap, blanket	Molded box, or sewed canvas	Self fixing

Table 1.7 Summary of Insulation Requirements for Equipment on Cold Service (In All Cases a Vapor Seal Is Required over the Insulation)

Type of Equipment	Form of Insulation	Finish	Insulation Supports
Storage tanks	Slab, block, or foamed in situ plastics	Corrugated or plain metal sheet or mastic	Self support cage, bands
Spheres	Slab, block, or foamed in situ plastics	Corrugated or plain metal sheet or mastic	Self support cage, bands
Spheres (LPG)	Block	Plain metal sheets, or mastic; reinforced mastic	Adhesive and bands
Storage tanks (double walled)	Loose fill (dry inert gas pressurized)	Plain metal sheets, or mastic; reinforced mastic	Adhesive and bands
Vessels, vertical	Slab, block, or foamed in situ plastics	Plain or reeded metal sheets, mastic	Self support cage or bands
Vessel, horizontal	Slab, block, or foamed in situ plastics	Plain or reeded metal sheets, mastic	Self support cage or bands
Vessel, sheltered	Slab, block, or foamed in situ plastics	Plain or reeded metal sheets, mastic	Self support cage or bands
Manhole covers, exchanger ends, valves and fittings	Slab, loose fill, or foamed in situ in removable or moldable box	Metal sheet or mastic	–
Piping, vertical	Slab sections, pre-insulated foamed in situ	Plain metal sheets, mastic	Rings 12" max. vertical spacing; bands
Piping, horizontal	Slab sections, pre-insulated foamed in situ	Plain metal sheets, mastic	Rings 12" max. vertical spacing; bands
Piping, sheltered	Slab sections, pre-insulated foamed in situ	Hard setting compound	Rings 12" max. vertical spacing; bands

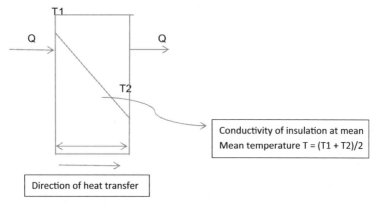

Figure 1.4 Schematic diagram of heat transfer through a single flat material.

1.5. HEAT TRANSFER THROUGH INSULATION

In this section the heat transfer equations that relate the heat flow through the insulation to the insulation thickness and to the thermal conductivity of insulation materials for flat and cylindrical insulation are presented. The symbols and units are as follows:

Q = heat flow in W/m

T_1 = temperature of hot surface in °C

T_2 = temperature of cold surface in °C

L = thickness of insulation in meter

λ = conductivity in W/mK

- Equation for heat transfer for one thickness of flat insulation:

$$Q = \frac{T_1 - T_2}{L/\lambda} \qquad (1.2)$$

Figure 1.4 shows a diagram of heat transfer through a single flat material.

Example 1.1

If the temperature of insulation is 148.89°C on one surface and 21.1°C on the other surface, heat conductivity at 85°C is 0.062 W/mK, and heat transfer is not to be over 94.577 W/m^2, what thickness of insulation is required?

As stated:

$T_1 = 148.89$°C

$T_2 = 21.1$°C

$$\lambda = 0.062 \text{ W/mK}$$
$$Q = 94.577 \text{ W/m}^2$$

From Eq. (1.2):

$$L = \lambda \left(\frac{T_1 - T_2}{Q} \right) = 0.062 \left(\frac{148.89 - 21.1}{94.577} \right) = 0.0838m$$

- Equation for heat transfer for two or more thicknesses of flat insulation:

$$Q = \frac{T_1 - T_S}{\left(\frac{L_1}{\lambda_1} + \frac{L_2}{\lambda_2} + \frac{L_3}{\lambda_3} \right)} \qquad (1.3)$$

where L_1, L_2, and L_3 are the thicknesses of the first, second, and third insulation material in meters and λ_1, λ_2, λ_3 are the conductivities of the first, second, and third insulation materials in W/mk, respectively. The factors are shown schematically in Figure 1.5.

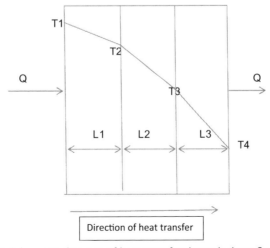

Figure 1.5 Schematic diagram of heat transfer through three flat materials.

Example 1.2

Find the thickness of two different insulations so that the heat transfer will be less than 315.26 W/m^2 when the temperature on one side is 760°C (T_1) and the temperature on the other side is 0°C (T_s). The conductivity of one insulation is 0.108 W/mK and that of the other is 0.072 W/mK. The maximum temperature of the insulation with the lower conductivity is 649°C (T_2).

$$Q = \frac{T_1 - T_S}{\left(\frac{L_1}{\lambda_1} + \frac{L_2}{\lambda_2}\right)} \Rightarrow \frac{L_1}{\lambda_1} + \frac{L_2}{\lambda_2} = \frac{T_1 - T_S}{Q} = \frac{760 - 0}{315.26} = 2.41$$

We see that $T_1 - T_2 = 760 - 649 = 111°C$ or less and Q is stated to be no more than 315.26 W/m^2:

$$L_1 = \left(\frac{T_1 - T_S}{Q}\right)\lambda_1 = \left(\frac{760 - 649}{315.26}\right)(0.108) = 0.038m$$

$$L_2 = \left(\frac{T_2 - T_S}{Q}\right)\lambda_2 = \left(\frac{649 - 0}{315.26}\right)(0.072) = 0.1482m$$

From thermal conductivity, we know:

λ_1 at mean temperature, $(T_1 + T_2)/2$

λ_2 at mean temperature, $(T_2 + T_3)/2$

λ_3 at mean temperature, $(T_3 + T_S)/2$

$$Q_t = \frac{T_1 - T_S}{\frac{L_1}{\lambda_1} + \frac{L_2}{\lambda_2} + \frac{L_3}{\lambda_3}} \tag{1.4}$$

- Equation for heat transfer through one thickness of cylindrical insulation:

$$Q_t = \frac{T_1 - T_2}{\left(\frac{r_2}{\lambda}\ln\left(\frac{r_2}{r_1}\right)\right)} \tag{1.5}$$

where

r_1 = inner radius of a single layer of cylindrical insulation in meters

r_2 = outer radius of a single layer of cylindrical insulation in meters

$$\text{Equivalent thickness} = r_2\ln\left(\frac{r_2}{r_1}\right) \tag{1.6}$$

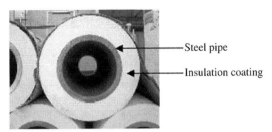

Figure 1.6 Insulation coating for a steel pipe. *(Source: Elsevier, N. Bouchonneau et al. / Journal of Petroleum Science and Engineering 73 (2010) 1–12).*

This is a more common method of calculating heat transfer in which all heat transfer is based on the outer area of the insulation and equivalent thickness has been used in the formula.

Figure 1.6 shows insulation coating for a steel pipe.

Example 1.3

Assume the following:

The temperature of the higher temperature side (inner) $T_1 = 232.2°C$.
The temperature of the lower (outer) side $T_2 = 37.7°C$.
The conductivity of the insulation $\lambda = 0.0577$ W/mK.
The inner radius $r_1 = 0.0381$ m.
Determine the thickness L and r_2, the outer radius of insulation:

$$Q = \frac{T_1 - T_2}{\dfrac{r_2 \ln\left(\frac{r_2}{r_1}\right)}{\lambda}} \tag{1.7}$$

From Eq. (1.7):

$$r_2 \ln\left(\frac{r_2}{r_1}\right) = \left(\frac{T_1 - T_2}{Q}\right)\lambda = \left(\frac{232.2 - 37.7}{220.7}\right)(0.0577) = 0.051$$

As r_2 is the unknown and appears twice in the remaining factor, it is difficult to compute directly.

However by looking at Table 1.8 for the values of $r_2 \ln\left(\frac{r_2}{r_1}\right)$, it can be determined that $r_2 = 0.0762$ m and insulation thickness will be

$$r_2 - r_1 = 0.0762 - 0.0381 = 0.0381 \text{ m}$$

- Equation for heat transfer through two or more thicknesses of cylindrical insulation.

The same form of equation presented to calculate the heat flow through one thickness of cylindrical insulation should also be used where the total

Table 1.8 Equivalent Thickness – Meters

Tube size, mm	Nominal Insulation Thickness, mm																			
	13	25	36	51	64	76	80	102	114	127	140	152	168	178	190	203	216	220	241	250
6.4	0.026	0.06	0.106	0.16	0.202	0.255	0.335	0.376	0.474	0.477	0.544	0.605	0.657	0.731	0.798	0.861	0.901	0.994	1.035	1.13
12.7	0.021	0.051	0.086	0.125	0.167	0.211	0.25	0.306	0.355	0.406	0.458	0.511	0.565	0.62	0.675	0.733	0.791	0.848	0.906	0.967
15.6	0.017	0.042	0.07	0.102	0.135	0.172	0.211	0.251	0.292	0.334	0.378	0.423	0.474	0.516	0.532	0.612	0.661	0.71	0.762	0.812
50.8	0.015	0.035	0.058	0.084	0.111	0.144	0.172	0.204	0.238	0.273	0.309	0.346	0.386	0.422	0.462	0.502	0.543	0.585	0.627	0.67
76.2	0.014	0.032	0.053	0.075	0.099	0.125	0.153	0.182	0.211	0.242	0.274	0.306	0.339	0.374	0.412	0.444	0.482	0.519	0.556	0.592
101.6	0.014	0.031	0.05	0.07	0.092	0.116	0.141	0.167	0.195	0.222	0.252	0.282	0.312	0.344	0.376	0.409	0.44	0.476	0.511	0.545
127.6	0.014	0.03	0.048	0.067	0.088	0.11	0.133	0.158	0.183	0.209	0.236	0.264	0.293	0.322	0.353	0.382	0.413	0.446	0.478	0.511
152.6	0.014	0.029	0.046	0.065	0.084	0.105	0.128	0.15	0.174	0.199	0.224	0.251	0.278	0.306	0.333	0.363	0.391	0.422	0.454	0.484
177.8	0.013	0.029	0.045	0.063	0.082	0.102	0.123	0.145	0.168	0.191	0.216	0.241	0.267	0.293	0.318	0.347	0.375	0.404	0.432	0.463
203.2	0.013	0.028	0.044	0.062	0.08	0.099	0.12	0.141	0.163	0.185	0.209	0.233	0.257	0.282	0.307	0.335	0.362	0.389	0.418	0.445
228.6	0.013	0.028	0.044	0.061	0.079	0.097	0.117	0.137	0.158	0.18	0.203	0.226	0.25	0.274	0.299	0.324	0.35	0.377	0.401	0.431
254	0.013	0.028	0.043	0.06	0.077	0.095	0.115	0.134	0.155	0.176	0.198	0.22	0.243	0.266	0.292	0.315	0.34	0.366	0.39	0.418
304.8	0.013	0.027	0.042	0.058	0.075	0.093	0.111	0.13	0.157	0.169	0.19	0.211	0.233	0.255	0.277	0.29	0.324	0.349	0.374	0.411
355.6	0.013	0.027	0.042	0.057	0.073	0.09	0.106	0.126	0.157	0.164	0.184	0.204	0.225	0.246	0.261	0.29	0.311	0.336	0.36	0.382
408.4	0.013	0.027	0.041	0.056	0.072	0.089	0.106	0.123	0.145	0.16	0.179	0.199	0.218	0.239	0.26	0.281	0.306	0.325	0.347	0.37
457.2	0.013	0.027	0.041	0.056	0.071	0.087	0.104	0.121	0.141	0.157	0.175	0.194	0.214	0.234	0.251	0.274	0.293	0.317	0.338	0.36
508	0.013	0.027	0.041	0.055	0.071	0.086	0.1	0.119	0.139	0.159	0.172	0.191	0.207	0.229	0.248	0.269	0.291	0.31	0.332	0.352
609.6	0.013	0.027	0.04	0.054	0.069	0.085	0.1	0.116	0.138	0.15	0.167	0.185	0.203	0.221	0.242	0.259	0.281	0.298	0.317	0.337
711.2	0.013	0.026	0.04	0.054	0.068	0.084	0.099	0.118	0.133	0.147	0.164	0.18	0.198	0.216	0.235	0.252	0.279	0.297	0.31	0.326
914.4	0.013	0.026	0.039	0.053	0.067	0.082	0.096	0.112	0.127	0.143	0.158	0.175	0.191	0.206	0.226	0.242	0.262	0.278	0.3	0.314
1016	0.013	0.026	0.039	0.053	0.067	0.081	0.096	0.111	0.126	0.141	0.157	0.173	0.189	0.205	0.223	0.239	0.26	0.274	0.299	0.309
1219.2	0.012	0.026	0.039	0.053	0.066	0.08	0.095	0.11	0.124	0.139	0.154	0.17	0.185	0.201	0.216	0.234	0.248	0.268	0.28	0.301
1422.4	0.012	0.026	0.039	0.052	0.066	0.08	0.094	0.108	0.122	0.138	0.152	0.168	0.182	0.198	0.214	0.23	0.245	0.262	0.276	0.295
1625.6	0.012	0.026	0.039	0.052	0.066	0.08	0.093	0.108	0.121	0.136	0.151	0.165	0.18	0.196	0.21	0.226	0.243	0.259	0.274	0.289
1928.8	0.012	0.026	0.039	0.052	0.065	0.079	0.093	0.107	0.121	0.135	0.15	0.164	0.179	0.194	0.209	0.224	0.239	0.255	0.27	0.286

thickness is composed of two or more thicknesses of different λ values. In all cases, however, the multiplier for loge is the outermost radius:

r_1 = inner radius of innermost insulation in meters

r_2 = outer radius of innermost insulation and inner radius of outermost insulation in meters

r_s = outer radius of outermost insulation in meters

T_1 = temperature of innermost surface in °C

T_2 = temperature of radius between inner- and outermost insulation in °C

T_s = temperature of outer surface in °C

λ_1 = conductivity of innermost insulation in W/mK

λ_2 = conductivity of outermost insulation in W/mK

Heat transfer through two different types of cylindrical insulation in W/m^2 is

$$Q = \frac{T_1 - T_s}{\left(\dfrac{r_s \ln\left(\dfrac{r_2}{r_1}\right)}{\lambda_1}\right) + \left(\dfrac{r_s \ln\left(\dfrac{r_s}{r_2}\right)}{\lambda_2}\right)} \tag{1.8}$$

Heat transfer Q through two different types of cylindrical insulation in W/m^2 is

$$Q = \frac{T_1 - T_s}{\left(\dfrac{r_s \ln\left(\dfrac{r_2}{r_1}\right)}{\lambda_1}\right) + \left(\dfrac{r_s \ln\left(\dfrac{r_s}{r_2}\right)}{\lambda_2}\right) + \left(\dfrac{r_s \ln\left(\dfrac{r_s}{r_3}\right)}{\lambda_3}\right)} \tag{1.9}$$

Note: Tabular values are those of the equivalent thickness

$$L_m = r_{2m} \ln\left(\frac{r_{2m}}{r_{1m}}\right) \tag{1.10}$$

for the nominal insulation thickness given

L_m = Equivalent thickness in meters

r_{1m} = Inside radius of insulation in meters

r_{2m} = Outside radius of insilation in meters

1.5.1. Correlation Approach for the Estimation of Thermal Insulation Thickness

Selection and determination of optimum thickness of insulation is of prime interest for many engineering applications. In this section, a simple method is developed to estimate the thickness of thermal insulation required to arrive at a desired heat flow or surface temperature for flat surfaces, ducts, and pipes. The proposed simple method covers the temperature difference between ambient and outside temperatures up to 250°C and the temperature drop through insulation up to 1000°C.

The proposed correlation calculates the thermal thickness up to 250 mm for flat surfaces and estimates the thermal thickness for ducts and pipes with outside diameters up to 2400 mm. The accuracy of the proposed method was found to be in excellent agreement with the reported data for a wide range of conditions where the average absolute deviation between reported data and the proposed method is around 3.25%. The method is based on basic fundamentals of heat transfer and reliable data. Therefore the formulated simple-to-use expression is justified and applicable to any industrial application.

- **Proposed method**

    The proposed method contains four steps.

1. Heat flow is estimated as a function of temperature difference between the outside surface and ambient temperature.

2. In the second step, the required thermal resistance is calculated as a function of the calculated heat flow in the first step and temperature drop through insulation.

3. The calculated thermal resistance is applied to estimate the thermal insulation thickness for flat surfaces according to insulation materials, typical conductivity values, and principal properties of common industrial insulations.

4. In this step, actual insulation thickness for ducts and pipes is estimated as a function of pipe/duct diameter and the required thickness calculated in step 3.

Equations (1.11) and (1.12) represent new equations to predict heat flow (Q) and the temperature difference (ΔT) between the outside surface and ambient temperatures, respectively, where the relevant coefficients have been reported in Table 1.9 for plain and coated fabric, dull metals, and

Table 1.9 Coefficients for Plain and Coated Fabrics, Dull Metals, and Unjacketed Surfaces

Factor	$Q = f(\Delta T)$	$\Delta T = f(Q)$
α	8.1476300369	$-2.050785108 \times 10^{-2}$
β	$1.0008706829 \times 10^{1}$	$9.3624934927 \times 10^{-2}$
γ	$3.270099068 \times 10^{-2}$	$-1.9785966213 \times 10^{-5}$
θ	$9.8114131125 \times 10^{-5}$	$2.3223257024 \times 10^{-9}$

Table 1.10 Coefficients for Bright Stainless Steel

Factor	$Q = f(\Delta T)$ for Bright Stainless Steel	$\Delta T = f(Q)$ for Bright Stainless Steel
α	4.7830053863	$-4.6237756384 \times 10^{-1}$
β	7.501603731	$1.3166512634 \times 10^{-1}$
γ	$1.2875229031 \times 10^{-2}$	$-2.5889539714 \times 10^{-5}$
θ	$2.2607312497 \times 10^{-5}$	$3.386990479 \times 10^{-9}$

unjacketed surfaces. Tables 1.10 and 1.11 cover tuned coefficients for bright stainless steel and aluminum metals, respectively.

$$Q = \alpha + \beta(\Delta T) + \gamma(\Delta T)^2 + \theta(\Delta T)^3 \tag{1.11}$$

$$\Delta T = \alpha + \beta Q + \gamma Q^2 + \theta Q^3 \tag{1.12}$$

- **Methodology for developing correlation**

    Equation (1.13) represents the simple equation in which four coefficients are used to correlate the heat flow as a function of required thermal resistance and temperature drop through insulation with the relevant coefficients shown in Table 1.12. This correlation has been developed based on more than 300 reliable data sets covering several cases and scenarios. As these data are widely accepted by industry and practice engineers in terms of accuracy for engineering calculations purposes, the formulated simple-to-use expression is justified and applicable to industrial applications.

    The factors A_1 through to D_4 are reported in Table 1.12.

$$\ln(Q) = a + \frac{b}{R} + \frac{c}{R^2} + \frac{d}{R^3} \tag{1.13}$$

where

$$a = A_1 + B_1(\Delta T) + C_1(\Delta T)^2 + D_1(\Delta T)^3 \tag{1.14}$$

Table 1.11 Coefficients for Bright Aluminum

Factor	$Q = f(\Delta T)$	$\Delta T = f(Q)$
α	$2.0737038264 \times 10^{-1}$	$1.3818480877 \times 10^{-1}$
β	4.707050285	$1.6112057171 \times 10^{-1}$
γ	$3.6827674245 \times 10^{-2}$	$-5.37963059307 \times 10^{-5}$
θ	$-6.0676449796 \times 10^{-5}$	$1.2181396248 \times 10^{-8}$

Table 1.12 Tuned Coefficients Used in Proposed Correlations (Eqs. 1.14–1.17)

Symbol	Coefficients for $\Delta T < 400°C$	Coefficients for $\Delta T > 400°C$
A_1	1.5384556496	1.00417011×10^1
B_1	$2.0855189351 \times 10^{-2}$	$-2.3990642064 \times 10^{-2}$
C_1	$-5.5714167184 \times 10^{-5}$	$3.4177418722 \times 10^{-5}$
D_1	$5.1423958307 \times 10^{-8}$	$-1.5189769392 \times 10^{-8}$
A_2	1.2529804561	$-1.1133812763 \times 10^1$
B_2	$1.8360054131 \times 10^{-4}$	$5.3672004787 \times 10^{-2}$
C_2	$2.6123146992 \times 10^{-6}$	$-7.0539620855 \times 10^{-5}$
D_2	$3.457810468 \times 10^{-9}$	$3.0944003705 \times 10^{-8}$
A_3	$-1.7994582993 \times 10^{-1}$	6.3515036069
B_3	$2.9931323873 \times 10^{-4}$	$-2.8350811015 \times 10^{-2}$
C_3	$-2.0629866707 \times 10^{-6}$	$3.7870341741 \times 10^{-5}$
D_3	$-6.831331144 \times 10^{-10}$	$-1.7142212602 \times 10^{-8}$
A_4	$8.9199903846 \times 10^{-3}$	-1.0596394373
B_4	$-1.956731750 \times 10^{-5}$	$4.694812062 \times 10^{-3}$
C_4	$7.3898608865 \times 10^{-8}$	$-6.4901664223 \times 10^{-6}$
D_4	$3.2512054476 \times 10^{-10}$	$3.074388661 \times 10^{-9}$

$$b = A_2 + B_2(\Delta T) + C_2(\Delta T)^2 + D_2(\Delta T)^3 \qquad (1.15)$$

$$c = A_3 + B_3(\Delta T) + C_3(\Delta T)^2 + D_3(\Delta T)^3 \qquad (1.16)$$

$$d = A_4 + B_4(\Delta T) + C_4(\Delta T)^2 + D_4(\Delta T)^3 \qquad (1.17)$$

In the next step, thermal resistance as a function of heat flow (W/m^2) and temperature drop through insulation can be estimated. The constants for determining thermal conductivity and unit heat-transfer rate for some common insulating materials are calculated based on the information provided in Table 1.13 and Eqs. (1.18) and (1.19); then one can multiply the calculated constants by the required insulation resistance (R) calculated by Eq. (1.13) (the value of R is less than 3) to obtain the required thickness (δ_f) for flat surfaces.

$$k = 0.1442(A' + B'T + C'T^2 + D'T^3) \qquad (1.18)$$

where

$$T = 1.8(°C) + 32 \qquad (1.19)$$

$$\delta_f = k \times R \qquad (1.20)$$

Table 1.13 Constants for Determining Thermal Conductivity and Unit Heat-Transfer Rate for Some Common Insulating Materials for Eq. (1.18)

Insulation	A'	B'	C'	D'	Temperature Range, °C
Calcium silicate ASTM C533-80 Class 1	0.2858	3.709×10^{-4}	–	–	0–350
	0.3504	5.196×10^{-4}	–	–	38–370
	0.651	3.437×10^{-4}	–	–	0–550
White fiberglass blankets with binder density $= 48\ kg/m^3$	0.2037	6.161	1.403×10^{-6}	-5×10^{-10}	10–425
White fiberglass blankets with binder density $= 96\ kg/m^3$	0.2125	−2.325	1.797×10^{-6}	-7.97×10^{-10}	10–480
Rigid fiberglass sheet ASTM C-547-77 Class 1	0.2391	9.192×10^{-4}	6.942×10^{-10}	–	-12–50
Rigid fiberglass sheet ASTM C-547-77 Class 2	0.2782	1.226×10^{-3}	–	–	3–95
Rigid fiberglass sheet ASTM C-612-77 Class 1	0.2537	3.051×10^{-4}	1.950×10^{-6}	–	-18–120
Rigid fiberglass sheet ASTM C-612-77 Class 3	0.2631	2.301×10^{-4}	1.614×10^{-6}	–	-18–135
Rigid fiberglass sheet density $= 64\ kg/m^3$	0.2113	3.857×10^{-4}	1.20×10^{-6}	–	-18–150
Rigid fiberglass sheet density $= 96\ kg/m^3$	0.1997	2.557×10^{-4}	9.048×10^{-7}	–	-18–150
Cellular glass foam ASTM C-552-79 Class 1	0.3488	5.038×10^{-4}	1.144×10^{-7}	7.172×10^{-10}	-185–260
Mineral wool Basaltic rock blanket, 144 kg/m³	0.2109	3.382×10^{-4}	5.495×10^{-7}	–	-18–425
Mineral wool Basaltic rock blanket, 192 kg/m³	0.2798	9.508×10^{-5}	6.478×10^{-7}	–	-18–425
Mineral wool Metallic slag block, 96 kg/m³	0.1076	5.714×10^{-4}	3.124×10^{-7}	–	-18–315
Mineral wool Metallic slag block, 288 kg/m³	0.3190	8.870×10^{-5}	2.174×10^{-7}	–	-18–650

(Continued)

Table 1.13 Constants for Determining Thermal Conductivity and Unit Heat-Transfer Rate for Some Common Insulating Materials for Eq. (1.18)—cont'd

Insulation	A'	B'	C'	D'	Temperature Range, °C
Mineral-wool-based cement	0.4245	6.293×10^{-4}	-1.638×10^{-7}	3.533×10^{-10}	-18–510
Preformed expanded perlite ASTM C-610-74	0.3843	3×10^{-4}	2.2381×10^{-7}	–	10–400
Expanded perlite-based cement	0.6912	5.435×10^{-4}	–	–	10–340
Expanded polystyrene block ASTM C-578-69 GR2	0.1711	2.76×10^{-4}	1.796×10^{-6}	-3.997×10^{-9}	-50–40
Polyurethane, 35.2 kg/m^3, aged 720 days at 25°C and 50% relative humidity	0.1662	-4.094×10^{-4}	-5.273×10^{-6}	2.534×10^{-8}	-50–0
Polyurethane, 35.2 kg/m^3 (85% closed cell)	0.1516	-3.37×10^{-4}	7.153×10^{-6}	-2.858×10^{-8}	0–50
Polyurethane, 35.2 kg/m^3, new polyurethane	0.1271	-2.49×10^{-4}	-7.962×10^{-7}	4.717×10^{-8}	-50–0
Polyurethane, 35.2 kg/m^3 (95% closed cell)	9.72×10^{-2}	7.813×10^{-4}	-7.152×10^{-6}	2.858×10^{-8}	0–50
Exfoliated vermiculite (insulating cement), Aislagreen	0.48	6×10^{-4}	–	–	-18–650
Exfoliated vermiculite (insulating cement), ASTM C-196-77	0.8474	5.071×10^{-4}	–	–	-18–650

In the next step the actual thickness for tubing and ducts is calculated by Eqs. (Eqs. 1.22–1.25) as a function of required insulation thickness (mm) for flat surfaces and outside diameter of tubing or duct. The tuned co-efficients for Eqs. (1.22–1.25) are reported in Table 1.14.

$$\ln(\delta_A) = a + \frac{b}{\delta_f} + \frac{c}{\delta_f^2} + \frac{d}{\delta_f^3} \tag{1.21}$$

where

$$a = A_a + \frac{B_a}{d} + \frac{C_a}{d^2} + \frac{D_a}{d^3} \tag{1.22}$$

$$b = A_b + \frac{B_b}{d} + \frac{C_b}{d^2} + \frac{D_b}{d^3} \tag{1.23}$$

$$c = A_c + \frac{B_c}{d} + \frac{C_c}{d^2} + \frac{D_c}{d^3} \tag{1.24}$$

$$d = A_d + \frac{B_d}{d} + \frac{C_d}{d^2} + \frac{D_d}{d^3} \tag{1.25}$$

where
A: tuned coefficient
B: tuned coefficient

Table 1.14 Tuned Coefficients Used in Proposed Correlations (Eqs. 1.22–1.25)

Symbol	For Diameter up to 200 mm	For 200 mm < Diameter < 2400 mm
A_a	5.4297244603	5.7722200946
B_a	$-2.4079209368 \times 10^1$	$-1.8130812838 \times 10^2$
C_a	2.6118609185×10^2	2.6544362945×10^4
D_a	$-8.9253215148 \times 10^2$	$-1.4268296841 \times 10^6$
A_b	$-1.1480830041 \times 10^2$	-1.296434321×10^2
B_b	5.7013386876×10^2	1.1243502457×10^4
C_b	$-8.4900426435 \times 10^3$	$-2.3778937967 \times 10^6$
D_b	3.2400753932×10^4	1.4658879342×10^8
A_c	1.8449381262×10^3	2.0975842835×10^3
B_c	$-9.5636463714 \times 10^3$	$-2.4344171539 \times 10^5$
C_c	1.6443383753×10^5	5.6957733228×10^7
D_c	$-6.6475305613 \times 10^5$	$-3.6210109262 \times 10^9$
A_d	$-1.0033101082 \times 10^4$	$-1.1484153918 \times 10^4$
B_d	5.2906504601×10^4	1.4912582138×10^6
C_d	$-9.7834592556 \times 10^5$	$-3.5828101148 \times 10^8$
D_d	4.0403726809×10^6	$2.2974736646 \times 10^{10}$

C: tuned coefficient

D: tuned coefficient

d: outside diameter of steel pipe and duct, mm

K: constants for determining thermal conductivity and unit heat-transfer rate for some common insulating materials

OD: outside diameter, mm

Q: heat flow (W/m^2)

R: thermal resistance coefficient

ΔT: temperature difference, °C

ΔT_f: temperature drop through surface air film, °C

T_m: mean temperature of insulation, °C

δ_f: flat insulation thickness, mm

δA: actual thickness for pipes and ducts, mm

- **Solved examples**

   Just to show the simplicity and make it clear for the reader and practice engineer, a couple of examples are shown below showing step-by-step calculations in a systematic way.

Example 1.4

A rectangular duct is operating at 230°C. The duct is finished with a silicone coated fabric. The ambient temperature is 27°C. We desire to maintain a surface temperature of 55°C. What thickness of cellular glass foam is required? What is the heat loss? Repeat the calculations and assume the same conditions as except that the surface to be insulated is a 100 mm OD duct.

Solution

$$T_h = 230°C$$

$$T_s = 55°C$$

$$T_a = 27°C$$

$$\Delta T_i = 230 - 55 = 175°C$$

$$\Delta T_f = 55 - 27 = 28°C$$

$$T_m = (230 + 55)/2 = 143$$

From Eq. (1.11) at $\Delta T_f = 28°C$, heat loss (Q) is 316.18 (W/m^2). Corresponding to a temperature drop through the insulation $\Delta T_i = 175°C$, assume $R_i = 0.53$.

From Eqs. (1.14) through (1.17):

$a = 3.7574676$
$b = 1.38364439$
$c = -0.1944061$
$d = 0.00950129$

From Eq. (1.13), one can estimate $Q = 311$ (W/m^2), which is in good agreement with 316.18 (W/m^2) and shows the assumed R_i is correct.

$$k = 0.1442\left(A' + B'T + C'T^2 + D'T^3\right) \tag{1.26}$$

From Table 1.13 ASTM C-552-79 Class 1 cellular glass foam has a "k" at $T_m = 143°C$:

$$k = 0.1442 \times \left[0.3488 + 5.038 \times 10^{-4} \times 290 + 1.144 \times 10^{-7} \times 290^2\right.$$

$$\left. + 7.172 \times 10^{-10} \times 290^3\right] = 0.52 \times 0.1442 = 0.075$$

Multiply the required insulation resistance R_i by k to obtain required thickness for flat surfaces (δ_f):

$$\delta_f = 0.58 \times 0.075 = 0.043m = 43mm$$

To calculate the thermal insulation thickness for a 100 mm OD duct, we obtain the following from Eqs. (1.22–1.25):

$a = 5.21415844$
$b = -109.923565$
$c = 1765.080293$
$d = -9597.830256$

The actual thickness = 32.8 mm from Eq. (1.21) for a 100 mm outside diameter duct. The heat loss of 320 W/m^2 of the outside insulation surface remains the same. The heat loss per linear m of outside duct surface (including insulation) is

$$OD = 100 + 2(32.8) = 165.6 \text{ mm}$$

$$\frac{\pi \times OD}{1000}Q = \left(\frac{168\pi}{1000}\right)320 = 166.5 \text{ W/m}$$

The calculated result is in good agreement with the reported data of 168 W/m with an average absolute deviation percent of 0.9%.

Example 1.5

A furnace is operating at 595°C. The outside surface is stainless steel. The ambient temperature is 24°C. We desire to limit the heat loss to 475 W/m^2. What thickness of mineral wool and cellular glass foam is required? What is the surface temperature?

Solution

$$T_h = 595°C$$

$$T_a = 24°C$$

$$Q = 475 \frac{W}{m^2}$$

For bright stainless steel, from Eq. (1.12) and at Q of 475 W/m², the temperature difference temperature between the outside surface temperature and ambient air temperature will be 56°C.

T_s is $(56 + 24) = 80$. In this case a combination of mineral wool (metallic slag block, 288 kg/m³) on the hot face backed by cellular glass foam (ASTM C 552-79 Class 1) is to be used. The temperature limit of cellular glass foam is 260°C. The inner face temperature between the two materials should be close to, but not above, this limit.

ΔT_i (min *real wood*) = 595 − 260 = 335

T_m (mineral wool) = (595 + 260)/2 = 428

K from Table 1.9 for mineral wool at 428°C (T_m) = 0.076

ΔT_i (cellular glass foam) (260 − 80) = 180°C

T_m (cellular glass foam) (260 + 80)/2 = 170°C

k from Table 1.9, cellular glass foam at 170°C (T_m) = 0.081

Through the insulation $\Delta T_i = 335°C$. Assume $R_i = 0.66$.

From Eqs. (1.14) to (1.17):

$a = 4.20572466651$

$b = 1.73765133586$

$c = -0.33687721969$

$d = 0.02288123919$

From Eq. (1.13), we have $Q = 466$ W/m², which shows acceptable agreement with 475 and shows the assumed R_i is correct.

Thickness of mineral wool required:

$$(0.66)(0.076) = 0.0501 \text{ m or } 50.1 \text{ mm}$$

The calculated result is in good agreement with the reported data of 52 mm with an average absolute deviation percent of 4%.

Similarly, cellular glass foam of $\Delta T_i = 180°C$ gives an insulation resistance of 0.39. The thickness of this cellular glass foam $(0.39)(0.081) = 0.032$ m or 32 mm.

- ## Summary of results

Figure 1.7 shows the temperature difference between the outside surface and ambient temperature as a function of heat flow (Q) for plain and coated fabrics, dull metals, unjacketed. Figure 1.8 illustrates the results of Eq. (1.13) in predicting the heat flow as a function of thermal resistance of insulation and temperature drop through insulation.

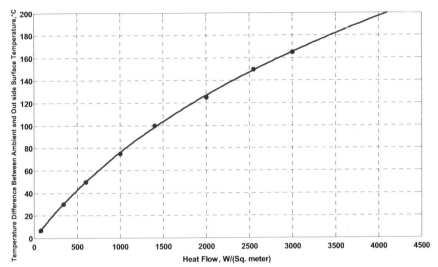

Figure 1.7 The results of Eq. (1.12) for prediction of temperature difference between the outside surface and ambient temperature as a function of heat flow (Q) for plain and coated fabrics, dull metals, unjacketed. *(Reprinted with permission from © Elsevier, A. Bahadori and H. B. Vuthaluru (2010). A Simple Method for the Estimation of Thermal Insulation Thickness, Applied Energy 87:613–619).*

Figures 1.9 and 1.10 show the results of Eq. (1.21) in predicting the actual insulation thickness for duct and tube as a function of outside diameter of pipe and ducts and required insulation thickness for flat surfaces. These graphs show good agreement between reported data and the estimated results from the proposed simple correlations.

Table 1.15 shows the accuracy of proposed correlation in comparison with the reported data where the average absolute deviation is around 3.25%. The correlations proposed in the present work are novel and unique, and are nonexistent in the literature.

This is expected to benefit in making design decisions leading to informed decisions for the determination and selection of optimum thickness of insulation for a given application in any process industry.

1.5.2. A Graphical Technique for the Estimation of Thermal Insulation Thickness

High fuel costs increase the need for insulation. A rule of thumb for estimating the thickness of insulation is to apply the thickness that produces a heat loss of 3% to 5% or less from the surface.

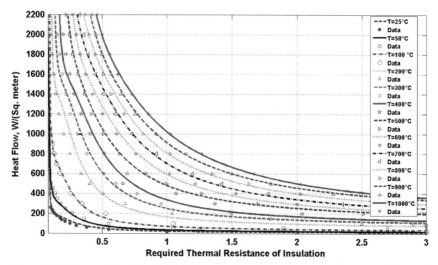

Figure 1.8 The results of Eq. (1.13) for prediction of heat flow as a function of thermal resistance of insulation and temperature drop through insulation in comparison with the data. *(Reprinted with permission from © Elsevier, A. Bahadori and H. B. Vuthaluru, 2010, A Simple Method for the Estimation of Thermal Insulation Thickness,* Applied Energy *87:613–619).*

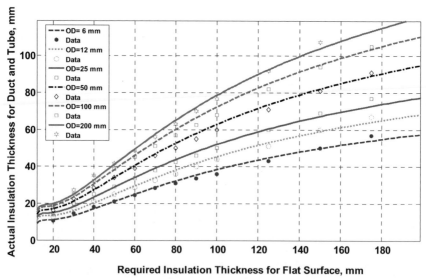

Figure 1.9 The results of Eq. (1.21) for prediction of actual insulation thickness for duct and tube as a function of outside diameter (less than 200 mm) and required insulation thickness for flat surfaces in comparison with the data. *(Reprinted with permission from © Elsevier, A. Bahadori and H. B. Vuthaluru (2010). A Simple Method for the Estimation of Thermal Insulation Thickness,* Applied Energy *87:613–619).*

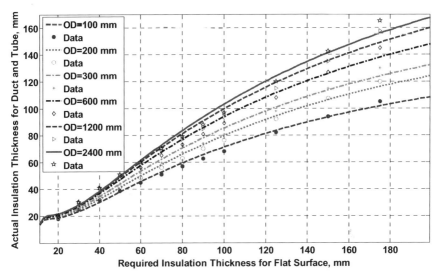

Figure 1.10 The results of Eq. (1.21) for prediction of actual insulation thickness for duct and tube as a function of outside diameter (between 100 mm and 2400 mm) and required insulation thickness for flat surfaces in comparison with the data. *(Reprinted with permission from © Elsevier, A. Bahadori and H. B. Vuthaluru (2010). A Simple Method for the Estimation of Thermal Insulation Thickness, Applied Energy 87:613–619).*

Table 1.15 Accuracy of Proposed Correlation for Predicting Heat Flow (W/m^2) in Comparison with the Literature Reported Data

Temperature Drop Through Insulation, °C	Thermal Resistance Coefficient	Heat Flow (W/m^2) Reported Data in the Literature	Heat Flow (W/m^2) Calculated Values	Absolute Deviation Percent (ADP)
25	0.112	240	234	2.5
50	0.1525	300	315	5
100	0.128	800	774	3.25
200	0.25	1000	1019	1.9
400	0.66	600	576	4
400	0.288	1400	1359	2.93
500	1.7	300	289	3.67
600	1.53	400	374	6.5
700	1.42	500	476	4.8
800	3.05	300	291	3
900	0.38	2400	2405	0.21
1000	0.41	2500	2532	1.28
Average absolute deviation percent (AADP)				3.25

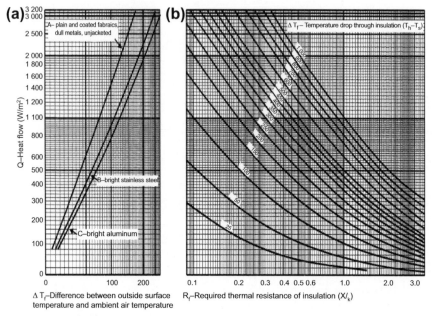

Figure 1.11 Calculation of heat flow for different insulators with varying thermal resistance. *(© Reprinted with permission from* Gas Processors and Suppliers Association Engineering Data Book, *Tulsa, OK, USA, 2004).*

Specific insulating materials and thicknesses for any large application should be determined with the assistance of the manufacturer. Figures 1.11 to 1.13 permit the rapid estimation of the thickness of thermal insulation required to give a desired heat flow or surface temperature when the hot face and ambient temperature are known. The method is based on elementary heat transfer theory and reliable experimental data.

The following examples illustrate the use of these graphs.

Example 1.6

A rectangular duct is operating at 230°C. The duct is finished with a silicone-coated fabric. The ambient temperature is 27°C. We want to maintain a surface temperature of 55°C. What thickness of cellular glass foam is required? What is the heat loss?

Solution

Using Figure 1.11:

$$T_h = 230, \; \Delta T_i = 230 - 55 = 175$$
$$T_s = 55, \; \Delta T_f = 55 - 27 = 28$$
$$T_a = 27, \; T_m = (230 + 55)/2 = 143$$

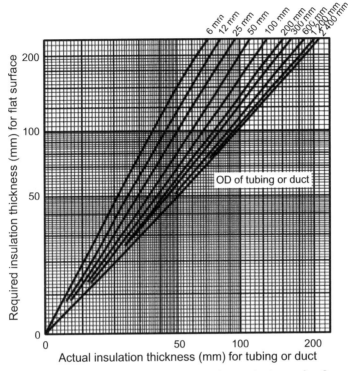

Figure 1.12 Relationship between required insulation thickness for flat vs. actual insulation thickness for tubing or duct. (© *Reprinted with permission from* Gas Processors and Suppliers Association Engineering Data Book, *Tulsa, OK, USA, 2004*).

In Figure 1.11 at at ΔT_f of 28, project vertically to curve A, then horizontally to the left to a heat loss (Q) of 320 W/m^2.

Project horizontally to the right along the 320 W/m^2 Q line to the point in Figure 1.11b corresponding to a temperature drop through the insulation (ΔT_i) of 175, then vertically downward to an insulation resistance (R_i) of 0.58.

From Table 1.13, ASTM C-552-79 Class 1 cellular glass foam has a k at $T_m = 143°C$:

$$k = 0.1442 \times [0.3488 + 5.038 \times 10^{-4} \times 290 + 1.144 \times 10^{-7} \times 290^2$$

$$+ 7.172 \times 10^{-10} \times 290^3] = 0.52 \times 0.1442 = 0.075$$

Multiply the required insulation resistance R_i by k to obtain the required thickness (X).

$$X = (0.58)(0.075) = 0.043 \text{ m} = 43 \text{ mm}$$

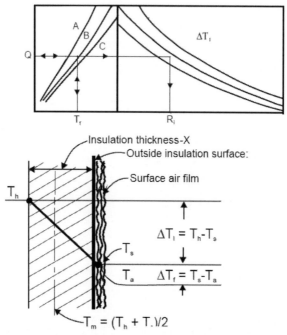

Figure 1.13 Details to illustrate how to use Figures 1.11 and 1.12. (© *Reprinted with permission from* Gas Processors and Suppliers Association Engineering Data Book, *Tulsa, OK, USA, 2004).*

Example 1.7

In the previous example, if the heat loss 320 W/m^2 is specified instead of a surface temperature of 55°C, the following procedure is used. Project a line horizontally on Figure 1.11a from a heat loss of 320 W/m^2 to curve A, then vertically downward to a ΔT_f of 28. The surface temperature = 27 + 28 = 55°C. The rest of the solution remains the same.

Example 1.8

Assume the same conditions as the previous example except that the surface to be insulated is a 100 mm OD duct.

After determining the required thickness of 43 mm for a flat surface, go to Figure 1.12. Project horizontally from 43 mm for a flat surface to the line representing a 100 mm OD duct then vertically to an actual thickness of 34 mm. The heat loss of 320 W/m^2 of the outside insulation surface remains the same. The heat loss per linear m of outside duct surface (including insulation) is

$$\frac{\pi \text{OD}}{1000}(Q) = \left(\frac{168 \cdot \pi}{1000}\right) 320 = 168.9 \text{W/linear m} \qquad (1.27)$$

Note: The insulation surface temperature on tubing and ducts in the horizontal position is generally higher than in the vertical position for the same heat flow. To correct for the horizontal position, multiply the ΔT_f for flat surfaces obtained from Table 1.13 by the following factors (H):

Q (W/m^2)	30–300	310–625	630–940	945 and up
H	1.35	1.2	1.10	15.0

Example 1.9

A furnace is operating at 595°C. The outside surface is stainless steel. The ambient temperature is 24°C. We want to limit the heat loss to 475 W/m^2. What thickness of mineral wool and cellular glass foam is required? What is the surface temperature?

Solution

$T_h = 595$

$T_a = 24$

$Q = 475 \text{ W/m}^2$

From Figure 1.11a at a Q of 475 W/m^2, project horizontally to curve B, then vertically to $\Delta T_f = 56$. T_s is $(56 + 24) = 80$. In this case a combination of mineral wool (metallic slag block, 288 kg/m^3) on the hot face backed by cellular glass foam (ASTM C 552-79 Class 1) is to be used. From Table 1.13, the temperature limit of cellular glass foam is 260°C. The inner face temperature between the two materials should be close to, but not above, this limit.

ΔT_i (mineral wool) $(595 - 260) = 335$

T_m (mineral wool) $(595 + 260)/2 = 428$

k of mineral wool at 428°C $(T_m) = 0.076$

ΔT_i (cellular glass foam) $(260 - 80) = 180$

T_m (cellular glass foam) $(260 + 80)/2 = 170$

k (from Table 1.13) of cellular glass foam at 170°C $(T_m) = 0.081$

Using Figure 1.11b, project horizontally along the 475 Q line to a ΔT_i (mineral wool) of 335°C, then vertically to an insulation resistance of 0.72. The thickness of mineral wool required is $(0.72 \times 0.076) = 0.055$ m or 55 mm.

Similarly, project horizontally along the 475 Q line to a ΔT_i (cellular glass foam) of 180°C then vertically to an insulation resistance of 0.39. Thickness of cellular glass foam $(0.39)(0.081) = 0.032$ m or 32 mm.

In the case of multiple layer construction Figure 1.12 should not be used to convert to a circular cross section. Figure 1.13 illustrates the details of how to use Figures 1.11a,b and 1.12.

1.6. ECONOMIC THICKNESS OF INSULATION

The required thickness of insulation for any specific application depends upon the characteristics of the insulating material and the purpose of the equipment. When the sole objective is to achieve the minimum total cost, the appropriate thickness is known as the economic thickness. If we need to calculate the economic thickness, the following additional information will be necessary:

- Cost of heat to be used for calculation purposes, e.g., US dollars per useful megajoule
- Evaluation period (working hours)
- Whether or not the cost of finish is to be included in the calculation.

Any increase in the amount of insulation applied raises the cost of the insulation but decreases the cost of heat lost. The sum of the two costs may be shown to lie on a curve with a rather flat region on either side of a minimum value representing the economic thickness.

Alternative systems should therefore be derived and compared. It should be noted that methods used for calculating economic thicknesses refer primarily to application over uniform types of surface, e.g., straight lengths of pipework. Thus they do not take account of insulating valves and bends, nor do they necessarily include the cost of providing for staggered joints or multiple layers.

Also the method of algebraic solution requires the assumption that the incremental cost of insulation can be related to the volume of insulating material, i.e., per cubic meter or per liner meter for given diameters of pipe.

1.7. METHODS OF OBTAINING ECONOMIC THICKNESS

The methods of obtaining economic thickness are given in the following sections.

1.7.1. Economic Thickness by Tabulation

The cost of insulation (including application and any normal finish) and of the heat lost are tabulated for a range of thickness of insulation (see Table 1.16). These costs are added for each thickness and the minimum total becomes apparent.

For the purpose of calculating the cost of heat lost, the surface coefficient can be taken to be 10.0 W/(m^2.K), since likely variations from this figure do

Table 1.16 Economic Thickness: Typical Method by Tabulation

Thickness of Insulation in mm	Rate of Heat Loss per Linear Meter per Second q_1 (in W/m)	Cost (Cent/m)		
		Heat Loss, Cost	Insulation + Labor + Ancillary Materials	Total
0	15018	2465335	0	2465335
50	624	102437	5139	107576
100	388	63695	9491	73186
160	288	47279	14850	62129
170	278	45636	15372	61008
180	269	44159	16506	60665
190	260	42681	17660	60341
200	252	41369	18444	59813
210	245	40220	19539	59759
220	239	39234	20633	59867
230	235	38250	21644	59894
240	228	37428	22532	59960
250	223	36608	23469	60077

not significantly affect the result. The insulating effect of finishing materials is ignored. The cost of heat lost is as follows:

- Cost of heat lost from flat surfaces:

$$5400\ Hq\ \ Y\ \ 10^{-6}\ \text{cent/m}^2 \tag{1.28}$$

- Cost of heat lost from cylindrical surfaces:

$$5400\ Hq_1\ \ Y\ \ 10^{-6}\ \text{cent/m}^2 \tag{1.29}$$

We know that

$$q_1 = 10^{-3}\pi d_o q \tag{1.30}$$

where:

d_o = outside diameter of pipe (m)
H = evaluation period in hours (working)
Y = cost of heat in cent/useful MJ
q = rate of heat loss through the insulating material per unit area of hot surface per second in W/m^2

This method is useful if the cost of increasing thickness of insulating material does not follow a uniform pattern, e.g., if there is a transition from a single layer to a double layer system. It is recognized that in particular instances allowances may be necessary for the capital or operating costs of the installation as a whole, but these are not considered in this book.

1.7.2. Economic Thickness by Algebraic Solution

• **Incremental cost derivation**

In the derivation of an algebraic expression for economic thickness, a term arises that is a function of the insulation cost, and for the equation given below, the term is represented by the symbol C, which is defined as the incremental cost of insulation.

It is important to realize that this is not the simple difference in applied cost between one thickness and the next higher one, but is more strictly interpreted as the derivative of the applied cost with respect to the volume of insulation. It should be noted that the additional cost of the finish and accessories resulting from the increasing thickness of insulation is included.

Within the context of this method of economic thickness calculation, therefore, the value of C should be obtained from a measurement or deduction of the slope of the curve for the plot of insulation cost against insulation thickness. However, in practice, such a graph is unlikely to exhibit a curve (unless forced smoothing of the plotted points is carried out) and the alternative approach is commonly adopted in these circumstances.

By this method, any one value of C can be estimated from the costs of two correspondingly successive thicknesses, with the understanding that the value applies to neither one, but may relate to a thickness about midway between them.

Sequential values obtained in this manner can still be very erratic and should be subsequently smoothed where the situation calls for an orderly progression of C values.

The equations for deriving the incremental cost in accordance with the two methods described in the previous paragraph are as follows:
From the slope of the cost curve:
1. Cylindrical surfaces:

$$C = \frac{s}{\pi d_n} \times 10^6 \qquad (1.31)$$

2. Flat surfaces:

$$C = S \times 10^3 \tag{1.32}$$

From the cost of two successive thicknesses:

This is determined from the change in cost divided by the corresponding change in volume on insulation:

1. Cylindrical surfaces:

$$C = \frac{(p_b - p_a)10^6}{\pi(d_o + L_b + L_a)(L_b - L_a)} \tag{1.33}$$

2. Flat surfaces:

$$C = \frac{(p_b - p_a)10^3}{(L_b - L_a)} \tag{1.34}$$

where

$S =$ slope of the curve of P against L

$d_n =$ the outermost diameter of the insulation (in mm)

$P =$ the installed cost of insulation (in S/m or S/m^2)

$L =$ the insulation thickness (in mm)

$d_o =$ the outside diameter of pipe (in mm)

$C =$ the incremental cost (in \$/m^3)

$H =$ evaluation period in hours (working)

$Y =$ cost of heat in cent/useful MJ

$q =$ rate of heat loss through the insulating material per unit area of hot surface per second in W/m^2

$\lambda =$ thermal conductivity of insulation material in W/mK

$\theta_1 =$ hot face temperature in $^\circ$C

$\theta_m =$ ambient air temperature in $^\circ$C

$\chi =$ economic thickness in mm

For the purpose of the calculation, the incremental cost C varies from 891 \$/m^3 (see Table 1.17). As an initial approximation, C is taken as the average of 698 \$/m^3.

1.7.3. Method of Calculation

If it is assumed that the cost per cubic meter of each increment through the insulation thickness is constant, it is possible to give a simple algebraic expression relating the total cost of the insulation, plus that of the heat loss in

Table 1.17 Incremental Cost of Insulation: Double Thickness

Outside Diameter of Steel Pipe (in mm)	Thickness (in mm)									Average Cost (in $/m³)
	Incremental Cost of Insulation (in m³):									
	120 to 140	140 to 160	160 to 180	180 to 200	200 to 220	220 to 240	240 to 260	260 to 280	280 to 300	
17.2	632	–	–	–	–	–	–	–	–	632
21.3	631	–	–	–	–	–	–	–	–	631
26.9	618	–	–	–	–	–	–	–	–	618
33.7	652	451	–	–	–	–	–	–	–	552
42.4	573	570	283	–	–	–	–	–	–	475
48.3	553	557	416	–	–	–	–	–	–	509
60.3	517	514	512	202	–	–	–	–	–	436
76.1	358	523	523	101	–	–	–	–	–	376
88.9	362	510	511	508	–	–	–	–	–	473
101.6	496	510	507	507	415	–	–	–	–	487
114.3	422	513	513	514	499	–	–	–	–	492
139.7	404	489	490	493	491	–	–	–	–	473
168.3	410	456	458	459	459	455	–	–	–	450
219.1	594	445	447	450	449	449	–	–	–	465
244.5	382	439	439	439	433	442	439	–	–	430
273	581	432	428	429	431	429	431	–	–	452
323.9	300	428	436	438	445	439	444	445	–	422
355.6	599	407	426	436	439	431	416	436	–	449
406.4	562	408	440	433	431	400	436	434	–	443
457	637	401	425	420	419	423	420	422	421	443
508	548	405	424	410	420	416	415	418	418	430
Over 508	552	–	–	–	–	–	–	–	–	552

an arbitrary period of time, to the thickness of insulation. This expression can then be used to find the thickness of insulation that gives the minimum total cost. The economic thickness obtained in this way is given by the following equation:

$$x = 5.76 \left[\frac{HY\lambda(\theta_1 - \theta_m)^{0.5}}{C} \right] \qquad (1.35)$$

The equation makes only an approximate allowance for the temperature difference between the atmosphere and the insulation surface, but this approximation has little effect on the calculated economic thicknesses and the method is applicable to all normal insulation systems. Appreciable inaccuracies occur when considering high cost insulation having a high thermal conductivity and in this case the method given in previous section has to be used.

In practice it may be found that the incremental cost of the insulation is not constant. The expression for economic thickness is given just above. When making a calculation it is first necessary to take an approximate value for C, which is used to calculate a value for the economic thickness. If the value of C taken is found not to be appropriate to the economic thickness obtained, a more appropriate value of C should be taken and the calculation repeated.

1.7.4. Examples of Calculation of Economic Thickness

To illustrate the application of the methods, the economic thickness for a typical case is deduced by each method:

- **Economic thickness by tabulation**
  Cost of heat, $Y = 1.14$ cent/useful MJ
  Evaluation period, $H = 40,000$ h
  Hot face temperature, $\theta_1 = +500°C$
  Ambient air temperature, $\theta_m = +20°C$
  Thermal conductivity of the insulation, $\lambda = 0.09$ W/(m.K)
  Pipe diameter (outside), $d_o = 210$ mm

In this example the lowest total cost tabulated occurs at a thickness of 100 mm; this is the economic thickness for the conditions indicated.

- **Economic thickness by algebraic solution**
  Cost of heat, $Y = 1.14$ cent/useful MJ
  Evaluation period, $H = 40,000$ h
  Hot face temperature, $\theta_1 = +500°C$

Ambient air temperature, $\theta_m = +20°C$
Thermal conductivity of the insulation, $\lambda = 0.09$ W/(m.K)
Pipe diameter (outside), $d_o = 219.1$ mm
Step 1
From Eq. (1.35),

$$x = 5.76 \left[\frac{40,000 \times 0.76 \times 0.09 \times (500 - 20)}{465} \right]^{0.5} = 306.1 \text{ mm} \qquad (1.36)$$

Therefore the economic thickness for a flat surface using insulation for which $C = 698$ $/m^3 is 306.1 mm.
Step 2
The equation

$$\chi = 0.5(d_o + 2L)\ln((1 + 2L)/d_o) \qquad (1.37)$$

is used to find the corresponding thickness L for a pipe of 219.1 mm outside diameter.

$$306.1 = 0.5(219.1 + 2L)\ln \left[\frac{1 + 2L}{219.1} \right]$$

The value of L cannot be obtained directly from this equation. It can however be obtained by trial and error. The result is $L = 192.4$ mm (approximately).
Step 3
If incremental costs are checked, it is found that for such a thickness the value of C is about 675 $/m^3. Recalculation with $C = 450$ gives $x = 311.7$ and $L = 194.9$ mm (approximately).

Checking incremental costs again gives a further value of 675 $/m^3 (approximately) for C, the same as the value previously taken. Further calculation is unnecessary. The economic thickness is therefore 195 mm.

Economic thicknesses for process pipework and equipment insulation for various ranges of operating temperature and varieties of thermal conductivity are given in Table 1.18. The method used to derive Table 1.18 was the algebraic solution method for economic thickness; the assumptions for the calculations are based on BS 5422, Section 7.

If the designer requires that the economic thickness be determined for the conditions other than those used in Table 1.18, the calculation shall be in accordance with the method indicated in the previous section.

Table 1.18 Economic Thickness of Insulation for Process Pipework and Equipment (These Thicknesses Are Sufficient for Personnel Protection)

Outside Diameter of Steel Pipe (in mm)	Hot Face Temperature at Mean Temperature (in °C) (with ambient still air at +20°C)														
	+100					+200					+300				
	Thermal Conductivity at Mean Temperature (in W/(m.K))														
	Thickness of Insulation (in mm):														
	0.02	0.03	0.04	0.05	0.06	0.03	0.04	0.05	0.06	0.07	0.03	0.04	0.05	0.06	0.07
17.2	28	31	35	38	41	45	49	52	56	59	52	57	61	66	70
21.3	29	32	37	40	43	46	50	54	58	62	55	60	65	70	74
26.9	31	35	39	43	46	50	54	59	63	67	59	64	69	74	78
33.7	33	36	40	44	48	52	56	61	65	69	61	66	72	77	82
42.4	36	40	45	49	53	56	61	67	72	77	67	73	79	84	90
48.3	38	42	47	51	55	59	64	70	75	80	70	77	82	88	95
60.3	41	45	50	55	59	63	69	75	81	86	76	82	89	96	102
76.1	42	47	52	57	62	67	73	79	85	90	78	86	94	101	107
88.9	44	49	54	59	64	70	76	82	89	94	83	90	98	105	112
101.6	45	50	56	62	66	73	79	85	91	97	85	93	101	109	116
114.3	46	52	57	63	68	76	80	87	93	99	87	95	103	111	118
139.7	49	54	60	66	71	78	84	92	99	105	94	102	110	118	125
168.3	52	58	64	70	76	83	90	98	105	111	101	107	117	126	134
219.1	54	60	67	74	80	87	95	104	112	119	105	114	124	133	142
244.5	55	62	69	76	82	89	98	106	115	122	108	117	127	137	146
273	56	64	71	78	84	94	100	110	118	126	113	120	132	142	151
323.9	58	66	73	80	86	94	104	114	123	132	115	123	135	145	154
355.6	59	67	74	81	88	97	107	116	125	134	116	125	137	147	156

(Continued)

Table 1.18 Economic Thickness of Insulation for Process Pipework and Equipment (These Thicknesses Are Sufficient for Personnel Protection)—cont'd

Hot Face Temperature at Mean Temperature (in °C) (with ambient still air at +20°C)

Thermal Conductivity at Mean Temperature (in W/(m.K))

Outside Diameter of Steel Pipe (in mm)	+100					+200					+300				
	0.02	0.03	0.04	0.05	0.06	0.03	0.04	0.05	0.06	0.07	0.03	0.04	0.05	0.06	0.07
	Thickness of Insulation (in mm):														
406.4	62	69	76	83	90	100	109	118	127	136	118	128	140	150	159
457	63	70	77	84	91	102	111	120	129	138	121	132	144	154	163
508	65	72	79	86	93	105	114	123	132	141	124	134	146	156	165
Over 508 and including flat surfaces	72	78	87	98	105	113	124	133	142	151	127	137	151	161	170

Hot Face Temperature at Mean Temperature (in °C) (with ambient still air at +20°C)

Thermal Conductivity at Mean Temperature (in W/(m-k))

Outside Diameter of Steel Pipe (in mm)	+400					+500					+600			
	0.04	0.05	0.06	0.07	0.08	0.05	0.06	0.07	0.08	0.09	0.07	0.08	0.09	0.1
	Thickness of Insulation (in mm):													
17.2	64	69	74	79	83	70	81	86	91	95	93	98	103	107
21.3	68	73	78	83	88	81	86	91	96	101	98	103	108	118
26.9	73	78	83	89	94	87	92	98	103	107	105	110	115	120
33.7	76	81	87	92	97	89	95	100	106	111	108	114	119	124

Outside Diameter of Steel Pipe (in mm)															
42.4	83	89	96	102	107	99	105	111	117	123	114	120	126	132	137
48.3	87	93	100	106	112	103	109	116	122	128	119	125	132	138	143
60.3	94	101	108	115	121	111	118	125	132	138	128	135	142	149	156
76.1	99	106	114	121	127	117	124	132	139	146	135	142	149	156	163
88.9	103	110	118	126	133	123	130	138	145	152	141	148	156	163	170
101.6	106	114	123	130	136	126	134	142	150	157	145	153	161	169	177
114.3	109	116	126	133	140	129	137	145	153	160	149	157	165	173	181
139.7	116	124	133	141	149	136	146	155	163	171	158	167	176	184	190
168.3	124	132	142	151	159	147	156	165	174	182	170	178	188	196	205
219.1	130	140	151	161	171	156	166	176	186	195	180	190	200	210	220
244.5	135	145	156	165	175	161	171	182	192	201	186	196	206	216	226
273	139	149	160	170	180	166	176	188	198	207	191	202	213	224	235
323.9	142	153	164	174	184	171	181	193	202	212	196	207	218	229	240
355.6	146	157	168	178	188	177	185	197	206	216	201	212	224	235	245
406.4	149	160	171	181	192	181	189	202	213	223	207	218	230	241	252
457	153	165	176	187	195	187	196	209	220	231	213	225	236	250	261
508	155	168	179	191	202	191	200	213	226	237	218	231	244	256	267
Over 508 and including flat surfaces	158	171	182	195	205	194	207	218	230	239	228	240	250	261	270

Hot Face Temperature at Mean Temperature (in °C) (with ambient still air at +20°C)

÷700

Outside Diameter of Steel Pipe (in mm)	Thickness of Insulation (in mm):				
	0.07	0.08	0.09	0.1	0.11
17.2	99	104	109	114	119
21.3	105	110	115	126	125
26.9	113	118	123	128	133

(Continued)

Table 1.18 Economic Thickness of Insulation for Process Pipework and Equipment (These Thicknesses Are Sufficient for Personnel Protection)—cont'd

Outside Diameter of Steel Pipe (in mm)	+700 Hot Face Temperature at Mean Temperature (in °C) (with ambient still air at +20°C)				
	Thickness of Insulation (in mm):				
	0.07	0.08	0.09	0.1	0.11
33.7	116	121	127	132	137
42.4	128	134	140	146	152
48.3	134	140	146	152	158
60.3	144	151	158	165	172
76.1	152	159	166	173	180
88.9	159	166	174	181	189
101.6	164	172	180	187	195
114.3	167	175	183	191	198
139.7	179	187	195	204	211
168.3	191	200	209	218	227
219.1	203	213	223	233	243
244.5	210	220	230	240	250
273	217	227	238	248	258
323.9	223	233	244	254	264
355.6	230	240	251	261	271
406.4	234	245	257	269	279
457	242	254	266	278	289
508	248	260	273	285	296
Over 508 and including flat surfaces	257	271	279	293	304

1.7.5. Simple Correlation for Estimation of Economic Thickness of Thermal Insulation

As noted above, where the sole objective of applying insulation to a portion of plant is to achieve the minimum total cost during a specific period (evaluation period), the appropriate thickness is usually termed as the economic thickness.

The principle is to find at what thickness further expenditure on insulation would not be justified by the additional financial savings on heat to be anticipated during the evaluation period. Although an increase in the amount of insulation applied will raise the initial installed cost, it will reduce the rate of heat loss through the insulation.

Therefore it is necessary to reduce the total cost during the evaluation period. In this work, a simple-to-use correlation, employing basic algebraic equations, which are simpler than current available models involving a large number of parameters that require more complicated and longer computations, is formulated to arrive at the economic thickness of thermal insulation suitable for process piping and equipment.

The correlation is a function of steel pipe diameter and thermal conductivity of insulation for surface temperatures at $100°C$, $300°C$, $500°C$ and $700°C$. A simple interpolation formula generalizes this correlation for a wide range of surface temperatures.

The proposed correlation covers pipeline diameter and surface temperature up to 0.5 m and $700°C$, respectively. The average absolute deviation percent of proposed correlation for estimating the economic thickness of the thermal insulator is 2%, demonstrating the excellent performance of the proposed simple correlation.

The required data to develop this method include the reliable and widely accepted by industry data for various optimum economic thicknesses of thermal insulation as a function of steel pipe and equipment diameter as well as thermal conductivity of insulation and surface temperature.

In this work, optimum economic thickness of thermal insulations are predicted rapidly as a function of steel pipe and equipment diameter and thermal conductivity of insulation by proposing simple correlation. The following methodology has been applied to develop simple correlation.

Equation (1.38) represents the proposed governing equation in which four coefficients are used to correlate the optimum economic thickness of thermal insulations (δ) as a function of steel pipe and equipment diameter (d) for various thermal conductivities of insulations (λ) where the relevant coefficients have been reported in Table 1.19.

Table 1.19 Tuned Coefficients Used in Eqs. (1.39) to (1.42) for Eq. (1.38) to Predict the Optimum Economic Thickness of Insulation (m)

Factors	Surface Temperature = 100°C	Surface Temperature = 300°C	Surface Temperature = 500°C	Surface Temperature = 700°C
A_1	-1.619063838	-1.4673416207	-1.3064287345	-1.5068329304
B_1	$-6.0440641629 \times 10^{-2}$	$-9.4579004057 \times 10^{-3}$	$3.7689223988 \times 10^{-2}$	$1.6368010607 \times 10^{-1}$
C_1	$1.2992412636 \times 10^{-3}$	$-9.0991682769 \times 10^{-4}$	$-5.7653162538 \times 10^{-3}$	$-1.9272634157 \times 10^{-2}$
D_1	$-1.0480516067 \times 10^{-5}$	$2.1036093111 \times 10^{-5}$	$1.5920740612 \times 10^{-4}$	$6.081932874 \times 10^{-4}$
A_2	$-5.675424778 \times 10^{-2}$	$5.6420129717 \times 10^{-3}$	$2.3856855469 \times 10^{-2}$	$2.1494907991 \times 10^{-1}$
B_2	$1.1266206576 \times 10^{-3}$	$-8.1324216389 \times 10^{-3}$	$-1.7666292295 \times 10^{-2}$	$-7.236908659 \times 10^{-2}$
C_2	$-4.5476251244 \times 10^{-5}$	$3.6013086233 \times 10^{-4}$	$1.2527015231 \times 10^{-3}$	$6.2521396767 \times 10^{-3}$
D_2	$5.4011484658 \times 10^{-7}$	$-5.0991959691 \times 10^{-6}$	$-2.9095395202 \times 10^{-5}$	$-1.7702985302 \times 10^{-4}$
A_3	$1.287145175 \times 10^{-3}$	$-1.0914548287 \times 10^{-3}$	$-2.2442814135 \times 10^{-3}$	$-1.2571952312 \times 10^{-2}$
B_3	$-3.1321987972 \times 10^{-5}$	$3.1336503528 \times 10^{-4}$	$7.879778678 \times 10^{-4}$	$3.6872340957 \times 10^{-3}$
C_3	$1.3585299744 \times 10^{-6}$	$-1.3704000608 \times 10^{-5}$	$-5.6411819564 \times 10^{-5}$	$-3.1830356934 \times 10^{-4}$
D_3	$-1.6951529528 \times 10^{-8}$	$1.9239332368 \times 10^{-7}$	$1.3264010427 \times 10^{-6}$	$9.0141402462 \times 10^{-6}$
A_4	$-1.1238180847 \times 10^{-5}$	$1.4969220576 \times 10^{-5}$	$3.3668468005 \times 10^{-5}$	$1.4981731667 \times 10^{-4}$
B_4	$3.574027836 \times 10^{-7}$	$-3.3674152566 \times 10^{-6}$	$-9.9103300792 \times 10^{-6}$	$-4.2562432156 \times 10^{-5}$
C_4	$-1.5140460085 \times 10^{-8}$	$1.4658761606 \times 10^{-7}$	$7.1509244179 \times 10^{-7}$	$3.6706099161 \times 10^{-6}$
D_4	$1.8649575376 \times 10^{-10}$	$-2.0517177043 \times 10^{-9}$	$-1.6973934089 \times 10^{-8}$	$-1.0392224103 \times 10^{-7}$

This correlation has been developed based on more than 1000 reliable data sets covering a wide range of cases and scenarios. As these data are widely accepted by industry and practice engineers in term of accuracy for the purpose of engineering calculations, the formulated simple-to-use expression is justified and applicable to any industrial application.

In brief, Eq. (1.38) represents a new correlation to predict the optimum economic thickness of insulation as a function of steel pipe diameter and the equipment's diameter.

$$\ln(\delta) = \alpha + \frac{\beta}{d} + \frac{\gamma}{d^2} \frac{\theta}{d^3} \tag{1.38}$$

In the above equation, δ refers to the optimum economic thickness of insulation (meters) as a function of steel pipe and/or equipment diameter (meters).

The factors α, β, γ, and θ take into account the thermal conductivity of insulation λ expressed in W/(m.K).

$$\alpha = A_1 + \frac{B_1}{\lambda} + \frac{C_1}{\lambda^2} + \frac{D_1}{\lambda^3} \tag{1.39}$$

$$\beta = A_2 + \frac{B_2}{\lambda} + \frac{C_3}{\lambda^2} + \frac{D_2}{\lambda^3} \tag{1.40}$$

$$\gamma = A_3 + \frac{B_3}{\lambda} + \frac{C_2}{\lambda^2} + \frac{D_3}{\lambda^3} \tag{1.41}$$

$$\theta = A_4 + \frac{B_4}{\lambda} + \frac{C_4}{\lambda^2} + \frac{D_4}{\lambda^3} \tag{1.42}$$

The factors A_1 through to D_4 correspond with the thermal conductivity of insulation λ in W/(m.K) and they are reported in Table 1.19. In this correlation ambient temperature is considered still air at $20°C$.

The following interpolation formula is recommended to extend the proposed method for other surface temperatures:

$$\delta = \delta_1 + (T - T_1) + \frac{\delta_2 - \delta_1}{T_2 - T_1} \tag{1.43}$$

where
A: tuned Coefficient
B: tuned Coefficient
C: tuned Coefficient
D: tuned Coefficient

λ: thermal conductivity of insulation, W/(m.K)

d: outside diameter of steel pipe and equipment, m

δ: economic insulation thickness, m

The correlation proposed in the present work is a simple and unique expression. This is expected to benefit the making of design decisions, which could lead to informed decisions on the economic thickness of thermal insulation suitable for process piping and equipment. In addition, we have selected an exponential function to develop the correlation, because these functions are smooth and well behaved (i.e., smooth and nonoscillatory) equations, which should allow for more accurate predictions.

To clarify these correlation applications, two solved examples are reported below:

Example 1.10

Calculate the optimum thickness of insulation for the following case:

Outside diameter = 0.25 m
Surface temperature = 100°C
Insulation thermal conductivity = 0.04 W/(mK)
Ambient temperature = 20°C

Solution

Use the second column of Table 1.19 as tuned coefficients for Eqs. (1.39–1.42):

$\alpha = -2.48181215$ (from Eq. 1.39)
$\beta = -4.85720938 \times 10^{-2}$ (from Eq. 1.40)
$\gamma = 1.08830906 \times 10^{-3}$ (from Eq. 1.41)
$\theta = -8.85190265 \times 10^{-6}$ (from Eq. 1.42)
$\delta = 7.0000415 \times 10^{-2}$m (from Eq. 1.38)

Example 1.11

Calculate the optimum thickness of insulation for the following case:

Outside diameter = 0.35 m
Surface temperature = 320°C
Insulation thermal conductivity = 0.04 W/(mK)
Ambient temperature = 20°C

Solution

Use the third column of Table 1.19 as tuned coefficients for Eqs. (1.39−1.42):

$\alpha = -1.94379819$ (from Eq. 1.39)

$\beta = -5.2261676 \times 10^{-2}$ (from Eq. 1.40)

$\gamma = 1.183816355 \times 10^{-3}$ (from Eq. 1.41)

$\theta = -9.6569899 \times 10^{-6}$ (from Eq. 1.42)

$\delta = 1.24471 \times 10^{-2}$ m   (from Eq. 1.38)

Use the fourth column of Table 1.19 as tuned coefficients for Eqs. (1.39−1.42):

$\alpha = 1.4799$ (from Eq. 1.39)

$\beta = -8.947755 \times 10^{-2}$ (from Eq. 1.40)

$\gamma = 2.9227943 \times 10^{-3}$ (from Eq. 1.41)

$\theta = -3.237472 \times 10^{-5}$ (from Eq. 1.42)

$\delta = 1.80423 \times 10^{-2}$ m   (from Eq. 1.38)

To calculate the insulation thickness in meters at 320°C, we must interpolate between previous results by Eq. (1.43). The final result will be:

$$\delta = 0.12447 + (320 - 300) \times \left(\frac{0.180423 - 0.124471}{500 - 300} \right) = 0.13m$$

So:

$$\delta = 0.13m$$

Figures 1.14 to 1.17 show the thickness of insulation as a function of outside diameter of pipe and insulation of varying thermal conductivity at 100°C, 300°C, 500°C, and 700°C, respectively. These graphs show good agreement between reported data and the estimated results from the proposed simple correlation. Table 1.20 shows that the average absolute deviation percent of the proposed correlation in comparison with typical data is 2.12%, demonstrating the excellent performance of proposed correlation. The proposed correlation covers pipeline diameter and temperature up to 0.5 meters and 700°C, respectively.

In summary, in this section, a simple correlation was formulated for the estimation of economic thickness of thermal insulation for process piping and equipment. The proposed simple correlation is a function of steel pipe diameter and thermal conductivity of insulation for surface temperatures at 100°C, 300°C, 500°C, and 700°C. A simple interpolation generalizes this correlation for a wide range of surface temperatures. Unlike complex mathematical approaches for estimating thermal insulation thickness, the

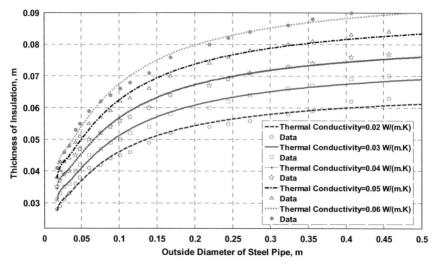

Figure 1.14 Comparison of predicted results from simple correlation with the reported data for surface temperature of 100°C. *(Reprinted with permission © Elsevier, Bahadori, A., and Vuthaluru, H.B (2010b). A Simple Correlation for Estimation of Economic Thickness of Thermal Insulation for Process Piping and Equipment, Applied Thermal Engineering 30:254–259).*

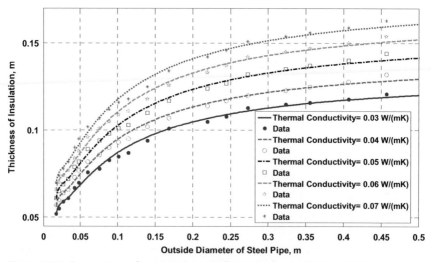

Figure 1.15 Comparison of predicted results from simple correlation with the reported data for surface temperature of 300°C. *(Reprinted with permission © Elsevier, Bahadori, A., and Vuthaluru, H.B (2010b). A Simple Correlation for Estimation of Economic Thickness of Thermal Insulation for Process Piping and Equipment, Applied Thermal Engineering, 30:254–259).*

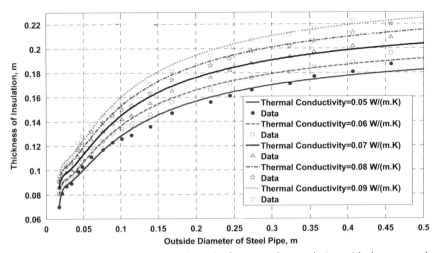

Figure 1.16 Comparison of predicted results from simple correlation with the reported data for surface temperature of 500°C. *(Reprinted with permission © Elsevier, Bahadori, A., and Vuthaluru, H.B (2010b). A Simple Correlation for Estimation of Economic Thickness of Thermal Insulation for Process Piping and Equipment, Applied Thermal Engineering, 30:254–259).*

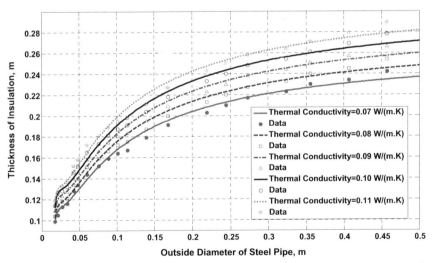

Figure 1.17 Comparison of predicted results from simple correlation with the reported data for surface temperature of 700°C. *(Reprinted with permission © Elsevier, Bahadori, A., and Vuthaluru, H.B (2010b). A Simple Correlation for Estimation of Economic Thickness of Thermal Insulation for Process Piping and Equipment, Applied Thermal Engineering, 30:254–259).*

Table 1.20 Accuracy of Proposed Correlation for Predicting the Economic Thickness of Thermal Insulation in Comparison with Data

Surface Temperature, °C	Diameter, m	Thermal Conductivity, W/(MK)	Calculated Economic Thickness of Thermal Insulation (m)	Economic Thickness of Thermal Insulation (data) (m)	Average Absolute Deviation Percent (AADP)
100	0.0269	0.02	0.03137	0.031	1.19
100	0.1016	0.04	0.0571	0.056	1.96
100	0.457	0.06	0.0894	0.091	1.75
300	0.0213	0.03	0.05667	0.055	3.03
300	0.0889	0.05	0.09875	0.098	0.77
300	0.4064	0.07	0.1576	0.159	0.88
500	0.0337	0.05	0.0909	0.089	2.13
500	0.1143	0.07	0.1513	0.145	4.34
500	0.508	0.09	0.2255	0.237	4.85
700	0.0172	0.07	0.0974	0.099	1.62
700	0.1683	0.09	0.214	0.209	2.39
700	0.3556	0.11	0.2697	0.271	0.48
Average Absolute Deviation Percent (AADP)					**2.12%**

proposed correlation is simple to use, employing basic algebraic equations that can easily and quickly be solved by a spreadsheet. In addition, the estimates are quite accurate, as evidenced by the comparisons with reliable data.

The average absolute deviation percent of the proposed correlation for estimating the economic thickness of the thermal insulators is 2.12%.

1.8. HOT INSULATION THICKNESS

When the base of insulation is conservation of energy, the generally accepted insulation thicknesses given in Table 1.21 for calcium silicate and mineral wool shall be used.

One of the previous sections gives the equations for heat transfer through flat and cylindrical insulation, which relates heat flow to the insulation thickness. If the designer wishes the insulation thickness for energy conservation to be calculated for specific conditions or different insulation materials these equations may be used.

Table 1.21 Hot Insulation Thickness for Heat Conservation: Calcium Silicate and Mineral Wool Slab, Centimeters

Nominal Pipe Size	Temperature °C								
	66 to 149	150 to 204	205 to 260	261 to 315	316 to 371	372 to 427	428 to 482	483 to 538	539 to 593
4 cm & under	3	3	4.3	4.4	4.7	6.4	6.7	6.8	8.7
5 cm	3	3	4.3	4.4	4.7	6.4	6.7	6.8	8.7
8 cm	3	3	4.3	4.4	6	6.4	7.9	8.1	10.3
10 cm	3	3	4.3	5.6	6	7.6	7.9	8.1	10.3
15 cm	3.8	4	5.5	6.7	7.1	10	10.4	10.7	13
20 cm	3.8	4	5.5	6.7	7.1	10	10.4	12.2	13
25 cm	3.8	4	6.5	7.9	9.3	11.4	11.9	13.6	14.5
30 cm	3.8	6	7.7	8.8	9.3	11.4	13.2	13.6	16
30 cm & over	5.8	6	7.7	8.8	10.6	12.7	14.7	15	17.3
Equipment	3.8	4	6.5	7.9	8.4	10	10.4	12.2	13

- **Insulation thickness for personnel protection**

  If economic thickness is used for hot insulation, the thicknesses given in Table 1.18 are sufficient for personnel protection. When the base of insulation is energy conservation, the thicknesses given in Table 1.22 for calcium silicate and Table 1.23 for mineral wool shall be used.

Table 1.22 Hot Insulation Thickness (mm) for Personnel Protection

Pipe Size (outside diameter), mm	To 300°C	Over 300°C
To 48.3	25	38
60.3	25	50
88.9	25	50
114.3	25	50
168.3	25	50
219.1	38	50
273.1	38	50
323.9	38	50
355.6 and over	50	50
Vessels, pumps, exchangers, etc.	38	50

Notes:
1. The above table is based on calcium silicate insulating material, with mean thermal conductivity of approximately 0.061 W/mK @ 93°C.
2. Insulation thicknesses stated here do not include the finishing or weatherproofing materials.

Table 1.23 Hot Insulation Thickness for Personnel Protection

Pipe Size (outside diameter), mm	To 300°C	Over 300°C
To 48.3	22	43
60.3	22	57
88.9	22	57
114.3	22	57
168.3	22	57
219.1	33	57
273.1	33	57
323.9	33	57
355.6 and over	44	57
Vessels, pumps, exchangers, etc.	33	57

Notes:
1. The above table is based on mineral wool insulating material.
2. Insulation thicknesses stated here do not include the finishing or weatherproofing materials.

1.9. COLD INSULATION THICKNESS

The required thickness of insulation for cold services to prevent condensation on the outer surface of insulating material shall be as given in Tables 1.24 and 1.25 for foam glass and polyurethane, respectively, under the stated conditions of ambient temperature, relative humidity, and emissivity of external surface.

If the ambient conditions are different from those stated in Tables 1.24 and 1.25, the thicknesses shall be calculated in accordance with the following section.

1.10. CALCULATION OF THICKNESS OF INSULATING MATERIAL REQUIRED TO PREVENT CONDENSATION

The limiting condition for the formation of condensation on the surface of an insulating material occurs when the surface temperature equals the dew point temperature, θ_d, i.e., when $\theta_2 = \theta_d$.

$$Q = \frac{\theta_1 - \theta_2}{R} = \frac{\theta_1 - \theta_m}{R + R_S} \tag{1.44}$$

Table 1.24 Cold Insulation Thickness for Energy Conservation/Condensation Prevention

Pipe or Equipment, Outside, mm	Normal Insulation Thickness, mm										
	25	38	50	64	75	89	100	114	125	140	150
	Minimum Temperature, °C										
To 33.4 mm	5	-12	-32	-54	-79	-107	-137	-176	-223	-	-
48.3	7	-7	-34	-54	-73	-98	-129	-159	-198	-259	-
60.3	7	-7	-21	-37	-57	-79	-104	-129	-159	-204	-251
88.9	10	-1	-15	-22	-46	-67	-84	-104	-134	-162	-196
114.3	10	-1	-12	-22	-40	-57	-73	-98	-121	-143	-171
168.3	13	4	-9	-21	-32	-48	-65	-82	-98	-118	-137
219.1	13	4	-7	-21	-32	-43	-57	-71	-87	-104	-123
273.1	13	4	-7	-15	-26	-37	-51	-65	-73	-93	-109
323.9	13	4	-4	-15	-23	-34	-48	-59	-73	-87	-104
355.8	16	7	-1	-12	-21	-32	-43	-54	-68	-82	-96
406.4	16	7	-1	-12	-21	-29	-40	-54	-65	-79	-93
457	16	7	-1	-9	-18	-29	-40	-51	-62	-76	-90
508	16	7	-1	-9	-18	-29	-37	-48	-62	-73	-87
610	16	7	-1	-9	-18	-26	-37	-48	-59	-71	-82
711	16	7	-1	-9	-18	-26	-34	-46	-57	-68	-79
813	16	7	-1	-9	-18	-26	-34	-46	-54	-65	-76
914	16	7	-1	-7	-15	-23	-34	-43	-54	-65	-76
1067	16	7	-2	-7	-15	-23	-32	-43	-51	-62	-73
1219	16	7	-2	-7	-15	-23	-32	-43	-51	-62	-71
1524	16	7	-2	-7	-15	-23	-32	-40	-48	-59	-71
1829	16	7	2	-7	-15	-23	-32	-40	-48	-57	-68
2438	16	7	2	-7	-15	-21	-29	-37	-48	-57	-65
3048	16	7	2	-7	-15	-21	-29	-37	-46	-57	-65
3658	16	7	2	-7	-15	-21	-29	-37	-46	-54	-65
Flat	16	10	2	-7	-12	-21	-26	-34	-43	-51	-59

Also,

$$Q = \frac{\theta_2 - \theta_m}{R_S} \tag{1.45}$$

Therefore,

$$\frac{\theta_1 - \theta_2}{R} = \frac{\theta_2 - \theta_m}{R_S} \tag{1.46}$$

Table 1.25 Thickness of Insulation for Cold Services

Normal Operating Temperature, °C	+21 to -18°C	-19 to -29°C	-30 to -40°C	-40 to -51°C	-52 to -73°C	-74 to -101°C	-102 to -129°C
Outside Diameter of Pipe, mm	Insulation Thickness, mm						
12.5	25	25	40	40	40	40	40
19	25	25	40	40	40	50	50
25.4	25	25	40	40	40	50	50
38	25	25	40	40	40	50	65
51	25	25	40	40	40	50	65
76	25	40	40	40	50	65	65
102	25	40	40	50	50	65	65
152	40	40	40	50	65	75	75
203	40	40	40	50	65	75	75
254	40	40	40	50	65	75	75
305	40	40	40	50	65	75	90
356	40	40	50	50	65	75	90
406	40	40	50	65	65	75	90
457	40	40	50	65	65	75	90
508	40	40	50	65	65	75	90
610	40	40	50	65	65	75	90
762	40	50	65	65	65	75	90
914	40	50	65	65	65	75	90
Over 914	50	65	65	75	75	90	100

$$\frac{\theta_1 - \theta_d}{R} = \frac{\theta_d - \theta_m}{R_S} \tag{1.47}$$

Because

$$\theta_m > \theta_d > \theta_1$$

$$\frac{\theta_d - \theta_1}{R} = \frac{\theta_m - \theta_1}{R_S} \quad \text{or} \quad \frac{R}{R_s} = \frac{\theta_d - \theta_1}{\theta_m - \theta_d} \tag{1.48}$$

$$R = 10^{-3}\left(\frac{d_o}{2\lambda}\right)\ln\left(\frac{d_1}{d_o}\right) \text{ and } R_s = \frac{d_o}{hd_1} \tag{1.49}$$

$$\frac{R}{R_S} = 10^{-3}\frac{hd_1}{2\lambda}\ln\left(\frac{d_1}{d_o}\right) \tag{1.50}$$

$$R = 10^{-3}\frac{L}{\lambda} \tag{1.51}$$

$$R_S = \frac{1}{h} \tag{1.52}$$

$$\frac{R}{R_S} = 10^{-3}\frac{h}{\lambda}L \tag{1.53}$$

$$\frac{R}{R_S} = 10^{-3}\frac{h}{\lambda}\chi \tag{1.54}$$

$$10^{-3}\frac{h}{\lambda}\chi = \frac{\theta_d - \theta_1}{\theta_m - \theta_d} \tag{1.55}$$

$$\chi = 10^{-3}\frac{\lambda}{h}\left(\frac{\theta_2 - \theta_1}{\theta_m - \theta_d}\right) \tag{1.56}$$

The symbols and units:
Q = rate of heat loss through the insulating material per unit area of hot surface per second in W/m^2
θ_1 = temperature of hot surface in °C

θ_2 = temperature of the exterior (surface temperature of the insulating material) in °C

θ_m = temperature of the ambient still air in °C

θ_d = dew point temperature in °C

R = thermal resistance of the insulation per square meter of hot surface in m^2K/W

R_s = thermal resistance of the insulation surface to air boundary in m^2K/W

λ = thermal conductivity of insulating material in W/mK

d_o = outside diameter of pipe or tube in mm

d_1 = outside diameter of the layer of insulating material in contact with the hot surface in mm

h = heat transfer surface coefficient per square meter of external surface of insulation in W/m^2K

L = overall thickness of insulation in mm

χ = provisional thickness adopted for calculation purposes in mm

1.10.1. Examples

Example 1.12

- **Calculate the minimum thickness to prevent condensation**
  A pipe of 60.3 mm outside diameter containing fluid at 0°C is insulated with material of thermal conductivity 0.039 W/(m.K), to prevent condensation when in air at +22°C and 85% relative humidity. Calculate the minimum insulation thickness needed for finishes of high emissivity.
  From Table 1.26 the dew point temperature is

  $$\theta_d = +19.4°C$$

Then consider the equation

$$x = \left(\frac{19.4 - 0}{22 - 19.4}\right)\left(\frac{0.039 \times 10^3}{h}\right) = \frac{291}{h}$$

An approximate value for the surface coefficient is now taken for an initial assessment of the thickness. Assume $h = 7.0$, then

$$x = \frac{291}{7.0} = 41.5 \; mm$$

Relative Humidity %

Ambient Temperature °C	50	55	60	65	70	75	80	85	90	95
					Dew Point Temperature °C					
-20	-27	-26	-25.2	-24.5	-23.7	-22.9	-22.3	-21.7	-21.1	-20.5
-15	-22.3	-21.3	-20.4	-19.6	-18.8	-18	-17.5	-16.7	-16.2	-15.6
-10	-17.6	-16.6	-15.7	-14.7	-13.9	-13.2	-12.5	-11.8	-11.2	-10.6
-8	-15.7	-14.7	-13.7	-12.8	-12	-11.3	-10.5	-9.8	-9.2	-8.6
-6	-13.9	-12.8	-11.8	-10.9	-10.1	-9.9	-8.6	-7.9	-7.2	-6.6
-4	-12	-10.9	-9.9	-9	-8.1	-7.4	-6.6	-5.9	-5.3	-4.6
-2	-10.1	-9	-8	-7.1	-6.2	-5.4	-4.6	-3.9	-3.3	-2.6
0	-8.1	-7.1	-6	-5.1	-4.2	-3.4	-2.7	-1.9	-1.3	-0.6
2	-6.5	-5.4	-4.4	-3.4	-2.6	-1.7	-1	-0.2	0.5	1.3
4	-4.9	-3.8	-2.7	-1.5	-0.9	0	0.9	1.7	2.5	3.3
6	-3.2	-2.1	-1	0.1	0.9	2	2.8	3.7	4.5	5.3
8	-1.5	-0.5	0.7	1.8	2.9	3.9	4.8	5.7	6.5	7.3
10	0.1	1.4	2.6	3.7	4.8	5.8	6.7	7.6	8.4	9.2
12	1.9	3.4	4.5	5.7	6.7	7.7	8.7	9.6	10.4	11.2
14	3.7	5.1	6.4	7.5	8.6	9.7	10.5	11.5	12.4	13.2
16	5.6	6.9	8.2	9.4	10.5	11.6	12.5	13.4	14.3	15.2
18	7.4	8.8	10.1	11.3	12.4	13.5	14.5	15.5	16.3	17.2
20	9.2	10.7	12	13.2	14.4	15.4	16.4	17.4	18.3	19.2
22	11	12.6	13.9	15.1	16.3	17.4	18.4	19.4	20.3	21.2
24	12.9	14.4	15.8	17	18.2	19.3	20.3	21.3	22.2	23.1
26	14.8	16.2	17.6	18.9	20.1	21.2	22.3	23.3	24.2	25.1
28	16.6	18.1	19.5	20.8	22	23.1	24.2	25.3	26.2	27.1
30	18.4	19.9	21.4	22.7	23.9	25.1	26.2	27.2	28.1	29.1
35	23	24.5	26	27.4	28.7	29.9	31	32.1	33.1	34.1
40	27.6	29.3	30.7	32.2	33.5	34.7	35.9	37	38	39
45	32.2	33.8	35.4	37	38.2	39.5	40.7	42	42.9	44
50	36.7	38.5	40.1	41.6	43	44.7	45.5	46.8	47.9	49

Note: The above table refers to the outer surface of the insulation.

From Eq. (1.19),

$$X = 0.5(d_o + 2L)\ln((1 + 2L)/d_o)$$

so $L = 30$ mm here

An approximate outer insulation diameter can now be determined, i.e., 120 mm, from which a better surface coefficient is selected. Thus a second approximation for h is 7.7:

$$x = \frac{291}{7.7} = 37.8 \ mm$$

From Eq. (1.19),

$$\chi = 0.5(d_o + 2L)\ln((1 + 2L)/d_o)$$

We have $L = 28$ mm (approximately).

Further approximations to the surface coefficient do not affect this result significantly and therefore the required thickness is not less than 28 mm.

Example 1.13

We want to check the outer surface temperature of a given thickness of insulation for comparison with the dew point temperature using the heat loss equations.

The user is frequently faced with a limited choice of insulation thicknesses. It is necessary to compare the outer surface temperature of a selected thickness with the dew point temperature, to determine whether or not condensation is likely to occur.

A pipe of 168.3 mm outside diameter carrying fluid at $0°C$ and located in air at $+26°C$ and 85% relative humidity is to be insulated with material of thermal conductivity 0.035 W/(m.K). Calculate the surface temperatures of insulation thicknesses of 30 mm and 40 mm for finishes of high emissivity and compare them with the dew point temperature.

From the heat transfer equation:

$$Q = \frac{\theta_1 - \theta_m}{R + R_s} \quad R = 10^{-3}\frac{d_o}{2\lambda}\ln\left(\frac{d_1}{d_o}\right) \quad R_s = \frac{d_o}{hd_1}$$

Also,

$$\theta_2 = \frac{qd_o}{hd_1} + \theta_m$$

By substitution:

$$\theta_2 = \frac{\theta_1 - \theta_m}{10^{-3}\frac{hd_1}{2\lambda}\ln\left(\frac{d_1}{d_o}\right) + 1} + \theta_m$$

The outer surface temperature of a 30 mm and a 40 mm insulation thickness is calculated as follows:

For 30 mm insulation thickness

The insulation outside diameter, $d_1 = 168.3 + 60 = 228.3$ mm. From Table 1.27 an approximate value of 7.2 is taken for h.

Therefore

$$\theta_2 = \frac{0.26}{10^{-3} \frac{7.2 \times 228.3}{2 \times 0.035} \ln\left(\frac{228.3}{168.3}\right) + 1} + 26 = +22.8 C$$

$\theta_m - \theta_2 = 26°C - 22.8°C = +3.2°C$. Thus, a second, closer approximation of 7.5 is taken for h, from which $\theta_2 = +22.9°C$.

From Table 1.26, the dew point $\theta_d = +23.2°C$ and therefore condensation would occur.

For 40 mm insulation thickness

The insulation outside diameter $d_1 = 168.3 + 80 = 248.3$ mm. From Table 1.27 an approximate value of 7.2 is taken for h, therefore

$$\theta_2 = \frac{0.26}{10^{-3} \frac{7.2 \times 248.3}{2 \times 0.035} \ln\left(\frac{248.3}{168.3}\right) + 1} + 26 = +23.6°C$$

$\theta_m - \theta_2 = 26°C - 23.6°C = +2.4°C$. Thus, a closer approximation of 7.4 is taken for h, from which $\theta_2 = +23.7°C$. The dew point $\theta_d = +23.3°C$ and therefore condensation is unlikely to occur.

Table 1.27 Variation of Outer Surface Coefficient with Temperature Difference Between Surface and Air for Various Outer Dimensions of Insulation

Outer Diameter Insulation (mm)	High Emissivity Surface				Low Emissivity Surface			
	Temperature Difference (K)							
	1	2	5	10	1	2	5	10
	Outer Surface Coefficient, k (in W/(m².K))							
40	8	8.4	9.1	9.7	3.4	3.9	4.7	5.4
60	7.6	8	8.7	9.3	3.1	3.5	4.2	4.9
100	7.3	7.7	8.3	8.8	2.7	3.1	3.8	4.4
200	7	7.4	7.9	8.4	2.4	2.8	3.4	4
Vertical flat surface	6.5	7	7.5	8	2	2.4	3	3.8

Notes: The above table refers to the outer surface of the insulation.
The values given in the above are typical for heat gain situations under conditions of natural convection.

1.11. HEAT-TRACED PIPE: INSULATION THICKNESS

Piping which is heat traced shall be insulated with oversize pipe insulation to include the tracer lines. Valves, flanges, unions, and tracer line loops shall not be insulated, unless specified on the piping spool drawings. The thickness of insulation shall be in accordance with Table 1.28.

The thicknesses given in the tables do not include the cleading or finishing, the insulation of which is neglected.

Where proved to be more economical, the insulation should consist of two or more layers of dissimilar approved materials provided their respective service temperature limits are appropriate for the duty. Multilayer insulation shall have the inner layers banded or taped to the pipe or equipment.

1.12. RELATIONSHIP OF SYSTEM REQUIREMENTS TO THE DESIGN OF INSULATION SYSTEMS AND TO THE PROPERTIES OF MATERIALS USED

For any specific set of installation requirements, the properties of a material determine its suitability. If there were only one, or a limited number of sets of installation requirements, selection of a material would be simple, and the need for all the various types of insulations and weather barriers reduced. However, this is not the case. Each installation must be considered, and its requirements evaluated to allow the selection of the best-suited material (or materials) for the individual installation under consideration.

Not only do the installation requirements change with the individual case, but the relative importance of the requirements also vary. Some of the possible variations of the properties required of individual materials, as the installation requirements change, will be treated in the short discussion that follows.

In the transportation phase, weight is always a factor. Although light density material is desirable, lightweight insulation is of greater importance in aircraft than in ships, trucks, or railroad cars. Conversely, when insulation is used in chemical and petroleum processing plants, where fire protection is required, a higher-density insulation is essential to obtain the low diffusivity necessary for fire protection.

Table 1.28 Heat-Traced Pipe: Insulation Thickness (mm)

Pipe Outside Diameter, mm	9.5 mm Tubing 6 Electric Cable		12.5 mm Tubing		19 mm Tubing & 21.3 mm Pipe		25 mm Tubing & 26.5 mm Pipe		33.4 mm Pipe	
	One Tracer	Two Tracer	One Tracer	Two Tracer	One Tracer	Two Tracer	One Tracer	Two Tracer	One Tracer	Two Tracer
21.3	25	31.75	31.75	31.75	50	50	–	–	–	–
26.7	31.75	31.75	31.75	38	50	50	–	–	–	–
33.4	38	38	38	50	50	63.5	–	–	–	–
48.3	50	50	63.5	63.5	63.5	75	–	–	–	–
60.3	63.5	63.5	75	75	75	75	75	90	90	101.5
73	75	75	75	90	90	90	90	101.5	101.5	127
88.9	90	90	90	101.5	101.5	101.5	101.5	127	127	127
114.3	127	127	127	127	127	127	127	152	152	152
168.3	178	178	178	178	178	178	203	203	203	203
219.1	229	229	229	229	229	229	255	255	255	255
273.1	279	279	279	279	279	279	305	305	356	356
323.9	356	356	356	356	356	356	356	356	380	380
355.9	380	380	380	380	380	380	380	406	406	406
406.4	432	432	432	432	432	432	432	457	457	457
457	483	483	483	483	483	483	483	508	508	508
508	533	533	533	533	533	533	533	559	559	559
560	584	584	584	584	584	584	584	610	610	610
610	635	635	635	635	635	635	635	660	660	660

Notes:
1. Double-traced lines calculated on 90° spacing.
2. 25 mm tubing refers to 1" tubing.

Where insulation may be contaminated by toxic or highly combustible chemicals, the most important single requirement is that it is completely nonabsorbent.

In a process where cyclic application is required, the single most important requirement is that the insulation be such that it can be quickly and economically removed and reapplied.

Other cyclic operations may require that the insulation have low mass and specific heat so that the operating temperature can be changed without an excessive amount of heat and time being lost in "heating up" the insulation.

Just the opposite property is desirable for cold storage and storage of materials that must be held at relatively constant temperatures. In these installations, in case of power failure, the insulation's retention of the existing temperature is of utmost importance.

Even on the same installation, the specific end use of insulation may dictate its properties. A high-temperature pipe may require very rigid, strong insulation to resist the mechanical stresses imposed upon it, but the insulation used in the expansion joint should be soft, fluffy, and resilient to cushion the movements of the adjacent rigid insulation.

On low-temperature installations, the single most important property is resistance to the passage of water vapor, or ability to construct the insulation system to be as near vaporproof as possible.

Economics always enters into the selection process. If a process is critical, the most important single consideration may be reliability. If conservation of heat or power is the deciding factor, the savings per year as compared to the installed cost is the most important factor. Again, and almost the opposite, when insulation is used for a temporary function such as holding the heat in while a lining is being heat cured, then the lowest possible installed cost would be decisive.

Thus, because of conflicting requirements, there can be no "all pur-pose" insulation. Nor is there a "perfect" insulation for each set of re-quirements. It is essential that the engineer evaluates the installation and determine which requirements must be fulfilled, and which are of lesser importance.

A list of installation requirements as related to properties of materials is presented in outline form in Tables 1.29 and 1.30 to assist the engineer in his calculations.

Table 1.29 Installation Requirements in the Fabrication Phase

Requirements	Insulation Properties	Adhesive or Bonding Cement Properties
Fabrication or operation To be able to be cut or ground to desired shape.	Rigid insulations Dimensional properties; size and shapes, straightness and smoothness of surfaces, squareness and trueness.	Preuse properties Shelf life freeze—thaw resistance; mixing or storing time.
Able to maintain true surface and surface suitable for application and retention together with adhesive or cements.	Cutting properties Hardness delamination characteristics, ease of cutting. Smoothness and trueness of cut surface with adhesive or cement.	Use properties Troweling or brushing characteristics, surface welting and bonding characteristics, wet adhesive strength, compatibility with surface, drying time, curing time, wet or dry shrinkage, toxicity (solvent, fumes, etc.) for life.
Pieces able to be handled without excessive breakage, dusting, or wear.	Handling properties Tensile strength, flexural strength, compressive strength.	Dried properties Adhesive strength, shear strength, temperature resistance (maximum and minimum), resistance to moisture (vapor and liquid), resistance to solvent acids, or caustics, flexibility and elasticity.
To be able to be assembled and bonded together accurately and tightly.	Assembly properties Surface suitability to spreading or brushing of cement of adhesive. Surface comparability to adhesive or cement, resistance to surface.	

(Continued)

Table 1.29 Installation Requirements in the Fabrication Phase—cont'd

Requirements	Insulation Properties	Adhesive or Bonding Cement Properties
	Shear on drying of cement or adhesive, original track of surface to bond to surface, absorbency.	
To be able to be handled in a short time after bonding assembly.	Flexible materials Cutting characteristics, tensile strength, bonding characteristics.	
To have relatively short setting, drying, or curing time so that fabricate and assembled pieces can be used in a short time.	Flexible materials Cutting characteristics, surface layer delamination.	

1.13. GENERAL APPLICATION OF THERMAL INSULATION

1.13.1. Conditions in Which Thermal Insulation Shall Be Applied

Thermal insulation shall be applied to the piping, vessels, and equipment with the following range of applications and conditions.

Piping, vessels, and equipment operating over 5°C shall not be insulated unless required for heat control or acoustical control. Piping, vessels, and equipment operating above 55°C shall be insulated if necessary for personnel protection.

All piping, vessels, and equipment operating below 5°C shall be insulated to control the fluid temperature, for the conservation of refrigeration, or to control surface condensation. Operating temperatures of this specification refer to the internal fluid or material temperature of the process system.

Piping, vessels, and equipment which are to be insulated shall be indicated on flow diagrams, pipeline lists, piping and spool drawings, and on the equipment. These drawings shall also indicate the insulation classification

Table 1.30 Installation Requirements in Application Phase

Requirements	Insulation Properties	Accessory Properties
Cutting and fitting.	Rigid insulations.	Adhesive mastics, bonding cements, or scalers (where required).
To be able to field cut, fit, and place into position efficiently and accurately.	Dimensional properties Size and shape, straightness and smoothness of surfaces. Squareness and trueness, within dimensional tolerances.	Brushing characteristics, troweling characteristics, wet adhesion, surface wetting, gap filling, bridging, sizing and sealing, drying time, curing time, shrinkage, compatibility with insulation surface, solvent toxicity.
	Cutting properties Ease of cutting, dusting, resistance to abrasion and cracking.	Pins and clips, tensile strength, pull resistance, clamping strength.
	Handling properties Tensile strength, resistance to breakage due to load or impact, flexural strength. Surface properties Smoothness, trueness, compatibility with cements or sealers which may be required.	Adhesives, wet and dry adhesive strength, shear strength.
	Nonrigid insulation Dimensional properties Trueness in width, length, and thickness.	
	Cutting properties Cutting characteristics, dusting.	

(Continued)

Table 1.30 Installation Requirements in Application Phase—cont'd

Requirements	Insulation Properties	Accessory Properties
Securement To be able to be secured in position by wires, bands, pins and clips, or adhesives, as required by the installation.	Strength properties Tear strength, resilience, flexibility. Rigid insulation Compressive strength, hardness, shear resistance tensile strength.	Tape: Tensile strength, flexibility, elasticity, peel back adhesive strength, shear strength.
Trowel application To be able to mix quickly and efficiently trowel into position insulating cements or mastics. To obtain good attachment to surface and obtain even thickness and smoothness of surface required by installation.	Mixing properties Mixing time, consistency. Application properties Wet adhesion, build, wet coverage, dry coverage, shrinkage wet to dry, trowel ability, corrosiveness to substrate metal, drying time.	
Sprayed application To be able to be applied to insulation on a surface by spray equipment in a fast, efficient manner, with good adhesion to surface. To be able to obtain a reasonably smooth, even surface without sags, hills, and valleys, suitable for weather barrier or coverings. To be able to obtain density and related thermal efficiency.	During spraying Wet adhesion, build rebound, overspray, compaction, corrosion to substrate metal, drying time, shrinkage wet to dry, toxic dust or fibers. Temperature limits Ambient air. Dried properties Compressive strength, dimensional tolerance, surface evenness, and smoothness.	

To be able to obtain desired strengths for permanent installation.
To be able to make spray application with least amount of cost involved in cleanup or cleaning of spatter or overspray.

Poured applications
To be efficient, pour or ram fibrous, granular, or powder insulation into cavities, so that all voids are completely filled.

Note:
As poured insulation requires some receptacle, requirements in reference to weather vapor barrier or covering will not apply to this type of insulation system.

Sprayed foam insulations
During spraying
Wet adhesion, spray ratio, build, reaction time, curing time, expansion ratio, density, sag, toxic fumes, corrosiveness to substrate metal.
Limits of application
Condition of surface, temperature of surface, temperature of air, humidity of air, wind, flash point, fire point.
After application
Dry adhesion, compressive strength, density, tensile strength, dimensional stability, surface evenness and smoothness, bridging, elongation, flexibility.
Compaction, compressibility, resiliency, dusting, density, toxicity of dust.

(Continued)

Table 1.30 Installation Requirements in Application Phase—cont'd

Requirements	Insulation Properties	Accessory Properties

Application of weather barrier, vapor barrier, coverings, jackets
 	
To be able to cut, fit, form, seal, and apply jackets over flat surface insulation, vessel insulation, pipe and fitting insulation.
		Jackets
		  Cutting characteristics, tear resistance, forming characteristics, handleability.
		Jacket accessories
		Lap adhesive
		  Adhesion, shear strength, shrinkage, flexibility.
		Circumferential closures
		  Type and means of attachment, sealing compound flexibility.
		Longitudinal laps
		  Method of securing and sealing.

Mastic weather barriers To be able to brush, trowel, palm, or spray smooth, even water and weather barrier mastic or interior coating to obtain a smooth dried, film of even texture & thickness over entire surface.	All insulations Surface compatibility, surface suitability, smoothness, dryness, dust free, clean.	Mastics or coatings during application Mixing or stirring time, consistency, brushing characteristics, troweling characteristics, palming characteristics, wet adhesion, wet covering capacity, surface wetting, gap filling, bridging, sizing, and sealing, shrinkage wet to dry, drying time, curing time, solvent toxicity.
Physical To withstand dead loads, wear, impact, and mechanical damage, forces of	All insulation Compressive strength, flexural strength, shear strength, tensile strength,	Weather, vapor barrier, covering Impact resistance, indentation resistance,

expansion and contraction, and vibration.	flexibility, compaction resistance, abrasion resistance, vibration resistance.	tear resistance, abrasion resistance, flexure, elongation, tensile strength, adhesion, elasticity, shear strength.
Chemical Compatibility with metal to which it is applied, resistance to atmospheric and spillage contamination.	Alkalinity, acidity, inhibitors to Corrosion, acid resistance, caustic resistance, solvent resistance.	Acid resistance, caustic resistance, solvent resistance
Moisture Resistance to moisture in liquid or vapor form or both.	Absorptivity, adsorptivity, hygroscopicity, capillarity, vapor permeability	Absorptivity, adsorptivity, hygroscopicity, capillarity, vapor permeability (these properties relate to a system, rather than to a material, and must include the joints and seals of jackets or films). Dissimilar metals in jacket and substrate, producing a potential difference which causes galvanic current to flow. Water resistance, solar radiation resistance, temperature stability, impact resistance, contamination resistance, wind resistance, mold resistance, freeze–thaw resistance.
Weather resistance To resist solar radiation, rain, sleet, snow, wind, and atmospheric contamination, maximum and minimum temperature.	Note: If property is weather-protected, this function is one of the weather barrier.	Water resistance, solar radiation resistance, temperature stability, impact resistance, contamination resistance, wind resistance, mold resistance, freeze–thaw resistance.

(Continued)

Table 1.30 Installation Requirements in Application Phase—cont'd

Requirements	Insulation Properties	Accessory Properties

Safety
  To maintain standard of fire safety; personal protection from bombs (as controlled by insulation thickness vs. surface remittance).
	Hazard properties
	  Absorptivity (of combustible toxic liquids), absorptivity (of combustible or toxic vapor).
	Combustibility
	  Flash point, flame spread, fire point, self-ignition point, fuel contribution, smoke density, smoke toxicity.
		Note:
		  Deterioration of a weather barrier system caused by weather is a deterioration of the properties of the system.
		Combustibility
		  Flash point, flame spread, fire point, self-ignition point, fuel contribution, smoke density, smoke toxicity.

To protect from fire exposure	Protection properties Noncombustible, conductivity, diffusivity, density, specific heat, fire resistivity.	Resistance to flame

Thermal properties
  To resist service temperature.
	Maximum service temperature
	  Coefficient of expansion, shrinkage
		Maximum surface temperature
		  Coefficient of expansion (metals).

Miscellaneous To resist rodents, insects, termites, etc. (particularly in building and cold storage insulation).	Vermin resistance Mold resistance	Vermin resistance Mold resistance

which appears in the following. The specification of insulation such as thickness and material shall be shown on drawings and also by a table.

The insulation classification which may appear on flow diagrams and/or drawings shall be defined as following:

- **Ih**: for heat control for operating temperature above 5°C
- **ST**: for steam traced and insulated
- **STT**: for steam traced with heat transfer cement and insulated
- **STS**: for steam traced with spacers and insulated
- **ET**: for electric traced and insulated
- **ETT**: for electric traced with heat transfer cement and insulated
- **IS**: for personnel protection insulation
- **Iac**: for acoustic insulation
- **Ias**: for cycling or dual temperature service where temperatures fluctuate from 15°C to 320°C.
- **IC**: to conserve refrigeration, surface condensation, and control fluid temperatures for operating temperatures 5°C and below.
- **Fireproof**: for all services requiring fireproof-type insulation

1.13.2. Personnel Protection

To avoid the possibility of accident caused by the touching of either hot or cold surfaces by personnel, who may be unprepared for a sudden thermal shock, it is recommended that some reasonable protection shall be given.

Where operating temperature of piping, vessels, and equipment is above 55°C and insulation is not required for heat conservation, those portions of equipment or piping which present a hazard to operating personnel where guard rails or screens are not provided shall be insulated for personnel protection.

These insulated surfaces shall extend approximately 2 meters above the operating levels and 1 meter from the edge of platforms, walkways, and ladders. The extent of personnel protection insulation shall be clearly indicated on piping drawings. Valves, flanges, and unions shall not be insulated for personnel protection.

Where insulation is not permitted for economic purposes or where dissipation of heat is desirable or there is the advantage of noninsulated surfaces remaining free and visible for inspection, guards or shields shall be installed for personnel protection in lieu of insulation.

The extent of personnel protection, acoustical control, or fireproofing required on piping, vessels, and equipment shall be indicated on drawings.

Contact with very cold surfaces will also result in thermal shock or skin damage and personnel protection may be required for temperatures of approximately –10°C and below.

1.14. CHARACTERISTICS OF INSULATING AND ACCESSORY MATERIALS

1.14.1. Consideration of Characteristics of Insulating Materials

- **Health hazards**

  Every effort should be made to avoid the use of asbestos or asbestos-containing materials in the insulation system. In those cases where the use of such material is unavoidable due to the lack of a technically acceptable substitute or any other means, the prior agreement of the user to their use shall be obtained. Safety regulations on the use and application of asbestos or asbestos-containing materials shall be observed.

  Reference shall be made to the appropriate instructions from manufacturers for health and safety. Certain finishing cements are strongly alkaline when wet and may cause skin irritation. Gloves shall be worn when handling these materials.

  Chemical fumes from the components of some insulating materials, e.g., foamed ureaformaldehyde, phenolformaldehyde resins, isocyanates, and polyurethanes can be toxic or cause bronchial irritation, sometimes with persistent sensitization effects. When materials of these types are sprayed, new hazards can arise because nonvolatile components are formed into respirable aerosols. Suitable respiratory protective equipment shall be provided and, particularly in confined spaces, air-fed hoods or respirators are required for many materials, especially isocyanates and epoxy-resin compounds.

  Contact with certain resinous coatings, adhesives, and epoxy-resin components can cause dermatitis, which may occur after brief contact, or long exposure leading to allergy. Associated solvents may cause damage to the eyes and skin irritation. Goggles and gloves shall be worn when necessary. Barrier cream and the provision of adequate washing facilities will be of assistance. Special respiratory hazards can exist when these materials are sprayed.

  Many fibrous insulating materials can cause skin irritation. Normally this is associated with the coarser fibers, so that the same materials of smaller fiber diameter may not cause this type of irritation, although they may then have other undesirable effects. (See BS 4275.) The use of barrier creams and the provision of adequate washing facilities will help to reduce the incidence of skin irritation, but tight clothing, e.g., around the neck and at the wrists, may trap fibers and thus increase local irritation.

Respirable dust particles from insulating material containing asbestos, silica, etc., may enter the bronchial passages and cause persistent lung disease. Because of this risk to health where these materials are manipulated, e.g., mixed, handled, or removed, adequate protective equipment shall be provided and used. Insulating materials maintained in good order in quiescent use are likely to present only a low degree of risk.

The risks to health associated with the removal of thermal insulation containing asbestos are particularly well known, especially when crocidolite (blue asbestos) is encountered.

Before any stripping work is carried out it is necessary to determine what type of insulating material is involved. Although the presence of crocidolite is sometimes discernible by its rich lavender-blue color, it is not correct to rely on this test alone. It is always necessary to obtain positive analysis by a responsible laboratory.

Where the asbestos regulations, if any, apply and where crocidolite is identified, written notice has to be given to the inspector for the district at least 28 days prior to commencement of the work, or such shorter notice as the inspector may agree to accept.

- **Substitute material**

  The asbestos or any product containing asbestos materials shall not be applied by means of a spray process. When required for use of an existing plant or installation, the selection of material shall be confirmed with the local management in order that account may be taken of preferences for particular materials.

- **Thermal conductivity**

  A low thermal conductivity is desirable to achieve a maximum resistance to heat transfer. Therefore, for any given heat loss, a material of low thermal conductivity will be thinner than an alternative material of high conductivity. This is of particular advantage for pipes because thinner layers of insulations reduce the surface area emitting heat and also reduce the outer surface that will require protection.

  The thermal conductivity of most insulating materials varies with temperature and with bulk density, so that both these factors shall be considered. It is the normal practice of manufacturers to provide tables or graphs showing thermal conductivities for each of their standard products at a range of hot-face temperatures together with the relevant cold-face temperatures used for the tests. The information shall also include the bulk density for each material tested.

Although, under service conditions for insulating materials at elevated temperatures, the outer exposed surface may reach a temperature above that quoted for the test, the figures for any calculation of heat flow can normally be based on those published by the manufacturer. Provided that the outer surface temperature does not exceed about 50°C, the resultant error is likely to be less than 5%.

For very high cold-face temperatures, which could apply, for example, with the inner layer of composite insulation or with the inner layer of composite insulation or with materials exposed to high ambient temperatures on the exposed surface, special calculations are required (see BS 5970 for interface temperatures and BS 5422).

These may involve reference to mean temperatures, i.e., the arithmetic mean between the hot-face and cold-face temperatures of the test; to avoid confusion, either the corresponding hot-face or cold-face temperature shall also be quoted.

- **Physical forms**

Insulation is usually supplied in one of the following forms:

- Rigid boards, blocks, sheets, and preformed shapes such as pipe covering, curved segments, etc.
- Flexible boards, sheets
- Blankets
- Plastic composition
- Cement
- Loosefill
- Mastic.
- Sprayed or metallic, e.g., crimped foil

    Each of these will be made up of granular, fibrous, flake cellular, or reflective material (or a combination of these).

- **Forms of insulation**

Forms of insulations for application on piping and equipment may be as follows:

- For pipework on hot service, preformed rigid sections are preferred.
- On tanks and vessels, preformed slabs or mattresses may be used on hot service, but on cold duties, slabs only should normally be used. When

slabs are used on vessels subject to thermal movement the fixing ar-rangements shall allow for this.

- Loosefill, compatible with the insulation, may be used for filling interstices or for double skin removable cover and boxed items of equipment.
- Foamed-in-situ materials may be used where these can be shown to be economically advantageous. Samples of foamed materials should be taken during application for fire resistance (where required) and quality control purposes.
- Sprayed form shall not be used without the prior agreement of the company.
- Consideration may be given to the use of pre-insulated pipework where this is economically justified.
- **Bulk density**

   The bulk density for most thermal insulating materials falls within the range 10 kg/m^3 to 320 kg/m^3 and their effectiveness depends essentially on the large numbers of minute air or gas cells that they contain and that restrict transfer of heat by convection and radiation; at the same time, the limitation in areas of solid thermal bridges forms a good barrier to the passage of conducted heat.

   Low bulk density is normally associated with low thermal conductivity in the low and medium temperature ranges, but it may also result in low compressive strength, in which case the material will not be suitable for load bearing purposes. As a general rule, increase in bulk density is associated with reduction in the size of the contained air or gas cells and this tends to maintain effectiveness of insulation at the higher service temperature.

   Note: Some physical properties of thermal insulating materials, in particular thermal conductivity and strength, are dependent on the direction in which they are measured.

1.14.2. Suitability for Service Temperature

The temperature at which the material is used shall be within the range for which it will provide satisfactory long-term service under condition of normal usage.

For materials to be used at temperatures below about 10°C, attention shall be paid to the relevant limiting minimum as well as to the limiting maximum temperatures, and to the effects of possible excessive shrinkage, embrittlement, and porosity, in addition to resistance under conditions of occasional heating for defrosting purposes.

For materials to be used at elevated temperatures it will be necessary to consider the many factors that may result in deterioration under conditions of service.

These include linear shrinkage under heat, loss of compressive strength and of weight during heating, and the effects of vibration and possible self-heating phenomena.

In establishing maximum operating temperatures for preformed high-temperature insulation, particularly that in the form of pipe sections, the ability of insulation to withstand moderate loads and vibration while in service shall be considered. Some knowledge of the compressive strength both before and after heating may be of value.

Additionally, the effect of long-term heating should be studied. Where the compressive strength in service undergoes significant reduction, as in the case of banded mineral wool due to volatilization of the organic binders at temperatures above about 230°C, it may be necessary to support and protect the insulating material.

Metal cleading is convenient for this purpose; where additional strength is required the metal shall be supported from the pipe independently of the insulation, e.g., by insulated metal spiders or by suitable load bearing insulation. It is preferable that the appropriate tests shall be carried out by the manufacturer and the results used to compile a report on suitability for particular applications; judgment is then needed in selecting material for specific conditions of use.

1.14.3. Thermal Expansion

Most thermal insulating materials have a lower thermal expansion coefficient than metals. Differential thermal movement between insulated surface, insulation, and outer finish shall be considered.

1.14.4. Resistance to Compaction

Loosefill material and unbonded mattresses are liable to compact under the influence of vibration and thermal cycling.

1.14.5. Resistance to Water Vapor Penetration and to Water Absorption

While closed pore materials, e.g., cellular glass, may have appreciably low water-vapor permeability characteristics, open-pore and fibrous insulating materials can absorb considerable quantities of water that will adversely affect the thermal conductivity and the effectiveness of the insulation.

Insulation applied to cold surfaces has to be protected from water-vapor penetration that gives increased conductivity. If water is allowed to penetrate and freeze, the cells of the insulation may rupture and the material be permanently damaged. It is an advantage if the material itself is of low water-vapor permeability.

1.14.6. Mechanical Strength

In general, insulating materials are mechanically weak and strength normally decreases after heating. The finish applied frequently affords some protection against mechanical damage, but the insulation itself shall be strong enough to withstand reasonable handling-strength and abrasion resistance shall be related to the work and material in question, and have to be taken into consideration at the design stage.

1.14.7. Durability

The durability of an outdoor insulation system is important. For instance, shrinkage in sunlight or breakdown by frost may have serious consequences.

1.14.8. Fire and Explosion Hazards

Not all the thermal insulating materials in common use are nonflammable. Some of them, often used for refrigeration systems, are entirely of organic composition and thus may constitute a fire hazard, or they may emit smoke and toxic fumes. Designers of thermal insulation systems shall therefore consider the process conditions and the plant arrangement before deciding whether or not the proposed thermal insulating material might contribute to the spread of fire, however initiated, and they shall vary their choice of material accordingly.

While it is usually desirable that insulation shall be noncombustible, these may be areas where such materials are not necessary, in which case some relaxation could be considered; the relevant building and fire authorities shall be consulted. The fire hazard, where caused by quantity of combustible material, susceptibility to ignition, ease of surface spread of flame, or quantity of smoke or toxic gas produced in a fire may be influenced considerably by the protective coating and associated material, e.g., adhesives, vapor barriers, and sealants. Many insulating materials are free from risk, but if inadequately protected can absorbs quantities of oil, etc., that may ignite spontaneously.

Some insulating materials that contain organic bonding agents, although generally suitable for the anticipated service temperature, may in fact constitute a fire risk through a phenomenon of internal self-heating. Evidence of this self-heating may be a transient rise in temperature above the theoretical value for specific locations within the insulation system. This hazard may be accentuated if air can enter into material of low bulk density, or by convection current induced with insulated vertical pipe work. The phenomenon is complex one, being associated with local concentration of organic bonding material, the thickness of insulation, and the temperature of the insulated surface together with its orientation.

If excessive, this internal rise in temperature may constitute a fire hazard, particularly if the surrounding atmosphere is flammable or if there are flammable materials in the immediate vicinity.

In flammable atmospheres, particularly with long runs of insulated pipe work, it may be advisable to take the precaution of connecting external metal cleading to earth in order to avoid possible build-up of static electricity.

In addition to the above the following recommendations shall be considered when selecting insulation materials. Porous insulation materials, such as calcium silicate, will have low ignition temperature when impregnated with oil (for instance mineral oil ignites at 200–300°C and higher class alcohol at 100–150°C). Rock wool containing over 0.3 percent mineral oil or over 0.5 percent resin-treated glass fiber felt or organic matter shall not be used for oxygen gas facilities.

Cold insulation materials composed of perlite or silica gel mixed with fine aluminum powder shall not be used for liquefied oxygen facilities, such as air separators. Organic cold insulation materials, if used, shall be weatherproofed with incombustible or fire-resistant materials.

In certain areas of high fire (or explosion) risk, e.g., where powerful oxidizing agents are handled, the insulation shall not contain any organic matter.

1.14.9. Chemical Resistance

Thermal insulation used in process plants shall not react with chemicals present and considering the properties of the fluids to be handled, selection shall be made from among chemically resistant materials that will not cause any hazard if they absorb the fluid.

1.14.10. Optimum Life

The required life of the insulation system shall be considered because this affects the annual cost and hence the economic thickness. If the plant has only a short life, a cheap insulation system may be adequate; if the plant has a longer life, a more expensive insulation system with longer life may be the more economic.

When the technical requirements of the application have been met, the total cost (as distinct from the initial cost) during the life of the installation is the prime consideration.

1.14.11. Optimum Heat Capacity

As the heat capacity (alternatively thermal capacity) of an insulating material varies according to its bulk density, it is preferred that this thermal property shall be expressed in terms of heat capacity per unit mass (alternatively specific heat capacity). In SI units this value is expressed in $J/(kgK)$. It should be noted that the value will vary with mean temperature even though it is given in terms of unit increment of one Kelvin.

Note: An insulating material of low thermal capacity will absorb relatively small quantities of heat with increase of temperature and consequently, under fluctuating temperature conditions, it will be associated with rapid heating and cooling. Conversely, a material of high heat capacity will tend to impart thermal stability to an insulated system.

For certain applications it may be convenient to refer to heat capacity per unit volume, $J/(m^3 K)$, but this value will be typical only for a representative product, i.e., at a specific bulk density.

1.14.12. Freedom from Objectionable Odor

It is of particular importance that insulating materials for use in canteens or buildings in which food is processed or served, shall be free from objectionable odor.

1.14.13. Maintenance Requirements

Maintenance costs can be significant in the total cost of an insulating system. These costs can be minimized by correct selection of materials and finishes and by attention to detail in the layout of the system. Particular attention shall be given to the accessibility of removable sections, valves, etc., and the protective systems used in inaccessible locations such as ducts.

1.14.14. Thermal Expansions

Most thermal insulating materials have a lower thermal expansion coefficient than metals. Differential thermal movement between the insulated surface, insulation, and outer finish shall be considered.

1.14.15. Resistance to Compaction

Loosefill material and unbonded mattresses are liable to compact under the influence of vibration and thermal cycling.

1.14.16. Resistance to Vermin and Fungus

The resistance of insulation to vermin, insects, and fungal growth can be important, particularly in cold stores for goods. Insulation surfaces likely to become wet shall not be finished with materials that may be attacked by these agencies. Finishing with nonabsorptive materials is desirable for severe cases.

1.15. CONSIDERATION OF CHARACTERISTICS OF ACCESSORY MATERIALS

1.15.1. Vapor Barriers

Condensation of water will occur on surfaces at temperatures below the atmospheric dew point, due to water vapor drawn towards the cold surface as a result of the difference in partial vapor pressure between the air at ambient temperature and the temperature of the cold surface. Unless it is possible for this moisture to be re-evaporated, it can be absorbed into any permeable insulating material that may be applied to the cold surface, thus increasing the thermal conductivity of that material, with impairment of its effectiveness.

The purpose of the vapor barrier is to reduce, and if possible to prevent, the ingress of water vapor into the insulating material, Thus, the barrier shall always be applied to the warmer surface of the material. It may take the form of a coating or sheet material resistant to the passage of water vapor, i.e., of low permeability, and the sealing of joints and overlaps shall be effective. Insulating materials that consist substantially of closed cells possess an inherent resistance to the passage of water vapor, but open–cell insulants and loosefill porous materials are readily permeable to water vapor.

Even with materials that have good resistance to the transmission of water vapor, differential movement of plant and insulation can cause joints in the latter to open, thus allowing moisture to penetrate towards the underlying surface. Joint-sealing compounds also may fail to exclude water vapor completely, in which case the contained water or ice may form strongly conducting paths from the surface of the plant to the ambient air.

1.15.2. Vapor Barriers for Use over Insulation on Surfaces Below Dew Point

Since it is essential to prevent the deposition of moisture, with the consequent risk of ice formation within the insulating material in those zones that are below the freezing point, the use of an effective vapor barrier is an important technical requirement.

The deposition of moisture within the insulating material could lead to eventual saturation of the material, with possible resultant mechanical and physical deterioration within the insulation system, and the risk of corrosive attack on the insulated metal surface. It is essential that the barrier shall not be used as the exposed surface finish if it is likely to be damaged during service; it should be noted that even penetration by pinholes will impair the effectiveness of the vapor seal.

1.15.3. Vapor Barriers for Use over Insulation on Surfaces Above Dew Point

If the cold surface is at temperatures below the atmospheric dew point for short periods only, any deposited moisture may evaporate later, in which case the use of a vapor barrier may not be essential, or a surface breather coat may be adequate.

It should be noted that the atmospheric dew point is likely to vary from day to day in accordance with the relative humidity, so that account shall be taken of this likely variation when the requirements of vapor protection are under consideration.

1.15.4. Finishing Materials

The contribution of finishing materials to the thermal effectiveness of the insulation system is in general negligible, but usually a finish is desirable for one of the following reason:

• To give appropriate protection against mechanical damage.
• To assist in identifying the pipe or vessel. This may be achieved by painting, either with a characteristic color or by means of colored bands

at intervals. This identification also may be used to indicate the direction of flow for the fluid content.

- To protect against water, snow, airborne deposits, sunlight, or ozone.
- To give protection against excessive moisture, corrosive vapor, etc., in the atmosphere.
- To give protection against spillage of oils and other flammable liquids.
- To improve appearance or to provide a surface that can be cleaned easily.
- To retard, or if possible to prevent, the spread of flame.
- To protect against chemical attack, vermin, and mold growth.
- To resist against heat in the case of metallic jacketing.

1.15.5. Securing Materials

Insulation systems may be secured by means of adhesives, by mechanical means, or by a combination of both.

- **Adhesives**

    Adhesion depends on molecular forces and, because the surfaces in thermal insulation systems are often rough and irregular, adhesives used for these systems require special gap-filling properties. Compatibility with the surfaces involved is also important and particular insulating materials may require specific adhesives. Fluid primers may be required to assist penetration and wetting, particularly when it is necessary to bond or to consolidate friable surfaces.

- **Classification according to use**

    Adhesives for insulation may be classified as follows:

    - **Insulation bonding adhesives**

        These adhesives are used for securing preformed slabs and sections, or flexible insulating materials to themselves and to structures such as equipment and ducts. Generally the substances in this group are of relatively heavy consistency and have good gap-filling characteristics, i.e., are of high build and have high solids content (see Table 1.31).

    - **So-called "lagging" adhesives**

        These adhesives are used for bonding, sizing, and coating surface-finishing fabrics over insulated pipework and equipment. Usually these are water-based polyvinyl acetate (PVA) and PVA copolymer emulsions.

    - **Facing and film-attachment adhesives**

        These adhesives are used for attaching flexible laminates, foils, and plastic film to thermal insulation, and for bonding the overlaps of these materials.

Table 1.31 Insulation Bonding Adhesives for Preformed Sections and Slabs

Mineral fiber (glass, rock, slag)	Styrene-butadiene rubber (SRB) emulsion and solutions, neoprene-phenolics, bitumen/rubber emulsion and solutions, natural rubber solutions, alkyd solutions, latex rubber/hydraulic cement
Polystyrene foam	Some SBR solutions, SBR emulsions, bitumen/rubber emulsions, rubber/alcohol solutions, polyurethanes, latex rubber/hydraulic cement
Polyurethane foam	SBR solutions, neoprene-phenolics, filled alkyd solutions, polyurethanes, bitumen/rubber emulsions, hot-melt bitumen
Polyisocyanurate foam	SBR solutions, neoprene-phenolics, filled alkyd solutions, polyurethanes, bitumen/rubber emulsions, hot-melt bitumens
Phenolic foam	SBR solutions, bitumen/rubber emulsions, filled alkyd solutions, latex rubber/hydraulic cement, polyurethane, hot-melt bitumens
Baked cork	Natural rubber solutions or lattices, SBR emulsion or solutions, hot-melt bitumens
Expanded polyvinyl chloride (PVC)	Neoprene-phenolic, nitrilephenolic, epoxies, hot-melt bitumens, polyurethanes
Cellular glass	Latex rubber/hydraulic cement, polyurethanes, epoxies, bitumen/rubber solutions and emulsions, hotmelt bitumens

Note: The adhesive groups shown here are indicative only and in no way preclude selection of alternative materials recommended by the adhesive or insulant manufacturer. Final selection should be made only after full consideration of all the factors outlined in this section together with any special chemical resistance or other requirements.

- **Application characteristics**

Important properties that affect use are:

- Build, i.e., ability to resist sagging, etc.
- Consistency, i.e., suitability for brushing, spraying, etc.
- Coverage
- Solids content (influencing the ease with which solvents or water are lost)
- Degree of absorption with porous insulating materials

Other properties that may require consideration are:

- Flash point
- Toxicity
- Temperature and humidity range (for application)

- Range of bonding times
- Curing times
- Stability in storage
- Possibility of corrosive attack
- Danger from attack by solvents

(Slow loss of water can give rise to corrosion; alkaline adhesives, e.g., sodium silicate, can attack aluminum and glass fiber; organic foams, especially polystyrene, are susceptible to attack by certain solvents.)

- **Service properties**

Properties affecting the behavior of the adhesive after installation are as follows:

- **Temperature limits**

    There are limiting temperatures between which an adhesive will remain effective without significant loss of bond strength. Most adhesives for thermal insulation have service temperature limits within the range -50°C to +120°C, although many special adhesive/insulation combinations can be used outside this range.

- **Adhesive strength**

    Because of the low bulk density and loose bonding of many insulating materials, high adhesive strength is not usually the most important requirement. However, the adhesive strength has to exceed the cohesive strength of the insulating material by a safe margin under all conditions of service. For materials of higher bulk density and strength, stronger adhesive will be required, and these have to be adequate to carry the required load with an ample margin of safety.

- **Mechanism of curing**
    - **Solvent-based adhesives**

        The mechanism of curing is by the evaporation of organic or aqueous solvents or by the absorption of water into porous adherents with subsequent evaporation.

    - **Chemically curing adhesives**

        The group of adhesives generally set by a chemically activated cross-linking process.

    - **Hot melt adhesives**

        These are applied as hot viscous liquids and form a bond by solidification on cooling to ambient temperatures.

- **Types of adhesives**

Following is the list of some adhesives as guidance only. For some information on properties, reference is made to BS 5970:
- Bitumens
- Epoxy resins
- Natural rubber
- Neoprene phenolic
- Nitrite phenolic
- Polyvinyl acetate
- Reclaimed rubber
- Styrene-butadiene-based (SBR) cement
- Alkyd resins
- Rubber latex/hydraulic cement
- Polyurethanes

1.15.6. Mechanical Securements

The securing materials under this heading generally can be classified as welded attachments, bolted fittings, or banding and wire securements. Care is required to avoid bimetallic contact between metals of appreciably different electrochemical properties.

Welded attachments are used mainly on the vertical faces of piping, or vertical and downward-facing vessels and equipment. They can be in the form of cleats, angles, pads, studs, bolts, nuts, etc., that will provide support for bolted fittings and permanent datum positions relative to each other.

Bolted fittings are used mainly in conjunction with welded attachments and they can be in the form of angle rings, flat rings, etc.; they provide the horizontal support for insulating and cleading materials.

Banding and wire securements are used mainly to hold materials firmly to the plant to be insulated; they may be of metal, fabric, plastics strip, etc.

1.16. SELECTION OF INSULATION AND ACCESSORY MATERIALS

Materials for insulation, fastening, jacketing, and finishing shall be selected from the materials described below unless otherwise agreed by the company. For more information on insulation materials reference is made to BS 5422.

1.16.1. Thermal Insulating Materials

Before deciding on the insulating materials to be used for any specific purposes the factors in Table 1.32 shall be considered.

Thermal insulation material(s) shall be selected based on Table 1.33.

Miscellaneous insulation materials shall be the following:

- Expansion or contraction joint material shall be loose mineral wool fiber with temperature limits of -101°C to 650°C.
- Hydraulic-setting thermal insulation cement shall have asbestos-free mineral fiber.
- Steam tracer lead insulation tubing should be comprised of an inner layer of braided asbestos tubing, a middle layer of glass fiber insulation, and an outer layer of braided asbestos tubing with a factory-applied weather coat mastic.

Typical characteristics for a number of thermal insulating materials available for use mainly for temperatures higher than ambient temperatures are indicated in Tables 1.34 and 1.35 and the corresponding characteristics for typical materials particularly suitable for the temperatures lower than ambient temperatures are indicated in Tables 1.36 and 1.37. It should be noted that maximum service temperature of some lower temperature insulation materials are such that they can also be used for elevated temperatures.

Table 1.32 Thermal Insulating Material Factors

Hot Insulation	Cold Insulation
Cold-face temperature (min. & max.)	Cold-face temperature (min. & max.)
Hot-face temperature (max. & min.)	Warm-face temperature (max. & min.)
Ambient temperature	Ambient temperature and humidity
Thermal conductivity	Thermal conductivity (aged)
Thickness of insulation required	Thickness of insulation required
Mechanical strength	Mechanical strength
Health hazards	Health hazards
Fire hazards	Fire hazards
Thermal movement (expansion)	Thermal movement (contraction).
Permeability of insulating material with need for protection	Vapor sealing of system
Protective covering and finish	Protective covering and finish
Cost (including that for application and finish)	Cost (including that for application and finish)

Table 1.33 Recommended Thermal Insulation Materials

Commodity	Service	Recommended Material		Form of Thermal Insulation	Temperature Limit °C
Thermal insulation	Hot (higher than ambient temperature)	Mineral fiber	Rock/slag	1-Blanket	600
				2-Preformed pipe section	600
				3-Preformed board/slab	540
			Glass	1-Blanket	500
				2-Preformed pipe section	450
				3-Preformed board/slab	450
		Calcium silicate		Preformed block and pipe section	650
		Ceramic fiber		Blanket	1150
		Cellular glass/foam glass		Block and pipe section	−268 to 427
	Cold (lower than ambient temperature)	Polyurethane and isocyanurate		1-Spray applied rigid cellular polyurethane	90
				2-Unfaced preformed rigid cellular polyurethane block and pipe section	−40 to 107
		Baked cork		Block and pipe section	80
		Cellular glass/foam glass		Block and pipe section	−268 to 427

Table 1.34 Typical Insulating Materials for Use at Temperatures Higher Than Ambient

Material	Physical Forms	Type	Approximate Maximum Operating Temperature	Normal Bulk Density kg/m³
Aluminium silicate	Loosefill (Loosefill materials should be packed to densities to suit the application and thermal conductivity required.)	Fibrous	1260	80 to 100
	Spray-applied	Fibrous	1260	150 to 250
	Blanket felt	Fibrous	1260	50 to 150
	Paper	Fibrous	1260	200
	Ropes	Fibrous	1260	100 to 150
Calcium silicate	Plastic composition	Granular	1000	160 to 320
	Slabs/lags	Granular	1010	160 to 320
	Preformed pipe sections	Granular	800	190 to 260
Cellular glass	Slabs and pipe sections	Cellular	400	130 to 160
Insulating compositions	Plastic composition	Granular	1000	250 to 1100
Magnesia	Plastic composition	Granular	310	180 to 220
	Preformed slabs	Granular	310	180 to 220
	Preformed pipe sections	Granular	310	180 to 220
Mineral wool (glass)	Rigid slabs/lags	Fibrous	230	15 to 100
	Preformed pipe sections	Fibrous	400	60 to 100
	Flexible mats	Fibrous	230	15 to 35
	Lamella mat	Fibrous	230	20 to 40
Mineral wool (rock)	Loose wool	Fibrous	850	20 to 150
	Resin-bonded slabs/lags	Fibrous	750	20 to 200

Material	Form			
Mineral wool /cement/binder	Inorganically bonded	Fibrous	1100	80 to 150
	Slabs/lags	Fibrous	230	Up to 45
	Preformed pipe sections	Fibrous	850	90 to 130
	Flexible mats	Fibrous	850	30 to 48
	Wired mattresses	Fibrous	230	140
	Lamella mat	Fibrous	650	240 to 250
	Sprayed	Fibrous	650	100 to 300
Perlite	Loosefill (Loosefill materials should be packed to densities to suit the application and thermal conductivity required.) Preformed slabs and sections	Granular	870	40 to 150
Phenolic rigid foam	Preformed slabs and pipe sections and foamed in situ	Cellular	120	35 to 200
Polyurethane rigid foam	Preformed slabs and pipe sections, sprayed, and foamed in situ	Cellular	110	30 to 160
Polyethylene	Preformed pipe sections	Cellular	80	30 to 40
Polyisocyanurate rigid foam	Preformed slabs and pipe sections and foamed in situ	Cellular	140	30 to 60
Polyurethane flexible foam	Blanket, slab, and pipe sections	Cellular	70	30 to 65

(Continued)

Table 1.34 Typical Insulating Materials for Use at Temperatures Higher Than Ambient—cont'd

Material	Physical Forms	Type	Approximate Maximum Operating Temperature	Normal Bulk Density kg/m³
Silica, microporous	Loosefill (Loosefill materials should be packed to densities to suit the application and thermal conductivity required.)	Fibrous	1000	50
	Blanket felt	Fibrous	1000	50 to 500
	Cloth tapes	Fibrous	1000	500 to 800
	Opacified aerogel, slatted	Granular	950	200 to 300
Stainless steel	Plain and dimpled foil	Reflective	760	–
Synthetic rubber	Flexible preformed pipe sections, flexible slabs, and rolls	Granular	105	60 to 100
Vermiculite	Loosefill (Loosefill materials should be packed to densities to suit the application and thermal conductivity required.) Preformed slabs and sections	Granular	1100	50 to 150
Vermiculite/cement		Granular	1100	320
Vermiculite/sodium silicate		Granular	1100	450

Note: Temperatures shown are maximum continuous operating temperatures depending on product type. Reference should always be made to manufacturers' literature before specifying. The limiting temperatures of any facing material should also be checked.

Table 1.35 Hot Insulation Material Selection Criteria

Type	Fire Resistivity	Thermal Conductivity	Water-proofness	Non-corrosiveness	Mechanical Strength	Chemical Resistivity	Physical Safety	Ease of Repair
Rock wool	Superior	Superior	Inferior	Superior	Inferior	Superior	Superior (may cause irritation to the skin)	Superior
Glass wool	Superior	Superior	Inferior	Superior	Inferior	Superior	Superior (may cause irritation to the skin)	Superior
Ceramic wool	Highly superior	Superior	Inferior	Superior	Inferior	Superior	Superior	Superior
Hair felt		Superior	Inferior	Superior	Inferior	Inferior	Superior	Highly superior
Perlite	Highly superior	Superior	Inferior	Superior (if water is absorbed, aluminium will be attacked)	Superior	Superior	Superior	Superior
Diatomaceous earth	Highly superior	Inferior	Inferior	Superior	Inferior	Superior	Superior	Inferior
Calcium silicate	Highly superior	Superior	Inferior	Superior	Highly superior	Superior	Superior	Superior
Cellular glass	Highly superior	Highly superior	Highly superior	Superior	Superior	Superior	Superior	Inferior
Foamed hard urethane	Inferior	Highly superior	Superior	Superior (HCl may form if water is absorbed)	Superior	Superior	Superior (poisonous gas is formed when subject to fire)	Superior

Table 1.36 Typical Insulating Materials for Use at Temperatures Below Ambient

Material	Physical Forms	Type	Approximate Maximum Operating Temperature	Normal Bulk Density kg/m³
Phenolic foam	Preformed slabs and pipe sections	Cellular	120	35 to 200
Polystyrene foam extruded	Preformed slabs, preformed pipe sections	Cellular	80	15 to 30
Baked cork	Preformed slabs, preformed pipe sections	Cellular	60	100 to 200
Granulated baked cork	Loosefill (Loosefill materials should be packed to densities to suit the application and thermal conductivity required.)	Granular	60	80 to 150
PVC expanded	Preformed slabs	Granular	65	60 to 120
Polyurethane rigid foam	Slabs, pipe sections, sprayed	Cellular	110	30 to 160
Polyurethane flexible foam	Slabs, pipe sections	Cellular	70	30 to 65
Polyisocyanurate rigid foam	Slabs, pipe sections, sprayed	Cellular	140	30 to 60
Cellular glass	Rigid slabs, rigid pipe sections	Cellular	400	120 to 160
Silica aerogel	Loosefill (Loosefill materials should be packed to densities to suit the application and thermal conductivity required.)	Granular	980	50 to 150
Perlite	Loosefill (Loosefill materials should be packed to densities to suit the application and thermal conductivity required)	Granular	870	40 to 150
Synthetic rubber	Flexible slabs, flexible pipe sections	Cellular	105	60 to 100
Polyethylene	Preformed pipe sections	Cellular	80	30 to 40

Notes:
1. Care is necessary in interpreting the maximum operating temperatures in this table. These are based on manufacturers' claimed figures and may require modification (possibly considerable) in practice.
2. For fibrous materials, the compressive strength will vary with bulk density and with the binder.

Table 1.37 Cold Insulation Material Selection Criteria

Type	Fire Resistivity	Thermal Conductivity	Moisture-proofness	Heat Resistivity	Mechanical Strength	Chemical Resistivity	Non-corrosiveness (Inertness)	Physical Safety (Odor & Dust)	Ease of Repair
Perlite	Highly superior	Inferior (0.05)	Inferior	Highly superior	Inferior	Highly superior	Superior	Inferior (dust)	Inferior
Glass wool	Highly superior	Superior (0.031–0.83)	Inferior	Highly superior	Inferior	Superior	Superior	Superior (may cause irritation to the skin)	Superior
Rock wool	Highly superior	Inferior (0.045–0.054)	Inferior	Highly superior	Inferior	Superior	Superior	Superior (may cause irritation to the skin)	Superior
Hair felt	Inferior	Inferior (0.041)	Inferior	Highly superior	Inferior	Inferior	Highly superior	Highly superior	Highly superior
Cellular glass	Highly superior	Inferior (0.047)	Highly superior	Highly superior	Superior	Superior	Highly superior	Superior	Inferior
Foamed polystyrene	Inferior (while combustible, self-extinguishable)	Superior (0.033–0.037)	Superior	Inferior	Superior	Inferior	Superior	Superior (poisonous gas will form when subject to fire)	Superior
Foamed phenol	Inferior (while combustible, self-extinguishable)	Superior (0.029–0.031)	Superior	Inferior	Superior	Superior	Superior	Superior (poisonous gas will form when subject to fire)	Superior
Foamed hard urethane	Inferior (while combustible, self-extinguishable)	Inferior (0.02–0.026)	Superior	Inferior	Superior	Superior	Superior	Superior (poisonous gas will form when subject to fire)	Superior
Foamed vinylchloride	Inferior (while combustible, self-extinguishable)	Superior (0.031–0.035)	Superior	Superior	Highly superior	Superior	Inferior	Superior (poisonous gas will form when subject to fire)	Superior

1.16.2. Summary of Insulation Types

The four basic types of thermal insulating material are fibrous, cellular, granular, and reflective. These materials differ in many characteristics. Refer to Table 1.25 for a description of these materials, typical conductivity values, and principal properties of common industrial insulations.

- **Uses**

   Principal uses of insulation are for personnel protection, process temperature control, prevention of condensation, and conservation of energy.

- **Personnel protection**

   Personnel protection is accomplished by the application of insulation of proper thickness where the surface temperature should be limited to approximately 150°F or as specified by applicable codes or local standards.

- **Process temperature control**

   Insulation thickness is specified in this case to help control the temperature of the process fluid. Electrical, steam, hot process fluid, hot oil, and glycol-water tracing are used to add heat to the process line to balance the heat loss. The insulation thickness must be matched to the energy input to achieve the desired result. Freeze protection is another use for insulation. This includes fluids that have higher viscosities or freeze points.

- **Condensation**

   Insulation thickness must be sufficient to keep the outside surface of the insulation above the dew point of the surrounding air. Moisture condensation on a cool surface in contact with warmer humid air must be prevented because of the deterioration of the insulation. In addition to the required thickness of insulation, a vapor tight membrane must be properly applied to the insulation. As a rule, insulation thickness for condensation control is much greater than the thickness required for conservation of energy.

- **Refrigerated tank insulation systems**

   Low-temperature insulation is required for both spherical and flat-bottomed cylindrical refrigerated tanks. Two types of insulation systems are commonly used for low-temperature service—single wall and double wall.

   In the single wall system, the vessel wall is designed to withstand the design service conditions of the liquid to be stored. The outer surface of this wall is then covered with a suitable insulating material such as rigid polyurethane foam. An aluminum jacket is then installed to provide protection against the elements and physical damage. It is extremely important that the insulation be sealed with a good vapor barrier to minimize air leakage and thereby reduce the quantity of water that may migrate into the insulation. Such moisture migration can ultimately damage the insulation.

The welded steel plate outer shell of a double wall system provides containment and vapor protection for the insulation material, generally perlite. The outer wall also provides protection against fires at temperatures up to 320°C. Double wall tanks are considered in storing products at temperatures below −35°C.

This system minimizes heat leak, which generally means lower operating and maintenance costs. As an added safety feature, the outer wall is completely sealed and therefore permits the insulation space to be continually purged with an appropriate inert gas, which keeps the insulation isolated from outside humid air. Tables 1.13 and 1.38 provide a range of typical thermal conductivities for various types of insulating and tank shell materials.

1.17. FASTENING MATERIALS

1.17.1. Banding

Bands where used under cleading (jacketing) for securing rigid preformed insulation on pipe, vessels, and tanks shall be hot-dip galvanized mild steel. Bands for securing metal jacket on pipe, vessels, and tanks shall be stainless steel 18-8 Cr–Ni AISI type 302 or 304.

Tape for securing pipe foam insulation used in low-temperature services shall be glass filament reinforced pressure sensitive type. The type for sealing contraction joints in cold insulation systems shall be foil-to-Kraft tape.

1.17.2. Miscellaneous Fastening Materials

Seals for bands shall be 18-8 Cr–Ni stainless steel AISI type 302 or 304. Wire to secure insulation shall be soft annealed galvanized mild steel for hot insulation and stainless steel 18-8 Cr–Ni AISI type 302 or 304 for cold insulation.

Breather springs shall be 18-8 Cr–Ni stainless steel AISI type 302 or 304. Sheet metal screws shall be hard aluminum alloy for aluminum jackets and 18-8 Cr–Ni stainless steel AISI type 302 or 304 for steel jackets.

Table 1.38 Summary of Specifications for Low-Temperature and Cryogenic Steels (Values Are Maximum Unless Otherwise Specified or a Range Is Given)

ASTM/AISI	Lowest Usual Service Temperature, ↑°C	Thermal Conductivity, W/(m.°C)					Example Applications	
		+95	+20	−45	−100	−195	Uses	Liquids Stored
Carbon Steels								
A333 Grades 1 & 6	−45	51	50				Welded pressure vessels and storage tanks, when weight and strength are not critical; refrigeration and transport equipment	Butane, isobutane, sulphur dioxide, ammonia, propane, propylene
A334 Grades 1 & 6	−45	51	50					
A442 Grades 55 & 60	−45	51	50					
A516 Grades, 55, 60, 65 & 70 (to ASTM A300 Specifications)	−45	51	50					
A537 (with modifications)	−45	51	50					
Alloy Steels								
A517 Grade F	−45		38				Highly stressed pressure vessels; tank trucks for handling LP gases	LP gases
A203 Grades A&B – 2 1/4% Ni	−60	42	39	36	34	21	Tanks, vessels, and piping for liquid propane	Propane
A333 Grade 7 – 2 1/4% Ni	−60	42	39	36	34	21		
A334 Grade 7 – 2 1/4% Ni	−60	42	39	36	34	21		

Alloy Steels

							Application	Products
A203 Grades D&E – 3 1/2% Ni	-100	39	37	33	31	18	Land-based storage of liquid propane, carbon dioxide, acetylene, ethane, and ethylene	Propane, carbon dioxide, acetylene, ethane, ethylene
A333 Grade 3 – 3 1/2% Ni	-100	39	37	33	31	18		
A334 Grade 3- 3 1/2% Ni	-100	39	37	33	31	18		

Stainless Steels

AISI – 300 Series (type 301)	-100	16	15	14	12	8		

Alloy Steels

							Application	Products
A333 Grade 8 – 9% Ni	-195	30	27	26	23	13	Large tonnage oxygen producing equipment; transportation and storage of methane, oxygen, nitrogen, and argon	Ethylene, methane, oxygen, carbon monoxide, nitrogen, LNG, argon
A334 Grade 8 – 9% Ni	-195	30	27	26	23	13		
A353 Grade 8 – 9% Ni	-195	30	27	26	23	13		
AMSE Code Case 1308 – 9% Ni								

Stainless Steels

							Application	Products
AISI – 300 Series (Type 302)	-195	16	15	13	12	8		
AISI – 300 (Type 304)	-270	16	17	13	13	8	In petrochemical, nuclear, missile, and other areas where purity of product is essential; handling liquid hydrogen rocket fuel	Hydrogen, helium

Wire mesh poultry netting shall be galvanized carbon steel. "S" clips shall be aluminum in the case of aluminum sheet and 18-8 Cr–Ni stainless steel AISI type 302 or 304 in the case of steel sheet. Clips and quick release toggles shall be 18-8 Cr–Ni stainless steel AISI type 302 or 304 or hard aluminum.

Support rings shall be carbon steel. Expanded metal as reinforcement shall be bitumen coated or galvanized mild steel or 18-8 Cr–Ni stainless steel AISI type 302 or 304.

1.18. JACKETING MATERIALS

Jacketing materials shall be one of the following recommended metallic alloys, the application of which depends on the conditions:

- Aluminum
- Aluminized steel
- Galvanized steel
- Coated steel
- Stainless steel

Note that the melting point of aluminum is considerably lower than other recommended jacketing materials.

The combination of jacketing and fastening materials for an insulation system shall be according to Table 1.39.

1.19. FINISHING MATERIALS

Joint sealer shall be a nonsetting, nonshrinking permanently flexible compound.

Table 1.39 Combination of Jacketing and Fastening Materials

Jacketing Material	Bands	Self-Tapping Screws	Rivets	Clips and Quick-Release Toggles
Aluminum	Stainless steel	Hard aluminum or stainless steel	Hard aluminum	Stainless steel or Aluminum
Aluminized steel and galvanized steel	Stainless steel	Stainless steel	Stainless steel	Stainless steel
Stainless steel	Stainless steel	Stainless steel	Stainless steel	Stainless steel

Flashing compound shall be high-temperature asphalt mastic with nonasbestos fiber or filler for weatherproofing applications.

Weatherproofing mastic for high-temperature insulation systems shall be emulsion-type polymeric protective coating: acrylic or polyvinyl acetate or a combination of both.

Vapor barrier or weatherproofing mastic for low-temperature insulation shall be nonasphaltic elastomeric mastic suitable for application over polyurethane or foamglass or other cold insulation.

Adhesive for cellular glass to a sphere shall be a two-part adhesive compound of latex rubber/hydraulic cement. Properties for some insulation outer (jacketing and finishing) materials are indicated in Table 1.40.

1.20. RAPID ESTIMATION OF HEAT LOSSES FROM OIL AND GAS PROCESS PIPING AND EQUIPMENT SURFACES

The most important part of the energy strategy in any process industry is energy saving. In this section, a simple-to-use predictive tool that is easier than existing approaches and suitable for process engineers, with fewer computations, is presented for the rapid estimation of heat losses in terms of the wind velocity and the temperature difference between the process piping and equipment surfaces and the surrounding air.

The tool developed in this study can be of immense practical value, as it allows the engineers and scientists to execute a quick check on the heat losses for air in contact with walls or surfaces without opting for any experimental measurements. The results can be used in follow-up calculations to determine heat losses from the process piping and equipment surfaces at various conditions. In particular, practice engineers would find the predictive tool to be user-friendly with transparent calculations involving no complex expressions.

Several rigorous studies have been reported on the combined effect of convection and surface radiation in the literature. However, there is an essential need to have a simple-to-use predictive tool for an accurate estimation of the combination convection and radiation film coefficients for air in contact with vertical walls or surfaces to give the combined heat transfer coefficient for heat loss calculations in various cases, in terms of the wind velocity and the temperature difference between the process piping and equipment surfaces and the surrounding air.

In view of this necessity, our efforts have been directed at formulating a simple-to-use predictive tool that can help practice engineers and applied

Table 1.40 Insulation Outer (Coverture) Material Selection Criteria

Classification	Type	Fire Resistivity	Chemical Resistivity	Required Mechanical Strength	Heat Resistivity	Water Resistivity
Weatherproofing and indoor use outer covering	Galvanized steel sheet	Highly superior	Superior	Highly superior	Highly superior (approx. 600°C)	Highly superior
	Prepainted galvanized steel sheet	Highly superior	Highly superior	Highly superior	Superior (approx. 200°C)	Highly superior
	Stainless steel sheet	Highly superior	Highly superior	Highly superior	Highly superior (approx. 800°C)	Highly superior
	Aluminum sheet	Inferior	Inferior	Highly superior	Highly superior (approx. 480°C)	Highly superior
Waterproofing	Asphalt emulsion type mastic	Inferior	Superior	Superior	Inferior (approx. 80°C)	Highly superior
	Resin (or rubber) emulsion type mastic	Superior	Superior	Superior	Inferior (approx. 80°C)	Highly superior
Indoor outer covering	Hard cement	Highly superior	Superior	Highly superior	Highly superior (approx. 500°C)	Superior
	Asbestos cloth	Superior	Inferior	Inferior	Superior (Heat resistivity depends on asbestos content: Be careful of this.)	Superior
	Glass cloth	Highly superior	Inferior	Inferior	Highly superior (approx. 300°C)	Superior
	Canvas	Inferior	Inferior	Inferior	Inferior (approx. 80°C)	Superior
	Vinyl tape; adhesive tape	Inferior	Inferior	Inferior	Inferior (approx. 60°C)	Highly superior

Classification	Type	Moisture Resistivity	Physical Safety	Corrosion Resistivity	Adaptability for Use with Insulation Material	Ease of Repair	Limitation
Weatherproofing and indoor use outer covering	Galvanized steel sheet	Inferior	Highly superior	Inferior	Highly superior	Inferior	Usable for both cold and hot insulation in outdoor and indoor areas
	Prepainted galvanized steel sheet	Inferior	Highly superior	Superior	Highly superior	Inferior	
	stainless steel sheet	Inferior	Highly superior	Highly superior	Highly superior	Inferior	
	Aluminum sheet	Inferior	Highly superior	Inferior	Superior (May be corroded by alkaline matter in insulation: be careful of this point.)	Inferior	
Waterproofing	Asphalt emulsion type mastic	Superior	Superior	Superior	Superior (water migrates from these hydrous materials to the insulation, impairing the insulation characteristics; fiber-state insulation is easily affected and cracked by pressure)	Highly superior	Hot insulation outdoor use
	Resin (or rubber) emulsion type mastic	Superior	Superior	Superior	Superior	Highly superior	

(*Continued*)

Table 1.40 Insulation Outer (Coverture) Material Selection Criteria—cont'd

Classification	Type	Moisture Resistivity	Physical Safety	Corrosion Resistivity	Adaptability for Use with Insulation Material	Ease of Repair	Limitation
Indoor outer covering	Hard cement	Inferior	Inferior	Superior	Superior	Highly superior	Hot insulation indoor use
	Asbestos cloth	Inferior	Inferior	Highly superior	Superior (surface irregularities will easily occur when fibrous insulation is wrapped with cloth or tape)	superior	Both hot and cold insulation; dust preventive measures required for cold insulation
	Glass cloth	Inferior	Highly superior	Highly superior	Superior (must be patched with cardboard or other material to provide good form)	Highly superior	indoor use
	Canvas	Inferior	Highly superior	Highly superior	Superior	Highly superior	
	Vinyl tape; adhesive tape	Inferior	Highly superior	Highly superior	Superior	Highly superior	

researchers. The proposed tool appears to be superior owing to its accuracy and simple background, wherein the relevant coefficients can be retuned quickly if more data are available in the future. Examples shown for the benefit of engineers clearly demonstrate the usefulness of the proposed predictive tool. The present study discusses the formulation of a simple correlation, which can be of significant importance for practice engineers.

1.20.1. Development of Simple Predictive Tool

Equation (1.57) calculates a coefficient or ΔT, which is the temperature difference between the surface and the surrounding air ($^\circ$C).

$$\Delta T = T_s - T_a \tag{1.57}$$

The required data to develop the first correlation include the reliable data for various values of wind velocity, and the temperature difference between the surface and the surrounding air.

Equation (1.58) presents a new correlation in which four coefficients are used to correlate the combination convection and radiation film coefficients for air in contact with the process piping and equipment surfaces and the temperature difference between the surface and the surrounding air values.

$$\ln(h_{cr}) = a + b(\Delta T) + c(\Delta T)^2 + d(\Delta T)^3 \tag{1.58}$$

where

$$a = A_1 + B_1 v + C_1 v^2 + D_1 v^3 \tag{1.59}$$

$$b = A_2 + B_2 v + C_2 v^2 + D_2 v^3 \tag{1.60}$$

$$c = A_3 + B_3 v + C_3 v^2 + D_3 v^3 \tag{1.61}$$

$$d = A_4 + B_4 v + C_4 v^2 + D_4 v^3 \tag{1.62}$$

The tuned coefficients used in Eqs. (1.59) to (1.62) are given in Table 1.41 to cover the reported data with wind velocity variations up to 20 m/s and temperature gradients (temperature of surface less temperature of air) up to 280°C.

Equation (1.63) calculates heat losses from equipment surfaces that occur primarily by radiation and convection.

$$Q = h^{cr}(A^o)(T^s - T^a) \tag{1.63}$$

where

A, B, C, and D: coefficients

Table 1.41 Tuned Coefficients Used in Eqs. (1.59) to (1.62)

Coefficient	Value
A_1	2.18201771352
B_1	$2.65054426648 \times 10^{-1}$
C_1	$-1.895691067097 \times 10^{-2}$
D_1	$4.641521705558 \times 10^{-4}$
A_2	$6.616737105648 \times 10^{-3}$
B_2	$-1.119534124965 \times 10^{-3}$
C_2	$9.485846901915 \times 10^{-5}$
D_2	$-2.43542487435 \times 10^{-6}$
A_3	$-1.581032525858 \times 10^{-5}$
B_3	$3.699760242622 \times 10^{-6}$
C_3	$-3.222348926402 \times 10^{-7}$
D_3	$8.774496064317 \times 10^{-9}$
A_4	$2.1240407483723 \times 10^{-8}$
B_4	$-5.7213183786357 \times 10^{-9}$
C_4	$5.3052858118086 \times 10^{-10}$
D_4	$-1.572916658647 \times 10^{-11}$

A^o: outside area, m^2

D: outside diameter

h^{cr}: combined convection and radiation heat transfer coefficient, $(\text{W}/(\text{m}°\text{C}))$

Q: heat loss, W/(linear meter)

T^s: surface temperature, °C

T^a: surrounding air temperature, °C

v: wind velocity, m/s

ΔT: temperature difference between the surface and the surrounding air (°C)

Figure 1.18 illustrates the results of the proposed correlation for predicting combination convection and radiation film coefficients for air in contact with walls or surfaces in comparison with some typical data obtained from the literature. Figure 1.19 demonstrates the performance of the proposed predictive tool for a wide range of conditions. As can be seen, the results of the new proposed correlation are accurate and acceptable. This graph also demonstrates the excellent performance of the proposed correlation. An example is given below to demonstrate the simplicity of the proposed predictive tool for the estimation of combination convection and radiation film coefficients for air in contact with vertical walls or surfaces.

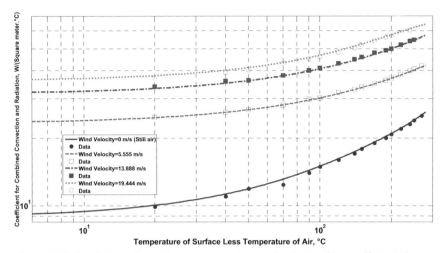

Figure 1.18 Prediction of combination convection and radiation film coefficients for air in contact with vertical walls or surfaces in comparison with some typical data. *(Reprinted with permission from © Crambeth Allen, A. Bahadori and H.B. Vuthaluru, (2010). Estimation of Heat Losses from Process Piping and Equipment,* Petroleum Technology Quarterly, *15(3):121–123).*

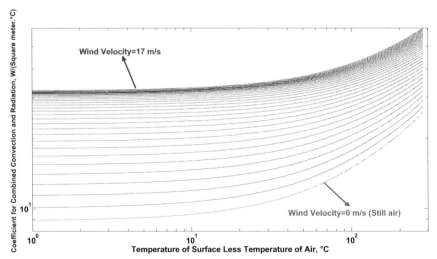

Figure 1.19 Performance of proposed predictive tool for estimation of combination convection and radiation film coefficients for air in contact with vertical walls or surfaces. *(Reprinted with permission from © Crambeth Allen, A. Bahadori and H.B. Vuthaluru (2010). Estimation of Heat Losses from Process Piping and Equipment,* Petroleum Technology Quarterly, *15(3):121–123).*

Example 1.14

How much heat can be saved per linear meter by covering a 200 mm NPS Sch 40 steam header, carrying 100 kPa (ga) steam at 120°C, with a 25 mm thick layer of block insulation? Assume ambient conditions are -1°C with a 24 km/hr wind. For the insulated pipe, assume the outside surface of the insulation is at 10°C.

Solution

Using the proposed Equations (1.58) to (1.62), the heat loss from the bare pipe is:

$a = 3.2440444304$ (from Eq. 1.59)
$b = 2.64750085877 \times 10^{-3}$ (from Eq. 1.60)
$c = -2.8669570727 \times 10^{-6}$ (from Eq. 1.61)
$d = 2.016839208 \times 10^{-9}$ (from Eq. 1.62)
$h^{\sigma} = 33.9874 \ W/(m°C)$ (from Eq. 1.58)
$D^{\circ} = 0.219 \ m$
$T^{w} = 120°C$
$T^{a} = -1°C$
$L = 1 \ m$

$$Q = h^{\sigma}(A^{\circ})(T^{w} - T^{a}) = (33.2)(1)(\pi)(0.219)(120 - (-1))$$
$$= 2917 \ W/(per \ linear \ m) \quad\quad\quad (from \ Equation \ 1.63)$$

For the insulated pipe, assume the outside surface of the insulation is at 10°C. Then:

$a = 3.24404443$ (from Eq. 1.59)
$b = 2.647500858 \times 10^{-3}$ (from Eq. 1.60)
$c = -2.86695707 \times 10^{-6}$ (from Eq. 1.61)
$d = 2.016839208 \times 10^{-9}$ (from Eq. 1.62)
$h^{\sigma} = 26.3857$ (from Eq. 1.58)

$$Q = 26.3857(1)(\pi)(0.219 + 2 \times 0.025)[10 - (-1)] = 245 \ W/(per \ linear \ m)$$
$$(from \ Equation \ 1.63)$$

Heat saved $= 2917 - 245 = 2672 \ W/m$

In this section, an attempt has been made to formulate a novel and simple-to-use predictive tool for the prediction of heat loss rate for air in contact with process piping and equipment surfaces. This simple predictive tool gives the combined heat transfer coefficient, in terms of the wind velocity and the temperature difference between the surface and the surrounding air. The results can be used in follow up design calculations to determine the heat losses from equipment surfaces at various conditions. The predictive tool proposed in the present work is a simple and unique expression. In particular, mechanical and process engineers would find the proposed approach to be user-friendly as it involves no complex expressions and has transparent calculations, which is evident from the cited example.

1.21. CORROSION CONSIDERATIONS IN DESIGN AND APPLICATION OF THERMAL INSULATION

An insulated metal surface may be affected by corrosive attack if the insulating material is applied in a wet state or if it is allowed to become wet for prolonged periods after application. The corrosion may be specific or general. A typical example of a specific corrosive attack is that of stress corrosion cracking (SCC) of austenitic steel due to the action of water-soluble chlorides in the presence of moisture.

Most thermal insulating materials contain traces of water-soluble chlorides in the presence of moisture. Most thermal insulating materials contain chlorinated organic compounds that may form soluble chlorides when heated in the presence of water. Both can give rise to stress corrosion cracking in susceptible austenitic alloys, even if only trace quantities are present.

It is not practicable to indicate a safe upper limit for chloride contents as water can leach out soluble chloride from substantial volumes of wet insulating material and allow them to be concentrated at the junction layer with the metal surface. Additionally, ingress of water from external sources, e.g., rainwater, plant spillages, and water used for hosing down equipment, may contain sufficient chloride to be potentially dangerous. General corrosion may occur if wet insulation is allowed to remain in contact with carbon steel surfaces for prolonged periods, particularly if acidic products are present in the water or if they can be extracted from the insulating material itself. Due regard shall also be paid to possible corrosive attack on a variety of metals, e.g., those used as finishes, that may be in contact with wet insulating materials.

To minimize corrosion the following recommendations shall be considered when selecting insulation materials:

- All insulation materials shall be noncorrosive, whether wet or dry, and suitably inhibited for application to stainless steel surfaces.
- In inhibited condition, insulation materials containing more than 600 ppm chloride are not acceptable for use on stainless steel surfaces.
- In uninhibited condition, the amount of leachable chloride in the insulation material shall not exceed 10 ppm.
- The pH of all insulation materials shall be between 6 and 9 in the wet condition.
- Insulation composed of alkaline materials shall not be used on aluminum surfaces.

The proliferation in recent years of corrosion failures of both steel and stainless steel under insulation has caused this problem to be of great concern to the operators of petroleum-, gas-, and chemical-processing plants. Piping and vessels are insulated to conserve energy by keeping cold processes cold and hot processes hot.

Once a vessel is covered with insulation and operating satisfactorily, concern for the condition of the metal under insulation usually diminishes. Thus, corrosion of steel and SCC of stainless steel begins and develops insidiously, often with serious and costly consequences many years later. To deal with corrosion under insulation effectively and economically, a systems approach must be developed that considers the metal surface, temperatures, water, insulation, and design.

Corrosion of steel under insulation did not receive a great deal of attention from the corrosion engineering community until the late 1970s and early 1980s. Although considerable attention was being given to the SCC of stainless steel under insulation, steel vessels and piping were rarely mentioned. This situation existed because of the slow corrosion rate of the steel equipment, together with the fact that many large hydrocarbon-processing plants were built 10 to 20 years earlier.

The most significant turn of events, which fortunately revealed the corrosion problem to many, was the energy shortage and the resultant emphasis on energy conservation. The effort to conserve energy soon led to replacement of much of the 10- to 20-year-old insulation with more efficient systems. As the old insulations were removed, localized, often severe corrosion damage was found. Corrosion of steel and of stainless steel under thermal insulation is discussed separately, because the nature of the problem, aside from the presence of insulation, is different.

1.21.1. Corrosion of Steel under Insulation

Steel does not corrode simply because it is covered with insulation. Steel corrodes when it contacts water and a free supply of oxygen. The primary role of insulation in this type of corrosion is to produce an annular space in which water can collect on the metal surface and remain, with full access to oxygen (air).

The most active sites are those where the steel passes through the insulation and on horizontal metal shapes, for example, insulation support rings where water can collect.

The other major corrosion problem develops in situations where there are cycling temperatures that vary from below the dew point to above ambient. In this case, the classic wet/dry cycle occurs when the cold metal develops water condensation that is then baked off during the hot/dry cycle. The transition from cold/wet to hot/dry includes an interim period of damp/warm conditions with attendant high corrosion rates. In both cases, if the lines had operated constantly either cold or hot, no significant corrosion would have developed.

- **The mechanism**

  The corrosion rate of steel in water is largely controlled by two factors: temperature and availability of oxygen. In the absence of oxygen, steel corrosion is negligible. In an open system, the oxygen content decreases with increasing temperature to the point where corrosion decreases even though the temperature continues to increase (Figure 1.20). A number of

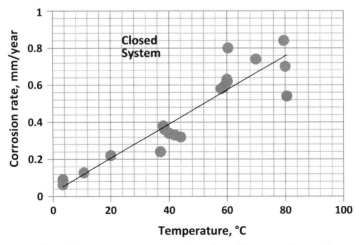

Figure 1.20 Effect of temperature on corrosion of steel in water. Data points are from actual plant measurements of corrosion under insulation.

recent case histories of corrosion under insulation have been reviewed, and the estimated corrosion rates have also been plotted in Figure 1.20. These filed data confirm that corrosion of steel under insulation increases steadily with increasing temperature, matching the curve for a closed system.

The problems of steel corrosion under insulation can be classified as equivalent to corrosion in a closed hot–water system. The thermal insulation does not corrode the steel, but, more correctly, forms an annular space where moisture collects.

The insulation also forms a barrier to the escape of water or water vapors, which hastens the normal corrosion rate of wet steel. In an attempt to determine which insulation contributes most and which contributes least to corrosion of steel, tests were conducted that compared 12 common insulation materials.

After a 1-year exposure to the elements, dousing with water, and daily heating with steam, it was concluded that, although the insulation materials that absorbed water developed higher corrosion rates than those that did not, the difference between the two was not great. Therefore, efforts to select a type of insulation to prevent corrosion would not be practical. Field observation and reports from the literature confirm that corrosion of steel occurs under any and all types of insulation.

- **Prevention**

  One attempt to prevent corrosion damage involves adding a corrosion inhibitor to the insulation. Inhibitors such as Na_2SiO_3 are added for this purpose. The practical effects of such an approach are questionable. In one set of laboratory tests, inhibited insulation material was submerged in water, then set out to dry. After several cycles of wetting and drying, the insulation was found to have no benefit over uninhibited insulation of the same brand.

  A corrosion inhibitor for insulation must be water soluble so that it can be dissolved and carried to the steel surface. If the insulation is water-proofed on the outside, water will enter through cracks and openings in the insulation system. This water is in the annulus and contacts only the inner surface of the insulation. The water–soluble inhibitor will soon be extracted and displaced by the water running down the vessel walls. Therefore, relying upon a consumable inhibitor for long–term protection may not be dependable.

  If water entry to the insulation/steel annulus is the problem, perhaps the solution lies in more effective waterproofing. Advancements in various metal jacketing, sealants, and mastics have improved the protection of insulation materials from water, weather, and man. However, these waterproofing systems probably do not prevent water entry, because, in order to seal and prevent the annulus form breathing with temperature changes, the seal would have to be equivalent to a pressure vessel.

Waterproofing systems are designed to keep the insulation dry. They are not capable of preventing water vapor and air from contacting the annular space. In fact, a waterproofing system keeps the water vapor in the annulus and effectively produces the hot-water closed system, with attendant high corrosion rates.

Therefore, selection of insulation, inhibiting the insulation, and waterproofing of insulation are not effective deterrents to corrosion. Instead, a high-quality protective coating, correctly applied to the steel before insulation, can offer long-term protection.

In the insulation industry, a primer is sometimes used, particularly under spray-on foam insulation materials. It should be understood that the purpose of the primer is to present a clean surface for bonding of the insulation. The primer-type paints are not intended to and will not prevent corrosion by hot water. The key to specifying a paint system is to remember that it will be exposed to hot water vapors, a very severe environment for paints.

The use of inorganic zinc paint as a primer, coupled with the hot water, will not cause accelerated corrosion of the steel. Zinc and the silicate binder are both dissolved by hot water. Even if conditions favored the zinc becoming cathodic to steel, the protective coating binder would dissolve and the coating would break down.

The advisability of using inorganic zinc under other protective coatings is not as clear. Some tests seem to indicate good performance. Others show that inhibited primers with topcoats achieve maximum performance in hot-water systems. Therefore, the main criterion for a protective coating under insulation is that it resist hot water and water vapors. This severe service will also require high-quality surface preparation.

When the oxidation of carbon steel is involved, various methods are available for depositing protective films on the surface of iron and steel, e.g., galvanizing, sheradizing, parkerizing, aluminzing, etc., and these can improve the resistance to oxidation, although often their main application is for accessory materials. Attention should be paid to the possible effect of change of conditions in service.

Thus, at temperatures over about 65°C under conditions of high humidity, oxidation, with severe scaling and possible loss of mechanical strength, may occur when a metal is heated to relatively high temperatures in air, e.g., about 400°C and above for carbon steel. The condition may be aggravated by the incorrect use of thermal insulating material on the external surface of refractory-lined equipment. This may raise the metal temperature to dangerous levels. If it is essential to have both internal and external insulation, great care is necessary in considering the effect on the metal temperature.

The formation of white deposits of oxide on the surface of zinc and certain types of aluminum sheet, particularly under moist conditions, may be unsightly rather than dangerous. Both may be eliminated by cleaning, followed by painting as necessary or, in the case of aluminum, by the use of sheets that have received a chemical or electrolytic oxidation treatment.

Other forms of corrosion that can occur on carbon steel and the methods of prevention follow.

1.21.2. Acidic Corrosion

This type of attack is most prevalent with carbon steels, although many of the nonferrous metals are also vulnerable. A typical example is the condensation of acidic gases inside a metal flue when the temperature is allowed to fall below the dew point of the contained gas; firing with heavy fuel oils or sulfite residues can be particularly troublesome, and special attention should be paid to vulnerable areas of thin metal, e.g., metal expansion bellows. In most cases the use of adequate thickness of insulating material to prevent internal condensation will be an effective safety measure.

As most insulating materials absorb water in the event of unsatisfactory or damaged weatherproofing, there is a serious risk of corrosion from wet insulation. The risk is increased in the presence of chlorides, nitrates, and sulfates, which may be introduced with rainwater.

Acidic corrosion on the external surface of carbon steel can result from the decomposition of chlorinated organic compounds in certain types of foamed plastic insulating materials. Normally this occurs when the products are heated in the presence of moisture; the attack may be minimized if the base metal is suitably painted prior to the application of insulating material. It should be remembered that even refrigeration plants may be heated on occasion, e.g., for defrosting, and that consequent decomposition of chlorinated organic compounds may occur.

The leakage of acidic gases through areas of faulty welding or faulty joints can result in corrosive attack on insulated metal surfaces, and this is the more dangerous because it is hidden from view. Also, any sulfur residues, e.g., certain slag wools, can produce acidic corrosion in the presence of moisture.

Corrosive attack on unprotected underground pipework can be dangerous, and effective measures should be adopted to avoid this. Insulating material of low capillarity can retain local concentration of water inside an insulation system. Also, the sulfate-reducing action of soil bacteria under anaerobic conditions can produce corrosive sulfide compounds; this action is especially dangerous with pipework buried directly in the ground. Protective measures will include coating the metal surface with asphaltic compounds, and these should be reinforced with inorganic woven fabric or staple tissue.

For some buried pipework systems, cathodic protection, e.g., by the use of zinc or magnesium sacrificial anodes or impressed current at intervals, may be effective, but generally specialist advice should be sought for difficult installations, particularly if there is a danger of contamination by seawater.

1.21.3. Alkaline Corrosive Attack

Some nonferrous metals may be attacked by alkalis extracted from certain types of insulating material under moist conditions, e.g., those that contain appreciable amounts of sodium silicate. Similar effects may occur with copper, brass, etc., and with zinc coating on galvanized surfaces. Aluminum sheet used to protect calcium silicate is also vulnerable. Protective measures will include keeping water out of the system and painting the metal surface as appropriate or using a factory-applied moisture barrier.

1.21.4. Intergranular Corrosion

Normally this consists of localized attack at the grain boundaries of ferritic and some austenitic alloy steels, which arises from the internal migration and deposition of chromium carbide at a sensitizing temperature over about $400°C$. Where doubt exists, it is recommended that the accessories should be selected from "stabilized" or special "low carbon" alloy steels.

1.21.5. Bimetallic (Galvanic) Corrosion

It is well known that direct contact between many dissimilar metals in the presence of moisture, particularly in a marine environment or near the sea, can result in rapid corrosion of one of the metals. When two different metals are in electrical contact and are also bridged by water containing an electrolyte (e.g., water containing salt, acid, combustion product, etc.) current flows through the solution from the anodic or baser metal to the cathodic or nobler metal.

As a result, the nobler metal tends to be protected, but the baser metal may suffer greater corrosion. In the past, schedules of electrode potentials have been published which have been of value in drawing the attention of designers to the dangers of bimetallic corrosion. Such schedules can, however, be misleading since the potential difference between metals, although it is the prime driving force of the corrosion current, is not a reliable guide to the rate and form of corrosion suffered at any particular area of contact. In particular, statements claiming that specific differences of potential are safe or unsafe are unreliable. Thus the use of unsuitable metals for screws, rivets, welded attachments, and even bands, should receive careful attention when such items are to be selected.

Where there is a danger of bimetallic corrosion, the metals should be isolated from each other, e.g., by plastic washers, insulation tape, bitumen mastic, or an appropriate paint of adequate film thickness.

1.21.6. Stress Corrosion Cracking (SCC) of Stainless Steel under Insulation

During and after World War II, many new petrochemical plants were built in the United States. Many of these plants contained equipment built with the new austenitic stainless steels. These alloys, such as types 304, 316, and 347, were widely used to combat process corrosion and to maintain product purity. As in all chemical processes, a large percentage of this equipment was insulated for thermal efficiency.

During this period, corrosion and materials engineers discovered some of the limitations of these stainless "wonder metals." In addition to crevice corrosion and weld decay, the engineers encountered SCC.

As the insulation systems and their weather barrier coatings aged, more incidents of SCC under insulation began to occur in the chemical processing industry. In the following years, much work was done to investigate the problem and the countermeasures.

- **The mechanism**

  Soon after the widespread use of the 18-8 austenitic stainless steels, beginning in the late 1930s, it became clear that the Cl^- ion in water could be very damaging. In addition to causing localized corrosion, such as pitting and crevice corrosion, rapid failure was seen in the form of a fine network of transgranular cracking. This pattern of cracking is very destructive and is found on the surface as well as in cross section. The theoretical processes of chloride SCC of the 18-8 stainless steels are under investigation. However, it has been established that four conditions are necessary for SCC to develop:

- An 18-8 austenitic stainless steel
- The presence of residual or applied surface tensile stresses
- The presence of chlorides; bromide (Br) and fluoride (F) ions may also be involved
- The presence of an electrolyte (water).

When these conditions are present, the occurrence of SCC is highly probable. The stainless steel alloys susceptible to SCC are generally classified as the 18-8s. This includes the molybdenum containing grades (types 316 and 317), the carbon-stabilized grades (types 318, 321, and 347), and the low-carbon grades (types 304L and 316L). Many variations of the basic 18-8 have been developed in order to combat SCC. These variations are higher in nickel, chromium, and molybdenum, and there are also grades with lower nickel and higher chromium (duplex stainless steels). All of these alloys are highly resistant to SCC and therefore not part of the problem of SCC under insulation.

For SCC to develop, sufficient tensile stress must be present in the material. If the tensile stress is eliminated or greatly reduced, cracking will not occur. The threshold stress required to develop cracking depends somewhat on the severity of the cracking medium. Most mill products, such as sheet plates, pipes, and tubing, contain enough residual tensile stresses from processing to develop cracks without external stresses. When the austenitic stainless steels are cold formed and welded, additional stresses are imposed. As the total stress increases, SCC becomes more severe.

The Cl^- ion is damaging to the passive protective layer on the 18-8 stainless steels. Once the passive layer is penetrated, localized corrosion cells become active. Under the proper set of circumstances, SCC can lead to failure in only a few days or weeks. Sodium chloride, because of its high solubility and widespread presence, is the most common culprit. This neutral salt is the most common, but not the most aggressive. Chloride salts of the weak bases and light metals, such as $LiCl$, $MgCl_2$, and $AlCl_3$, can rapidly crack the 18-8 stainless steels under the right conditions of temperature and moisture content.

The concentration of chlorides necessary to initiate SCC is difficult to ascertain. Researchers have developed cracking in solutions with remarkably low levels of chlorides ($<$ 10 ppm). The situation of chlorides under insulation is unique and ultimately depends on the concentration of chlorides deposited on the external surface of the metal. The concentrating mechanism is discussed in some detail in the following paragraphs.

Therefore, the amount of chlorides present becomes a debatable issue. If any chlorides are detected, there will probably be some localized sites of high concentration.

The most important condition affecting chloride concentration is the temperature of the metal surface. Temperature has a dual effect: first, elevated temperatures will cause water evaporation, which in turn concentrates the chloride salts, and, second, as the temperature increases, the rate of the corrosion reaction increases.

Finally, chlorides have been discussed because they are the most common and aggressive of the halogen family. Bromides and fluorides may also cause SCC, but are less common and probably less aggressive.

Water is the fourth necessary ingredient in SCC. Because SCC is an electrochemical reaction, it requires an electrolyte. As water penetrates the insulation system, it plays a key role at the metal surface, depending on the equipment operating conditions.

Examination of the phenomenon of corrosion of steel under insulation provides a better appreciation of the widespread intrusion of water. In effect, water must be expected to enter the metal/insulation annulus at joints or breaks in the insulation and its protective coating. The water then condenses or wets the metal surface, or if it is too hot, the water is vaporized.

This water vapor (steam) penetrates the entire insulation system and settles into places where it can recondense. Because the outer surface of the insulation is designed to keep water out, it also serves to keep water in. The thermal insulation does not have to be in poor condition or constantly water soaked. A common practice in chemical plants is to turn on the fire protection water systems on a regular basis. This deluges the equipment with chloride-bearing water. Some seacoast locations use seawater for the fire protection water.

Hot food-processing equipment is regularly washed with tap water, which contains chlorides. All insulation system water barriers eventually develop defects. As the vessel/insulation system breathes, moist air contacts the metal surface. From the insulation standpoint, the outer covering acts as a weather barrier to protect the physical integrity of the insulation material. The outer coverings are not intended, nor can they be expected, to maintain an air- and watertight system.

- **Prevention of external SCC**

    As an understanding of external SCC and its mechanisms developed, selection of a preventive method became much easier. Table 1.42 summarizes the possible methods of prevention that apply to each causative agent. As can be seen, application of a suitable protective coating system is generally the most economical method, although other methods are included and may be practical under certain circumstances.

    The critical step then becomes the implementation of whichever preventive method has been selected. Assuming a protective coating has been chosen, it is necessary to convince project or operating managers of the benefits of painting stainless steel. Although painting stainless steel may at first seem unusual, experienced field personnel will usually understand the need for preventive measures.

    Application of any protective coating requires a good specification and inspection. The use of manufacturer application guidelines and the knowledge of an experienced inspector are indispensable in producing an acceptable protective coating.

    Fortunately, protective coatings work very well on stainless steel because the metallic substrate does not oxidize or rust. Therefore, coating adhesion failures are very rare. The main objective is to achieve a continuous coating at a reasonable cost that will resist the hot-water environment encountered under insulation.

1.21.7. Stress Corrosion Attack on Other Metals

Many alloys, when subjected to the combined effects of stress and corrosion under conditions specific to the individual alloy, can develop cracking, although the incidence may not be frequent under normal operating conditions.

Carbon steel (mild steel) can be attacked by soluble nitrates present as contamination in water from external sources, under conditions of mild acidity and temperatures above about 80°C.

Commercial thermal insulating materials do not contain nitrates, but these may be present in natural waters. Thus, it is dangerous to allow water of any kind to penetrate and, above all, to saturate insulation systems where contaminants may concentrate at the metal surface. As a general rule, the danger of cracking from this type of attack is remote, but suitable painting of the surface should be adopted as a precaution where the consequences of fracture of the metal could be serious.

Table 1.42 Guidelines for Selecting External SCC Preventive Methods

Causative Agent	Preventive Method	Comments	Evaluation
Austenitic stainless steel	Change to SCC resistant alloy	Stainless steel alloys with > 30% Ni and the duplex stainless alloys are alternative choices, but cost considerably more and may not be readily available. Annealing at 1065°C (1950°F) followed by water quenching will distort and scale equipment severely.	Extra cost compared to other preventive methods makes this an unwise choice.
Tensile stress	Thermal treatment (anneal or stress relieve)	Stress relieving at 955°C (1750°F) and slow cooling will sensitize the grain structure and cause some warpage and scaling. *Note:* A stress-relieved vessel or pipe will be subjected to tensile stresses in assembly and under operating conditions may override the thermal treatment.	Generally not practical for piping and vessels; may be used for small individual components.
	Shot peen	Shot peening converts the surface stresses to compressive stress and is a proven SCC preventive method. It is a delicate process requiring specific skills and experience and may be costly or difficult to apply in the field.	Should be considered. But may be more costly and difficult to obtain than other prevention methods.
Chlorides	Remove or eliminate Cl$^-$ ion	Because of their widespread occurrence highly soluble chloride salts are difficult to avoid or keep off equipment.	Not practical

Water	Apply barrier coating to stainless steel	Use of a protective coating on the stainless steel surface can prevent Cl^- contact with the alloy.	This is a practical and proven preventive method being used with success.
		Wrap stainless steel with aluminum foil, which serves as both a barrier coating and cathodic protection anode.	Extended life of the aluminum has not been determined.
	Improve waterproofing to prevent water entry	No type of coating, cementing, or wrapping of insulation can keep air and water from entering the insulation system, except for constructing an external pressure shell. *Note:* The application and maintenance of a weather barrier is important to good insulation performance and should have a high maintenance priority.	Not practical to expect a wrap or coating to keep water out.
	Apply barrier coating to stainless steel	A carefully selected protective coating can provide long-term protection for stainless steel equipment.	This is a practical and proven preventive method.
		Use of aluminum foil wrap as above. *Note:* Use of inorganic zinc primer or paint system is not safe due to the possibility of liquid-metal embitterment upon subsequent welding or exposure to extreme heat.	Limited use but with success.

Nonferrous metals, notably the aluminum–zinc, aluminum–zinc–magnesium alloys, and various copper–zinc alloys, may be sensitive to stress corrosion attack, particularly in the presence of moisture and of alkalis. Precautions should be taken to exclude moisture from the system and, where appropriate, the surface of the metal should receive a protective coating of suitable compound.

1.21.8. Attack by Liquid Metals

Although thermal insulation materials themselves do not contain zinc, this metal is commonly associated with securing and reinforcing materials, usually as a protective coating, e.g., galvanized carbon steel.

The melting point of zinc is 419°C and at temperatures above 450°C molten zinc will diffuse and penetrate stainless steel, although embitterment is unlikely to occur until a temperature of about 750°C is reached. The depth of penetration will depend on the time and temperature of the exposure, and the structure and composition of the steel. It will also depend on the degree of applied stress. In the absence of stress it is unlikely that very rapid embitterment will occur.

Below 750°C, the influence of stress upon the rate of diffusion and penetration is relatively small. At temperatures above 750°C, however, the tensile stress in the steel will bring about rapid penetration, and cracking may occur in a matter of minutes or even seconds in the area of contamination. Zinc is volatile at such high temperatures and, while the major hazard arises from molten zinc, there is evidence that embitterment can even be caused from the gaseous phase.

As it is thought that embitterment arises from interaction between zinc and nickel in the steel, it may be anticipated that the austenitic chromium–nickel family of stainless steels will be particularly susceptible to zinc embattlement.

Under no circumstances should galvanized steel be welded and, additionally, contact between galvanized accessories, e.g., galvanized wire netting, and steels that are intended for use at service temperatures above about 350°C should be rigorously avoided.

Paints containing metallic zinc should not be used as surface protection for insulated austenitic steel surfaces in areas of high fire risk where the temperature can exceed 350°C.

Embrittlement also can occur if molten zinc is allowed to drop on to heated alloy steel surfaces, e.g., under conditions of external fire. This is of particular importance when the plant is under stressed conditions, particularly if the contents are flammable.

CHAPTER TWO

Thermal Insulation Installations

Contents

Alireza Bahadori, Thermal Insulation Handbook for the Oil, Gas, and Petrochemical Industries
© 2014 Elsevier Inc.
http://dx.doi.org/10.1016/B978-0-12-800010-6.00002-2

Next to raw materials, energy represents the largest single cost element of most manufacturing processes. Experts recognize thermal insulation as a valuable investment and a tremendous asset to energy conservation and process efficiency. Properly maintained insulation contributes significantly to the operating bottom line of oil, gas, and chemical plants, in turn helping to reduce CO_2 emissions, carbon footprint, and operating costs.

The insulation installation requirements are those requirements that insulation shall fulfill from the time it is manufactured until it is in service. After insulation is produced, it passes through several phases, i.e., it is shipped, stored, possibly fabricated, reshipped to the job, installed, and finally used in the service for which it was originally specified.

These phases that insulation shall pass through may differ depending upon the method of manufacture, type of material, and use, but in general they may be listed as follows:

1. Shipping
2. Storage
3. Fabrication
4. Application
5. Service

Each of these phases imposes a set of requirements upon the insulating material. It must meet these requirements in order to fulfill successfully its economic function. Because of a large number of requirements and also lack of reliable data on the requirements themselves, it is impossible to provide both an exact list of the requirements and of the properties whereby a precise selection of material can be determined. Therefore, this discussion must be kept general, for the final selection will be determined by engineering judgment.

Each "phase" includes one or more of the following:

- Mechanical requirements
- Chemical requirements
- Thermal requirements
- Moisture requirements
- Safety requirements
- Economic requirements

2.1. PHASES

2.1.1. Shipping

To be usable, insulation produced in one location and used in another must be capable of being transported. Rigid insulation must have a combination of compressive, tensile, and shear strengths to permit shipment without excessive breakage or excessively costly protective packaging. Nonrigid materials, such as blankets, shall have sufficient strength in length, width, and depth to resist delamination. Their binders shall withstand vibration without dusting or excessive fiber release. Fibrous materials must resist permanent compaction caused by either compressive forces or vibration.

Fill materials must also withstand excessive compression or compaction, even though packaged in bags. Practically all insulating materials must be packaged in cartons, bags, rolls, or some other type of acceptable shipping container. Proper selection of container and packaging influences the cost of handling and space requirements.

During loading, shipping, and unloading, materials may encounter water damage caused by rain or water vapor due to high humidity. The more highly absorbent or adsorbent the insulation is, the greater the hazard of excessive moisture pickup. Thus, the greater the adsorbency and absorbency, the greater the need for care in handling and weather-tight packaging.

2.1.2. Storage

Insulation is seldom used immediately upon receipt from the manufacturer. In most instances it is stored first by a distributor and later on the job. During these periods of storage it often receives more physical abuse than it will receive in service. Thus rigid materials must have sufficient strength to withstand such handling. Flexible materials must withstand the same type of handling without compaction, or pulling apart. Of course, all insulation should be dimensionally stable and not warp or shrink from moisture or age.

Few inorganic insulating materials are subject to chemical change during storage. However, some of the organic foam spray chemicals may be affected by shelf life or temperature. These must be stored and used within their limits, or they may become unusable. The same applies to some accessory materials such as adhesives, sealers, and weather barriers.

The problem of keeping absorbent insulations dry during this storage period is most important. Insulation that contains a large percentage of water may lose its strength, or chemically deteriorate, and will always lose its thermal efficiency. Improper warehousing and handling of absorbent insulation can be costly not only to the insulation itself, but its packaging may be adversely affected by water.

Water damage to cartons, with resultant loss of identification and damage to the contents, may be very expensive; needless to say, the more absorbent insulation is, the greater the need for its protection from water in any form.

During storage, safety to property and personnel must be considered. Any combustible insulation shall be stored with care. Fire will spread rapidly in such light density material. Sealers, weather/vapor barriers, or other liquid combustibles are always a potential fire hazard. Many insulations that do not support combustion by themselves will burn in the presence of flame. Any jacketing will produce toxic fumes when burning. Safe storage is, therefore, essential for these materials.

2.1.3. Fabrication

Advanced techniques have made it practical and economical to prefabricate, preform, or premold insulation shapes prior to their installation in the field. A rigid material requires sufficient mechanical strength to permit handling without breakage. It should have good cutting characteristics. Its cut surfaces shall be sufficiently smooth and free of dust in order that pieces can be cemented together into strong, integral finished shapes.

For ease of fabrication, the original blocks of insulation should be dimensionally true within acceptable tolerances. Out-of-square, nonparallel, and untrue planes in the original blocks will cause excessive waste of both material and labor. Materials that delaminate when cut also cause excessive labor and waste.

In most instances, materials that have good tensile strength in all directions are more suitable for cutting into compatible parts, and can then be bonded together into the finished shape.

Light density, fibrous materials generally do not lend themselves to this cutting and cementing type of fabrication. Fittings composed of these materials are formed most frequently by molding and heat-curing uncured batts or blankets into the desired shapes. Inorganic insulations are treated in a similar manner. Organic insulations may be molded into desired shapes. During fabrication, or molding, safety is an important consideration.

2.1.4. Application

Industrial applications impose additional requirements on the insulation materials that are important to a successful installation. The materials must have the necessary strength to resist an excessive amount of handling. Storage space at most of these installations is limited, and the materials may be moved many times before they are finally installed.

When secured by wires, materials must resist the tendency to crack into pieces due to cleavage along the wire, yet be of such texture that twisted wire ends can be embedded into their surface. Vessel block insulation and rigid pipe covering must withstand considerable force when pulled up to a tight fit by straps. The strapping tool will exert a 262–272 kg tensile pull with an insulation strap to draw the joints tight.

The dustier a material is the more difficult it becomes for field personnel to work with and install it. Dust, besides being a health hazard, irritates the eyes, makes scaffold boards slick, and causes a cleanup problem. In addition, dusty surfaces resist good bonding with insulation and finishing cements, and weather-barrier mastics.

The tendency of some insulating materials to absorb moisture adds to the cost of installation. Such installations must be protected from water, rain, and snow, both before and during their installation.

The need for weather protection before, during, and after installation until the weather barrier is installed adds considerably to costs. Trueness of dimensions of block pipe covering is essential to efficient field installation. If the ends of pipe covering are not square and true, the gaps must be plugged with insulating cement, or the end recut to fit.

2.1.5. Service

To perform its intended service, insulation must remain where it was installed. Its properties must fulfill all the requirements imposed upon it during its service life. The requirements of service phase is divided into various technical divisions in the following order:

2.1.5.1. Mechanical Requirements of Insulation on Equipment and Pipes

Insulation on equipment located in industrial plants is subject to external and internal physical forces. The external forces include bumps, persons walking upon the insulation, vibration, wind, or even explosions. Many insulations are expected to protect pipe and equipment from fire. Many fires are started by an explosion, and if the insulation is blown off of surrounding items and piping it cannot protect against fire.

The insulation shall be sufficiently strong to resist ordinary usage. This entails the compressive force of people walking on insulation, the force of vibration that causes abrasion between the insulation and the surface to which it is secured, and the force of wind or partial vacuums that tends to compress or pull insulation apart.

The forces most frequently damaging to insulation applications are those built into the system by poor design and application, which ignore the movement of vessels and pipes caused by thermal expansion (or contraction) of the metal.

Expansion and contraction of vessels and pipes can cause serious damage to thermal insulation and weather-barrier coverings so that the cracks may develop. Additional heat loss from these small–sized cracks would be quite small. The big difficulty is water entering the cracks, which makes the insulation wet and increases thermal conductivity.

2.1.5.2. Chemical Requirements

The insulation to be used shall not react to the chemical contained in the vessel or piping to which it is attached. Another factor to be considered is that the insulation shall be nonabsorptive when used on toxic processes.

A selected insulation shall not be chemically corrosive to the metal to which it is applied. Basically, insulation installed on steel shall be neutral, or slightly alkaline. That installed on aluminium shall be neutral or slightly acidic. Austenitic stainless steels are susceptible to stress corrosion cracking by chloride ion; therefore insulations with limited chloride content shall be used or resistant coating shall be applied on stainless steel surface.

2.1.5.3. Thermal Requirements

The thermal requirements of insulation are related to a number of properties of materials. These thermal properties of materials are as follows:

- Temperature limits
- Thermal shock resistance
- Thermal diffusivity
- Thermal specific heat
- Thermal conductivity

The temperature requirements of an installation are the maximum and minimum temperature to which the insulation will be subjected. Other factors such as duration of time the insulation is exposed to heat, mode of applying heat, and the time rate of raising or lowering the temperature should also be considered. The following questions must also be answered: Is it on only on one side, and is it continuous, intermittent, cyclic, or rapidly changing?

2.1.5.4. Moisture Requirements

Insulation, to be efficient, must be kept dry. Moisture in two forms, the liquid or the vapor state, can saturate the insulation. It can enter by various means, due to water pressure, vapor pressure, and gas or other simple leaks.

The protection required will be determined by the installation conditions. When insulated piping or equipment is submerged in water, a completely watertight outer shell is required around the insulation. Such systems must be encased in metal, fabricated by welded sections, and/or flanged with jacketed flanges, so that the outer jacket is sufficiently tight and strong to withstand the water pressure.

Underground installations where insulation systems may not be directly subjected to definite water pressure must still be protected to prevent the entry of water. Such protection must not only prevent the said entry, but it must also resist corrosion to remain watertight. The corrosion may be ordinary rust, chemical corrosion, or galvanic corrosion.

If the function of the two previously mentioned examples is the insulation of hot surfaces, vapor migration will be of no concern. However, if the installation is for low-temperature service, water vapor migration will be an additional problem. Many conduit systems may be watertight and not be vapor-tight. Practically, it is relatively easy to construct watertight systems, but almost impossible to construct vapor-tight ones. For this reason, it is suggested that, wherever possible, the cold equipment and piping be installed above grade.

The moisture protection necessary on equipment and piping located above grade is determined by whether the location is indoors or outdoors, and the operating temperature above or below ambient.

High-temperature insulated equipment and piping located indoors and not subjected to rain, snow, or sleet does not require water or moisture protection. A word of caution: the insulation must be adequately protected from moisture due to possible spillage or the washing down of vessels.

Low-temperature insulated equipment and piping located indoors must be protected against moisture vapor. High-temperature insulated equipment and piping located outdoors must be protected from liquid moisture in the form of rain, sleet, or snow.

Low-temperature insulated equipment and piping located outdoors must be protected both from liquid water, and also moisture in the vapor phase. These moisture requirements are summarized in Table 2.1.

2.1.5.5. Safety Requirements

The major safety requirements may be separated into the following divisions:

- Safe surface temperature for personnel protection or ignition
- Safety from radioactivity or chemical reaction.
- Safety from toxic conditions
- Fire protection.

When hot equipment or piping is located where its insulation may be touched by personnel, the insulation shall be so designed that its thermal resistance, surface emittance, and jacket conductivity together create a condition that will not produce skin burns when its outer surface is accidentally touched.

Table 2.1 Moisture Requirements

Service		Moisture Protection Required		
Temperature	Location	Water Pressure	Water	Vapor
High temp.	Underwater	Required	Required	Not required
Cyclic temp.	Underwater	Required	Required	Required
Low temp.	Underwater	Required	Required	Required
High temp.	Below grade	It depends if the condition of high-pressure water or water may exist	Required	Not required
Cyclic temp.	Below grade	It depends if the condition of high-pressure water or water may exist	Required	Required
Low temp.	Below grade	It depends if the condition of high-pressure water or water may exist	Required	Required
High temp.	Above grade indoors	Not required	It depends if the condition of high-pressure water or water may exist	Not required
Cyclic temp.	Above grade indoors	Not required	It depends if the condition of high-pressure water or water may exist	Required
Low temp.	Above grade indoors	Not required	It depends if the condition of high-pressure water or water may exist	Required
High temp.	Above grade outdoors	Not required	Required	Not required
Cyclic temp.	Above grade outdoors	Not required	Required	Required
Low temp.	Above grade outdoors	Not required	Required	Required

Where combustible materials are present and changes of leakage or spillage prevail, a suitable low-surface temperature of insulated vessels and piping may be very important in preventing the ignition of these materials. When this condition prevails, it becomes necessary that all hot surfaces, such as valves, flanges, and metal projections, be insulated so as to maintain all exposed surfaces below the maximum allowable surface temperature. Some areas that cannot be insulated, such as valve bonnets and stuffing boxes, may require shielding to prevent contact with the chemical.

As previously mentioned, noncombustible but absorptive insulation may be a fire hazard in two ways. First, in the case of combustible material leaks, the insulation's tremendous internal surface area can assist in causing rapid oxidation, which will result in spontaneous combustion. Second, also in the case of leaks, any absorbent can hold large quantities of combustibles and, if an accidental fire occurs, the saturated insulation will feed and spread it. However, properly used insulation can be used as fire protection.

An installation on equipment and piping must first consider the combustibility of the insulation, including the speed at which the insulation will spread the fire, or the amount of heat and smoke it will contribute. Second, the time rate at which temperature will pass through the insulation from the fire side to the vessel or pipe must be thought of.

2.1.5.6. Economic Requirements

The economic value of the insulation depends upon the installation requirements, and as this has been previously discussed, a simple listing of the factors to consider is given.

- Heat transmission allowable as determined by the process.
- Savings in boil-off or vaporization of the product.
- Temperature that must be maintained in the process.
- Fire protection required by the process. Insulation as related to savings in additional fire protection such as spray heads, larger safety valves, etc.
- Savings in investment in heat or refrigeration equipment.
- Savings in heat or refrigeration energy.

2.2. INSTALLATION SYSTEM DETAILS

The success of an insulation system depends upon individual consideration of design and insulation for the many details of the system. In all insulation design, thermal "short circuits" should be avoided or minimized.

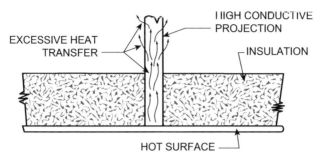

Figure 2.1 Thermal "short circuits" (excessive heat transfer through void).

These thermal short circuits are most frequently the result of metal or other high conductive materials through the insulation. This is illustrated in Figure 2.1.

The illustration shows the excessive heat transfer from a hot surface to a lower ambient temperature. If the condition were reversed, with the temperature of the surface lower than ambient, the same excessive heat transfer would exist, but the heat flow as shown would be reversed.

2.2.1. Voids in Insulation

Some voids in insulation are the result of insulation design and specifications. Typical of voids in the insulation system, which are a result of design, are those specifications that state that flanges and valves are not to be insulated. Voids that are not deliberately designed in, but develop in, an insulation system are most often the result of expansion of the substrate to which the insulation is installed, shrinkage of the insulation, or poor design or workmanship.

2.2.2. Projection Through Insulation

Metal projections through insulation should be avoided, if possible. Metals of two different temperatures should be isolated by insulation where it is practical.

Where it is necessary to have high conductive materials, such as metal, project through insulation, the heat transfer can be minimized by the following methods:

- Keeping the cross section of the projection to a minimum
- Use of a material of the lowest possible thermal conductivity

- Providing a long flow path through insulation
- Providing a thermal barrier in some phase of the projection connections or bearings

The use of stainless steel, which, besides having a lower conductivity than carbon steel, also has higher strength, can reduce any thermal transfer to one-fourth that of carbon steel. The length of the flow path of projection through insulation can be extended by insulating the projection beyond the basic insulation installed on a pipe or vessel.

Great emphasis has been placed on voids and projections through insulation on low-temperature service. The reason for this is that these voids and projections will cause more rapid insulation failure on such service than on high-temperature service.

One area where voids and projections through the insulation completely nullify the purpose of insulation is where that insulation is used for fire protection of vessels and piping. In such a case a heavy projection through the insulation, which would be exposed to the fire, would transfer the fire heat through the insulation at that point and fire protection would not be achieved. This is also true of voids in insulation. During fire the insulation will shrink and cause voids at the butt joints. For this reason, when fire protection is a critical function of insulation, double-layer, broken-joint construction shall be used.

2.2.3. Traced Piping and Vessel

The tracers are used where it is necessary to add or remove heat from piping or equipment. If the additional temperature to be supplied by the tracers is small then they may simply be fastened to the pipe or vessel and the air space between the insulation and metal substrate will distribute the heat over the pipe or vessel surface. However where high temperature or close tolerance of temperature is required, then the tracers shall be thermally bonded to the pipe or vessel.

For efficient operation of tracing systems they must be installed correctly. The metal of the system shall resist corrosion. Where necessary, the tracer must be thermally bonded by heat transfer cement, or pre-molded strip, to the process pipe or equipment. The heat transfer cement must be suitable for the process pipe and tracer temperatures to which it is to be subjected and shall not cause corrosion or rusting of the tracer or piping.

2.2.4. Weather/Vapor Barrier Details

- **Vapor barrier**

    Because vapor barriers on the outer surface of insulation are subject to all the ravages of time, mechanical abuse, and stresses imposed upon the system, cracks, breaks, punctures, or tears are likely to occur and in this case a small leak will allow the entry of vapor into the entire system. This being true, the goal is to provide a system that is compartmentalized so that a break or puncture does not ruin the entire system. Application of such a system is not important for cellular glass as almost universal material for low-temperature insulation on pipes and vessels and may be eliminated but it is essential for efficient long service of permeable organic foam insulation.

    It is critical to understand what causes failure to some insulating materials over time. The primary cause for insulation material losing its thermal resistance is the water vapor or moisture condensation problem. Over time, insulation materials absorb water and settle or cause surrounding materials, for instance drywall, to be damaged. This water absorption can cause indoor air pollution and damage to a structure. Figure 2.2 illustrates the role of vapor barriers in the installation of a thermal insulation system.

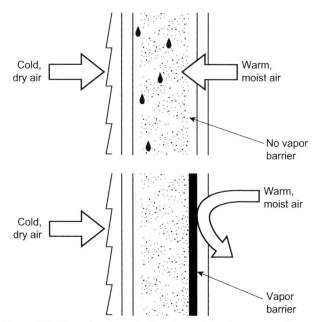

Figure 2.2 Vapor barrier roles for thermal insulation installations.

Air retarders block air and liquid water but they are not capable of blocking water vapor, which can condense inside insulation and cause damage to substances such as gypsum board and housewrap. But vapor barriers allow very little water vapor travel and are measured as per the ASTM C755 vapor retarder standard. In regards to residential applications, it is required that they cannot exceed 1 perm. The US perm is defined as 1 grain of water vapor per hour, per square foot, per inch of mercury and 1 grain = 64.79891 milligrams.

The major property a vapor barrier must have is its ability to retard the flow of vapor. This is its most important property, but its other properties, such as its ability to withstand mechanical abuse or weathering, may not be sufficient to fulfill the installation requirements. Where this is true, it may be necessary to provide a weather barrier over the vapor barrier. Although there are mastic materials that are formulated to serve both as vapor and weather barriers, in many low-temperature applications it is desirable to install a vapor barrier and protect it with a weather barrier.

The vapor sealing of cellular glass is important, even though it has greater resistance to vapor transmission than most materials used as vapor barriers. With proper joint sealing of cellular glass, and proper design to prevent its cracking or shearing due to expansion and contraction movement, there is no need to apply vapor-barrier material over its exterior surface. However it does require a weather barrier outdoors, and a finish indoors to protect it from ordinary weathering and mechanical abuse.

- **Weather barrier**

The prime function of the insulation weather barrier is to provide conditions in which the insulation can function to retard the heat flow. The weather-barrier covering jacket or finish forms the necessary protection around the insulation and for a successful installation, the details of application are of utmost importance. Regardless of the weather barrier used, it is ineffective if it is installed in a manner that will allow water to leak into the insulation. An example of proper method of installing a jacket that prevents water from leaking into insulation is shown in Figure 2.3.

Fitting and valve covers present a difficult problem in obtaining watertightness. Too frequently metal covers are installed by slipping the halves together without sealing the overlapping metal with mastic sealers. This type of construction will not provide a watertight installation, and only serves to hide the water as it soaks into the insulation with every rain.

Another common incorrect installation practice is the use of mastics as the weather barrier on valves and fittings while the straight pipe is jacketed with metal that projects over the extended mastic of the fittings, but with no sealing mastic used to prevent water from entering the overlap of the two. Nonsetting sealers must be used in these spots, as the expansion and contraction movement will break any sealer that sets and hardens.

When projections are so hot where they emerge from the insulation that suitable mastics cannot be found to seal this joint, then the projection should be insulated a sufficient distance outward to a location where its exit point through the insulation can be sealed.

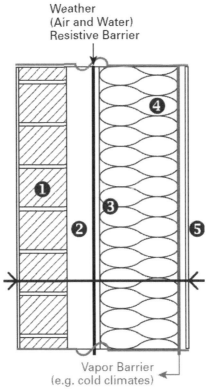

Weather (Air and Water) Resistive Barrier

Vapor Barrier (e.g. cold climates)

Figure 2.3 Proper sealing of steel stud cavity wall with fiberglass batt insulation in the stud cavity. 1. Exterior cladding 2. Air space (optional; recommended for water management) 3. Exterior sheathing 4. Cavity insulation 5. Interior sheathing.

Regardless of the type of weather barrier used, the inside corner of these projections should be caulked with a caulking compound of high solids content to eliminate the sharp inside corner. This same procedure shall be used at all projections through the insulation, such as where the pipe itself projects through the insulation, and where vessel nozzles, manholes, or skirts project through the insulation.

After the projections are properly insulated and caulked, the inside corners shall be weather-protected with a flexible mastic coating reinforced with extensible fiber cloth. If the weather barrier is mastic-reinforced with cloth, the weather-barrier system shall extend a minimum of 100 mm out over the projection.

If metal jacketing, felts, or plastic films are used as the basic weather-barrier system, then the flexible mastic with the extensible fiber cloth shall extend a minimum of 100 mm in each direction from the inside corner. Where projections are quite hot, the mastic used may be required to be high-temperature sealer instead of ordinary weather-barrier mastic.

2.3. APPLICATION PROCEDURE

The following application procedure, which involves consideration of many variables, must convey to the company, clearly and without equivocation, what the insulation expert intends to accomplish in the installation. Wherever necessary the procedure may be tailored to the specific job. The procedure must be written definitely and exactly to obtain the proper quality of material and assembled system.

We will not attempt to just present a single insulation application procedure, but will present a method for constructing a procedure, combined with a checklist from which the items for the individual specification may be selected.

The insulation application procedure may be broken down into the following major sections

There are general conditions that state what job conditions are to be expected, such as housing, storage, and responsibilities. This further acts as the specifications to the contract agreement between the contractor and his customer. As such, the general conditions section is a legal document on agreement, and not a technical matter that is out of place in application specification.

The scope of the procedure sets the boundaries within which the procedure is to be used. This section shall be sufficiently directive to indicate to the reader as to the general purpose of the detailed specifications that will follow, and should include a general classification of the material to be used, the boundary limits of facilities to be served, and the location. To illustrate, if the procedure is to be used to insulate the roof of a building, say so in the scope. Also state the conditions under which the insulation is to serve.

2.3.1. Application Specification

The general insulation specification to which individual specifications are referred shall have the following format:

- **General specification**
  - List of materials required
  - Preparation
  - Insulation application
  - Weather barrier or covering application

But each individual specification shall have the following format:

- **Individual specification**
  - Preparation
  - Insulation application
  - Weather barrier or covering application.

The preparation section shall state what must be done before the insulation is applied. In almost every instance, these are operations that must be performed by other craftsmen and possibly other contractors prior to the application of the insulation. The specification must state what these operations are and place the responsibility for them correctly, so as to prevent conflict and delay.

The insulation application section shall spell out the intent as to the quality of the installation, and what is expected from the finished product of assembled insulation and accessories. Of course, no specification can detail each and every piece of the installation, but sufficient typical details must be provided such that the applicators understand what is expected. For this reason, it is extremely helpful to include drawings in specifications to illustrate typical assemblies. Most workpeople, when shown a drawing, can get the idea of what is desired much more quickly than by reading a detailed description.

The weather barrier or covering application section may be made a part of the insulation application section, as the barrier is a part of the insulation system, but because these are installed last, it does lend itself to being a separate section, especially as a different set of materials is being used. Another advantage in making such a separate section is that, under various conditions, insulation installed as called for under the insulation application section may require one type or system of weather barriers in one location, and another in a different location. Conversely, one set of weather barrier application specifications may be suitable for use on many various types and forms of insulation.

An example of the preparation of application specification follows.

2.3.1.1. Application Specifications: Equipment and Piping
- **Preparation**

    Insulation supports, of the same metal as the vessel, projecting to a point 25.4 mm less than the thickness of the insulation, shall be welded or bolted to the vessels to support vertical insulation. They shall be located at the base of the insulation, at the top of the vessel, and at 457.2 mm centers above the base support. A support shall be located above each vessel flange, a sufficient distance above the flange bolts to allow for their easy removal. The bottom and top supports shall be slotted with 25.4 mm by 7.6 mm slots for attachment of straps or wire.

    Pipe above 8.9 mm diameter NPS shall be supplied with insulation supports welded or bolted just above lower elbow of risers. Projection of this support shall be a minimum of 38 mm. On long vertical runs of pipe, additional supports shall be installed on 457.2 mm centers.

    Flat surfaces and large bottom heads shall be equipped with rectangular punched welding pins to support the insulation. These pins shall be welded to surface in an approved manner. On the top surfaces the pins shall be located using a 0.61 m centers-diamond pattern, on the sides using a 0.457 m centers-diamond pattern, and on bottom surfaces using a 0.305 m centers-diamond pattern.

    All surfaces shall be clean and dry. Where metal surfaces are to be painted to prevent corrosion, the paint shall be completely dry prior to the installation of the insulation. Where heat tracing is required it shall be installed prior to application of the insulation.
- **Equipment**
  - **Application to curved surfaces**

      On equipment operating at over 260°C and 1.83 m and larger in diameter, a one-inch cushion blanket shall be applied over the entire surface before application of the calcium silicate curved segmental insulation. The blanket shall not be compressed, and shall be secured with a minimum amount of stainless steel wire.

On all equipment, calcium silicate shall be molded, or cut, into curved segments to fit the vessel surface, with a one-inch space allowed for cushioning blanket, where used. Curved block shall be applied to vessel sides in staggered position with all joints tightly butted. Insulation shall be secured with straps on 0.229 m centers. Multiple layers shall be applied so that the butt joints of one layer do not coincide with those of the other.

All layers of insulation shall be secured with straps where the contour of the vessel permits firm attachment. Insulation applied to irregular surfaces, where use of straps is impractical, shall be secured with wire.

- **Application to flat surfaces**

  Block insulation shall be applied to flat surfaces in staggered position with all joints tightly butted. Blocks shall be secured with wire attached to welding pins welded to the surface. Multiple layers shall be applied so that the butt joints of one layer do not coincide with those of any other layer. After the specified thickness of insulation has been applied, wire netting shall be stretched tightly over block and secured with tie wires.

- **Application to vessel heads**

  Vessel heads shall be insulated with preformed block (and cushion blanket where required). Insulation on top heads up to 3.658 m diameter shall be secured with straps. Top heads larger than 3.658 m diameter, all bottom heads, and vertical heads shall have insulation secured with wire attached to rectangular pins located.

- **Application to vessel flanges**

  All vessel flanges shall be insulated unless specifically excepted. The flange cover shall be a preformed-slip type flange cover of the same thickness as the specified insulation thickness. Flange cover shall be of step construction, so designed that it functions as an insulation expansion joint. The upper and lower sections of the cover shall be secured to the curved side wall insulation with straps and skewers. The midsection shall be secured at its bottom so that it may slide in respect to the upper section.

- **Application of expansion joints**

  Insulation expansion joints shall be provided on 4.572 m centers. The insulation support shall be 25 mm above the termination of the insulation below. This void shall be packed tightly with fibrous glass wool. Slip sleeves of stainless steel sheet shall cover this opening and protect it from the weather.

Where a vessel flange occurs on a vessel, expansion shall be considered as taking place, and be compensated for by the flange cover at that location.

- **Application to manholes and nozzles**

  All manholes, blind nozzles, and connecting piping flanges shall be insulated (unless specifically excepted) with precut oversize covers. Thickness shall be the same as specified for the vessel insulation. These covers shall be attached to the vessel insulation by wire, straps, and skewers in such a manner that movement of the vessel does not cause them to break loose.

- **Application to vessel legs**

  Channel legs for vessels shall be insulated with blocks that shall extend over channel flanges and down to fireproofing or footer. The thickness of insulation shall be a minimum of 13 mm more than that specified for the vessel but in no case less than 50 mm.

  Vessel insulation shall be butted firmly against the tank leg flange. Re-entrant space between the channel flanges and over the channel flange between body insulation shall be filled with block. Leg insulation shall be secured with straps and wire.

- **Application to vessel skirts**

  Skirts of vertical vessels shall be insulated down a minimum of four times the insulation thickness specified for the vessel from their junction points with the vessels. The thickness of insulation shall be the same as specified for the vessel. Where required for fire protection, the outside and inside of skirts shall be insulated entirely, with a minimum of 50 mm of block insulation precut to fit the skirt curvature. No cushion blanket shall be used under this block insulation.

  The outer insulation on the skirt shall be secured with straps. The inner insulation on the skirt shall be secured with wire attached to rectangular pins welded to the inside of the skirt on 25 mm centers.

- **Surface finish**

  Regular cylindrical surfaces, limited area flat surfaces, and preformed covers do not require any insulation or finishing cement surfacing. Any slight voids may be pointed up with insulating cement troweled flush with adjacent surfaces.

Large flat surfaces and irregular surfaces over which wire netting is installed shall be given a 12.5 mm thick, smooth and uniformly applied trowel coat of hydraulic setting insulating cement. The weather barrier shall be as designated in the insulation schedule. Materials and application shall be as called for under that designation.

- **Piping**
  - **Application to straight pipe**

    Vertical pipe over 76 mm NPS shall have insulation supported by an insulation support, welded or bolted to pipe directly above the lowest pipe fitting. Additional insulation supports shall be located 457 cm on center above the bottom support. An insulation support shall also be installed above each valve or pair of line flanges located in the vertical run of the pipe.

    Insulation shall be section up to 324 mm diameter and may be sectional or curved segments. Above this diameter insulation shall be applied in staggered joint construction. Multiple layers shall be installed so that the butt joints of one layer do not coincide with that of another.

    Securement of insulation shall be by wire up to 324 mm. Above this diameter the insulation shall be secured with straps, except that all inner layers shall be secured with wire. If metal jacketing is specified for the weather barrier and it is secured by straps, then all layers of insulation, regardless of size, shall be secured with wire.

  - **Expansion joints**

    Expansion joints in the insulation shall be installed every 457 cm of uninterrupted straight pipe in both the horizontal and vertical. The insulation of single and each of multiple layers shall be terminated in a straight cut. A space of 25.4 mm shall be left between the insulation terminations. This void shall be packed tightly with glass wool blanket. The expansion joint shall be protected by stainless steel sleeves.

  - **Application to flanged fittings**

    All flanged valves and fittings, with the exception of ball and plug valves, shall be insulated with preformed covers in accordance with the dimensions given in ASTM Recommended Practice C-450, latest revision. Ball and plug valve covers shall be field fabricated of the proper sectional pipe insulation. These covers shall be secured in position with straps.

- **Application of welded fitting covers**

    All welded fittings over 89 mm diameter shall be insulated with preformed covers in accordance with ASTM Recommended Practice C-450, latest revision. These covers shall be secured in position by straps or wires, depending upon pipe diameter.

- **Application to small welded or screwed fittings**

    Welded fittings under 89 mm diameter and screwed fitting covers may be preformed or field fabricated and secured in position with wire. Fittings less than 48 mm diameter may be insulated with insulating cement installed to the specified thickness.

- **Finish**

    No insulating cement or finishing cement shall be used to cover any preformed pipe or fitting covers. Slight voids shall be pointed up with cement to bring flush with adjacent surfaces. Application to heat traced pipe (where required).

    Piping requiring tracing by tubing or electric conduit up to 16 mm diameter shall be insulated with oversize insulation. The size of preformed fittings for this traced piping is given in the "Traced" section of ASTM Recommended Practice C-450, latest revision. After tracing is installed and is thermally connected by heat transfer cement, the pipe and fittings shall be insulated as previously specified.

2.3.2. Weather Barrier Designation I: Metal Jacket Pipe

- **Application of metal jacket to insulated straight pipe**

    Jacket shall be installed by placing it around the pipe insulation and engaging the "Z" joint. The "Z" joint on horizontal piping shall be on the side of the insulation, with the open edge of the "Z" joint pointed down. The butt joints between adjacent jackets shall be sealed with a closure band. Closure band sealing compound shall be used to seal voids across the interior of closure bands where they lap over "Z" joint. Closure bands shall be secured in place with insulation strap.

    On pipe insulation less than 300 mm outside diameter an insulation strap shall be installed at the halfway point from each end to secure the entire assembly. On pipe insulation above 300 mm outside diameter it shall be secured in position by two straps. Straps shall be spaced 300 mm from each end. Weather barrier for fittings and irregular shapes of piping insulation is given in weather barrier designation II, which follows.

2.3.3. Weather Barrier Designation II: Reinforced Mastic

• **Application**

The surface of insulation shall be smooth, even, and free of voids, and in a relatively dry state. On outside installations, insulation shall be sloped for water drainage.

Sharp outside corners of insulation shall be rounded off to not less than a 25.4 mm radius. Inside corners shall be caulked with caulking mastic to obtain a minimum 25.4 mm radius inside corner.

A heavy fillet of heat-resistant sealer shall be installed around all metal that projects through the insulation. The sealer shall extend over the insulation and 152 mm over the projection. Insulation shall be protected from weather as soon as possible. However, the barrier shall not be applied when atmospheric temperature is below 0°C or when the temperature is expected to be as low as -3.9°C within the ensuing 24 hours.

Reinforcing cloth shall be bonded, taut and smooth, to the insulation surface with the weather-barrier mastic. All joints shall be overlapped a minimum of 51 mm. All inside and outside corners shall be rounded and overlapped with two layers of cloth. Cloth shall extend a minimum of 102 mm out on all projections through the insulation.

Mastic shall be troweled or palmed over the entire surface, pressing it through the mesh of the cloth to obtain a bond with the insulation. The weather-barrier mastic shall be carried out 154 mm onto metal, beyond termination of the insulation on supports, skirts, or other metal projections. Care must be taken that mastic completely seals the openings in the cloth. After the weather barrier has partly set, it shall be water brushed to a smooth, even surface. The combined thickness of the weather-barrier coating and the reinforcing cloth shall not be less than 1.6 mm when dry.

Drip on floors or concrete, or splatter on gages, valve stems, instruments, or other items must be immediately washed clean with water, then dried. Expansion joints in weather barriers over insulation expansion joints shall be constructed with expansion sleeves.

In this sample specification reference illustrations may be given for easier understanding of the intent of the specification. The use of drawings to illustrate specifications is highly recommended as being a very effective means of communicating information.

In the preparation of specifications, various means may be used to reduce the space and amount of written matter. For example, the properties of material may be presented in tabular form rather than by referring to each property and test method in written form. These methods of presentation are quite efficient and desirable.

Here it shall be reemphasized that the specification, as presented, was a sample of the items to be covered by a specification and a suggested format by which all these items can be presented to communicate the desires of the designer–engineer to the contractor and his personnel. The presentation given is not a recommendation of insulation application. Only the engineer who knows his installation requirements, the properties of materials, and how they should be applied is in a position to prepare application specifications for his installation.

2.4. EXTENT OF INSULATION

Hydrostatic pressure tests on pipe, vessels, and equipment shall, if possible, be completed before insulation is installed. If insulation is applied before testing, all welds, threads, and bolted joints shall be left exposed until completion of testing.

To facilitate regular inspection of welds, bolted joints, thickness measurements, etc., a removable portion of insulation and finishing material shall be provided in appropriate locations to be selected at the site by the employer. The junction between removable and permanent insulation shall be made readily discernible, e.g., by painting the end of permanent insulation or laying a suitable textile fabric over the end.

The removable cover shall be of the same basic material and thickness as the permanent insulation and of sufficient rigidity to withstand handling. Vertical vessels that have a marked decrease in temperature from bottom to top shall be insulated as follows: the lower half of the vessel shall be insulated for the bottom service temperature, and the upper half may be insulated for the vessel overall average temperature.

Projections beyond the normal insulation thickness, such as stiffener rings on vessels that are on integral parts of such equipment, shall be fully and independently insulated and finished in the same manner as the equipment and the cleading shall be arranged to allow for expansion of the vessel.

Pumps, compressors, and turbines shall not be insulated unless required for process control or for safety aspects.

Nameplates containing design and/or operating data, stampings, thermowell bosses, pressure tappings, and warning notices on heat exchangers with differential pressures shall not be insulated and the insulation shall be beveled back and sealed or a window shall be provided if necessary.

Seals vent chamber and drip pots in pipelines shall not be insulated unless specified otherwise by relevant standards.

The word "fitting" shall designate ells, tees, caps, reducers, meters, and stub-in connections. The installation of rings around nozzles of insulated vessels shall be avoided since this would interfere with the easy access for inspection measurements and hammer testing. Distance of 600 mm from their point of contact with the shell or to within 50 mm of the top of the fireproofing, where applicable should be considered for installation purpose.

2.5. HOT INSULATION

Vessel manways, handholes, weep hole nipples, sample connections, nozzles with blind flanges, exchanger tube sheets, flanges, etc. shall not be insulated on hot insulated vessel exchangers or equipment, but shall be insulated where required for acoustical control. If they are to be insulated, preformed insulation of a design permitting removal and replacement shall be applied.

Piping bends are usually insulated to the same specifications as the adjacent straight piping. Where preformed material is used it shall be cut in mitered-segment fashion and wired or strapped into position; alternatively, prefabricated or fully molded half-bends may be used if these are available. Plastic composition may be used to seal any gaps that may appear between mitered segments.

It is preferable that flanges, valves, and other fittings on hot piping above 300°C be insulated, but where hidden flange leakage may cause a possible fire or other hazard, e.g., with oil lines, or where repeated access will make it uneconomical, insulation may be omitted.

Valves and flanges in piping for viscous fluids, e.g., asphalt and paraffin wax service, and in all lines in services above 300°C shall be provided with removable insulation covers. Attention shall be paid to the insulation details to prevent leaking product into the line insulation. Drainage outlets should be provided to give visible indication of a possible valve or flange leak.

Bonnet and channel hangs on heat exchangers shall preferably be insulated by means of a removable double skin box. If the weight of the box exceeds 25 kg it shall be in two or more parts and the weight of each part be less than 25 kg.

For heat exchangers on hydrogen duty, tube–sheet and channel flanges shall not be insulated, but a simple removable galvanized sheet metal protecting shroud shall be placed over the bolts to protect them from the effect of thermal shock from rain storms. A suitable gap shall be left between the bolts and the shroud to allow adequate ventilation. Also flanges for hydrogen service shall never be insulated.

Steam traps and the outlet side of the steam trap piping shall not be insulated. Lines to steam traps shall be insulated. In the case of thermostatic type traps, 600 mm to 1000 mm of line before trap shall be left uninsulated.

Heat exchanger flanges, exchanger channels and shell covers, saddle support for horizontal equipment, equipment shell closure flanges, nozzles to which noninsulated piping is attached, tube unions, and tube and shell connectors on fin–tube exchangers also shall not be insulated unless specified otherwise or required for personnel protection.

The bodies of safety valves shall not be insulated. Steam condensate lines shall remain uninsulated except as required for personnel or freeze protection.

2.6. COLD INSULATION

All external surfaces shall be completely insulated except for pumps, which shall be insulated only if specified. Vessels and their skirts shall be thermally isolated from their holding down bolts and pipes shall also be thermally isolated from their supports.

Hardwood blocks or other suitable material shall be used for this purpose. Ladders and platforms shall also be isolated from the vessels to which they are attached. Where hardwood blocks or similar material is used, this shall be suitably fireproofed. Insulation shall be extended sufficiently to prevent frosting of adjacent parts. Flanged connections on cold insulated vessels, heat exchangers, equipment, or piping shall be insulated.

2.7. FINISHES

Insulation exposed to the weather or subject to mechanical damage should normally be protected with metal sheeting and shall be arranged to shell water. Consideration may be given to the use of plastic sheeting or mastic finishes in certain areas.

Within units processing hydrocarbons or other flammable materials, and on crude oil tanks and on tanks and vessels storing liquefied flammable gases, the cleading shall be galvanized or aluminized steel. Outside these areas and on the other pipes, tanks, and vessels, aluminum cleading may be used. In particularly corrosive or difficult situations, consideration may be given to the use of light gauge stainless steel.

2.7.1. Hot Services

When cleading is precluded by the shape of the equipment or when sheltered from the weather, rigid insulation finished with a hard-setting cement shall be used.

Hard-setting cement finishes shall only be used over flexible insulation in combination with well-supported expanded metal reinforcement. Heads of vessels should normally be finished with segmental metal cleading, which shall be overlapped and sealed to prevent moisture from entering under the vertical cleading.

2.7.2. Cold Services

Sealing materials must be compatible with the insulation. Where required as protection from mechanical damage or to reduce the fire risk, insulation shall be protected with metal cleading, which shall be secured by bands at approximately 500 mm centers. Self-tapping screws should not be used since these may penetrate the sealing materials. For rigid materials a joint sealer shall also be applied to all joints in the insulation material. Precautions shall be taken to prevent leakage of solvents onto vapor barriers.

2.8. METHODS OF INSULATION APPLICATION

Insulating materials shall be kept dry in storage and during erection, since wet insulation cannot always be dried out on site. Insulation removed from storage shall be used the same day and neither be returned to storage or left overnight on the job site. Cartons containing rigid sections or segments need to be stored end-up and be stacked no more than three high.

Bales containing slab insulation shall be stored flat, stacked not more than six high and preferably only four high if storage is likely to exceed a year.

Bales containing mattress insulation shall be stacked flat and not more than four high.

Metal sheet, e.g., aluminum or galvanized mild steel, may be delivered as single sheets or in bundles, dependent on the type of handling facilities in the stores.

Sheets or bundles should be stored under cover in a dry atmosphere under ambient temperatures. The first layer of sheets not exceeding 0.5 ton or the first bundle shall be stacked on a pallet board, and the second and subsequent layers or bundles shall be supported on timber spaced at not more than 600 mm center between each layer or bundle.

Sheet edges shall be examined at least monthly to see if any discoloration has taken place. If discoloration occurs, the faces of the sheet shall be examined and, if necessary, dried and restacked. When storage for longer than three months is contemplated, the supplier shall be consulted for recommendations on packaging the sheets prior to storage.

Ancillary fixing materials, bandings, screws, mastics, etc., need to be stored under cover in a dry atmosphere not below 5°C. It is important, particularly with two-part adhesives, mastics, or foam systems, to verify the shelf life.

Apart from certain loadbearing materials, most types of insulating materials require support or reinforcement when applied; they also must be secured to the surface to be insulated. For these reasons, it may be necessary to attach fixing accessories to the piping or equipment before application of the insulating materials is commenced. All insulating materials, however fixed, shall be in close contact with the surfaces to which they are applied, unless an air space is specially required.

Where the main insulation consists of preformed, or flexible material, all edges or ends shall be closely butted; for multilayer work all joints shall be staggered.

As a general rule insulation shall be carried out at ambient temperature. In certain cases it may be advisable to apply finishing material at operating temperatures.

Before any section of the insulation work on piping, vessels, or ductwork is commenced, all hangers, brackets, pipe clips, etc., shall be in position and the necessary acceptance tests for pressure/vacuum, etc., shall have been carried out.

Attention is drawn to the possible danger of skin irritation when using plastic compositions. Materials with a high free-lime content require particular care. The junction between removable and permanent insulation is to be so arranged as to be readily discerned, e.g., by painting the end of the permanent insulation or by laying a suitable textile fabric over it. Application of "in situ" foaming or spraying shall be subjected to the approval of the company.

2.8.1. Working Conditions

Equipment, paved areas, etc., adjacent to components being insulated shall be protected from dripping compounds and cements. Any damage resulting through failure to observe protective measures shall be repaired.

The work area shall be kept clean and free of debris resulting from insulating work; and on completion of work all coating, unused insulating material, scaffolding, etc., shall be removed. With regard to stripping old asbestos-containing insulation, see also BS 5970.

The old insulating material shall be examined initially by the company to determine whether asbestos is present and the contractor shall remove the insulating material so specified.

During removal of insulation that contains asbestos, it is essential that safety requirements are observed. Thus all operators who are engaged in stripping asbestos-containing materials have to wear protective clothing and approved respiratory equipment, which is available from the safety department. The area of the work has to be segregated by the provision of barriers in locations outside which the level of asbestos dust will not contravene asbestos regulations.

Wetting is not obligatory, provided that the respiratory equipment is adequate for the concentration of asbestos dust produced, and that, where appropriate, there is an adequate standard of separation between the working area and other parts of the site to prevent the scope of dust.

Preformed insulation that is easily accessible shall be removed dry and placed immediately in nonpermeable bags, which shall be tied at the neck. Plastic compositions, hard-setting compositions, or self-setting cements that require the use of saws or pneumatic tools for cutting shall preferably be wetted before removal. After the outer finish has been removed, the main insulating material can be wetted, either by spraying or by injection techniques. The use of water should be controlled in order to avoid the formation of slurry with consequent risk of injury on slippery floors. Asbestos waste material shall be enclosed in bags marked clearly for identification. Bags that contain asbestos waste shall be removed from the place of work for safe disposal.

2.8.2. Surface Preparation

Piping and equipment shall be clean, dry, and free from grease, dirt, loose rust, and scale. In special cases where ingress of rainwater or condensation of water vapor in the insulation can cause severe underlagging corrosion of carbon steel and low alloy equipment and piping, the surface shall be blast cleaned and painted.

The steel to be sand blasted shall be dry and operating conditions shall be such that condensation does not occur on it during work. When compressed air is used, this shall be dry and free from oil.

Weld defects and surface imperfections such as sharp edges shall be removed.

Blasting operations shall never be allowed in the vicinity of painting work or near a wet paint surface, or anywhere that blast abrasive, grit, or fallout shall impinge on a freshly painted surface, or on any uncovered primed surface.

Blast cleaning operations shall not be conducted on surfaces that will be wet after blasting and before coating and when the surface temperatures are less than 3°C above the dew point, when the relative humidity of the air is greater than 85%, or when the ambient temperature is below 3°C. Blast cleaning is permitted only during daylight.

After blast cleaning, the residual shot, grit, and dust shall be completely removed by any means as appropriated. Care shall be taken not to recontaminate the blast-cleaned surface.

The prepared blast-cleaned surface shall be completely primed the same day as blasted and before any visible rusting or deterioration of the surface occurs. No blasted surface shall stand overnight before coating. If such surfaces are not primed in accordance with the above they shall be reblasted. Care shall be taken not to contaminate blast-cleaned surfaces prior to painting.

Austenitic stainless steel vessels, equipment, and piping need special attention in coastal areas or where corrosive products, e.g., chlorine, chlorides, and hydrogen chloride, are being handled. In order to abate the effects of chlorides and moisture trapped from the atmosphere in such concentrations that contamination of the metallic surfaces under the insulation may cause stress-corrosion cracking, the external surfaces of the vessels, equipment, and piping used occasionally or operating in temperatures above 50°C shall be thoroughly cleaned and painted, or wrapped with aluminum foil.

2.8.3. Accessories

Accessories include attachments, insulation supports, securing devices, and reinforcements.

2.8.3.1. Attachments

Usually the term "attachment" is used for any anchor fitting fixed permanently to the surface to be insulated, usually by welding. Some types of plastics fittings, e.g., nylon clips, may be fixed to the surface by means of a suitable adhesive, subject to temperature limitations.

Attachments for welding may be flat or angle cleats, pipe bosses, threaded pillar nuts, washers or nuts welded "on edge," or studs of various kinds, e.g., round, flat, split pins, and fork studs. Their purpose is to serve either for the direct support of insulating materials or as fixtures to which insulation supports can be secured, e.g., by bolting. Carbon steel fittings should not be welded directly to alloy steels.

For those surfaces on which site-welding of attachments is not permissible, e.g., for certain types of alloy steel, or where subsequent internal temperature and pressure could be a hazard, it may be satisfactory to weld suitable metal pads in the appropriate locations. These shall be applied during manufacture of the equipment at the works, which would permit subsequent stress relieving; the attachments would then be joined to the pads at the site.

When the service conditions (including any abnormal conditions that may occur as a result of incorrect operation or accident) make it necessary for a pipe or portion of plant equipment to have post-weld heat treatment to avoid any risk of brittle fracture at subzero temperatures, of stress–corrosion cracking, or of other type of failure, either

• weld the attachments in place before final post-weld heat treatment; or
• attach the insulating material by means that do not involve welding.

For the choice and methods of attachment of plastics or metal clips by means of adhesive, the manufacturer's literature should be consulted. Frequently these clips are formed with an integral perforated flat base that permits penetration of the adhesive through to the upper surface. Self-adhesive insulation pins may also be used on flat smooth surfaces, such as galvanized metal or plastic ducting, provided that the surface is free from dust, and the manufacturer's weight and temperature limitations are not exceeded.

Typical attachments for gun welding would be plain pins of 3 mm diameter, end-fluxed studs of 10 mm diameter (plain or threaded), flat cleats 12 mm wide and 3 mm thick (or a similar angle welded on edge), threaded pillar nuts up to 12 mm diameter, etc. Split pins, fork studs, and similar fastenings follow the general rule for studs, as above, with a preference that the contact area should not exceed about 80 mm^2.

The use of hand-welding techniques may permit the application of attachments with large contact areas; with relatively long angles or cleats, it may be sufficient to use intermittent (stitch) welding.

If the difference in thermal expansion between the welded contact edge and the free edge is likely to be significant, distortion of the attachment shall be avoided by the introduction of saw cuts at intervals along the free edge; these cuts shall penetrate to a distance of about half the width of the attachment.

In order to avoid mechanical damage to the stud during transit or during erection of the plant, a thick threaded nut shall be welded on to the surface to be insulated, thus providing a fixture into which a threaded stud may be served at a later stage. The locations of studs or cleats will depend on the weight of insulation to be attached, as well as on the location of the surface, and on the degree of vibration to which the plant may be subjected under service conditions; for large flat surfaces, reasonable average spacing would be:

- Vertical surfaces: 450 mm square spacing
- Upward-facing surfaces: 600 mm or 750 mm square spacing
- Overhanging and downward-facing surfaces: 300 mm (maximum) square spacing

Alternate rows may be offset by 50% of the spacing, depending on the dimensions of the material used. Usually there should be a row of attachments parallel to each edge and to each stiffener or flange, at a distance of 75 mm to 150 mm away from the edge.

For large-radius curved surfaces, if welding is permitted, 450 mm to 600 mm uniform spacing is considered suitable, but this may be modified for vertical large cylindrical surfaces when the cleats are required to prevent downward movement of the insulating material. Cleats may not be required for horizontal cylindrical surfaces if it is possible to provide circumferential straps that can be tensioned over the insulation.

Welded attachments should preferably penetrate into the insulating material only to the minimum extent necessary to achieve effective support except in special circumstances that shall be approved by the company. In such cases penetration shall not be greater than about half the thickness of the insulating material. The cross-sectional area of the attachments shall be the minimum consistent with the required mechanical strength in order to avoid excessive transfer of heat (or cold) by metallic conduction.

It is important to remember that a welded attachment will be subjected to the same extent of thermal movement as will the insulated surface, with the resultant possibility of tearing the insulation or finish, unless care is taken to allow for this, e.g., by expansion or slip joints.

2.8.3.2. Insulation Supports

Supports for insulation and any associated cleading are required only to bear relatively light loading; they may consist of metallic flat bars, rings, part rings, varying lengths of angle, or studs.

Insulation supports should be installed on vertical vessels on 3700 mm centers (see Figures 2.4 and 2.5). Compressive strength and change in length caused by expansion or contraction of the vessel compared to the insulation change in length are two factors controlling the support centers. In some instances, the support distances are set by vessel flanges.

In all cases, a support shall be installed above a vessel flange to prevent the insulation from sliding down and resting on the flange studs. The support shall be a sufficient height above the flange to permit easy removal of the studs. On low-temperature vessels, the differential movements that occur at the contraction joint shall not exceed the dimensional flexibility of the caulking weather-vapor seal at these joints. Insulation on long runs of vertical pipe shall be supported in the same manner as that described for vertical equipment.

Except for vessels operating at very moderate temperatures, in which expansion and contraction of vessels is insignificant, the insulation shall be sufficiently free of friction with the vessel surface so that expansion and contraction movement of the vessel is independent of the insulation. To obtain this freedom of movement, insulation shall be installed slightly oversize.

Factory-attached, metal-jacketed insulation panels are supported by a vessel in several ways. The first panel side rests on the tank base, and successive panels are supported on "S" clips attached to the top of the installed panel. In addition, the metal jackets are supported by pins welded to the side of the vessel. The jackets in turn support the insulation cemented to it.

The panels are secured by clips or straps. Large horizontal vessels (over 1.8 meters in diameter) require supports to which securement straps, used to pull the bottom insulation into position, can be attached. Such supports are necessary because if straps were simply brought around the vessel insulation on top, the pull required to draw the bottom insulation into position would exceed the compressive limits of the insulation over which it passes.

The supports are so located that bands can be drawn to secure the bottom-third sector of insulation into position. Smaller horizontal vessels are

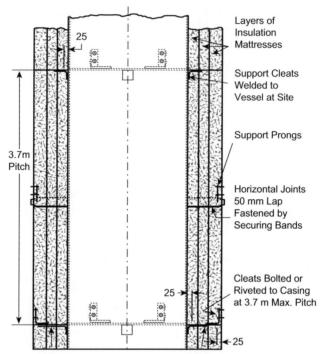

Figure 2.4 Typical method of insulating vertical vessels when welding is permissible on site.

their own support for the cylindrical insulation. The straps drawn around the insulation transmit the load of the lower insulation to that on top, which in turn is supported by the vessel. This is also true of all insulation installed on horizontal pipes.

The flat bars, rings, and angles normally will rest on stud or cleat attachments welded in the appropriate locations to the surface to be insulated (see Figures 2.6 and 2.7). Angle supports, where used, may be bolted to the welded attachments on the plant, or they may rest loosely on top of floating flat rings; in the latter arrangement short angle pieces may be used to secure external metal cleading to which they are fastened, either by bolts or by rivets.

Stud-type supports are used mainly for preformed or for spray-applied insulation, although they may serve as suspension points for the metal-mesh covering of flexible insulating mattresses. The studs may be in the form of attachments welded directly to the plant, e.g., split pins, fork studs, or plain studs, or as threaded studs screwed into nuts that themselves are welded to the surface to be insulated.

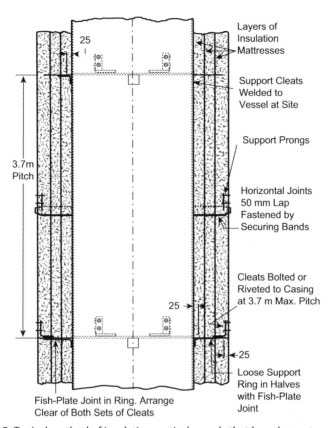

Figure 2.5 Typical method of insulating vertical vessels that have been stress relieved.

Alternatively, especially for vertical alloy-steel pipework, the studs may be welded radially to a ring that can be clamped around the pipe at the required vertical intervals. As these rings tend to slip downwards under service conditions, suitable support lugs or pads for the rings shall be welded on to the pipe at the manufacturer's works, to be followed by any necessary stress-relieving process.

As a general rule, studs shall not be greater than 10 mm in diameter, or of equivalent contact area, for gun welding. Angle cleats and flat bar normally will be about 5 mm to 10 mm thick, depending on the weight of insulation (and finish) to be supported; widths conveniently may be about 75 mm, varying according to the total thickness of insulation to be supported.

For anchorage of spray-applied mineral fibers and insulating concretes it is usual to provide Y-shaped studs of approximately 5 mm to 10 mm diameter for the main leg; these may be welded directly to the surface of the

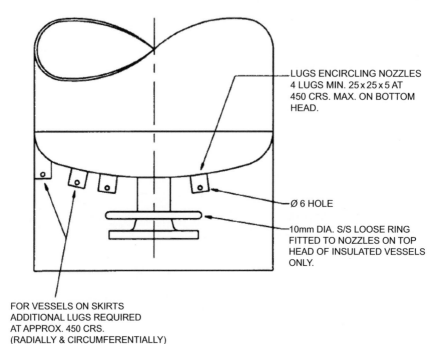

LUGS ENCIRCLING NOZZLES
4 LUGS MIN. 25 x 25 x 5 AT
450 CRS. MAX. ON BOTTOM
HEAD.

Ø 6 HOLE

10mm DIA. S/S LOOSE RING
FITTED TO NOZZLES ON TOP
HEAD OF INSULATED VESSELS
ONLY.

FOR VESSELS ON SKIRTS
ADDITIONAL LUGS REQUIRED
AT APPROX. 450 CRS.
(RADIALLY & CIRCUMFERENTIALLY)

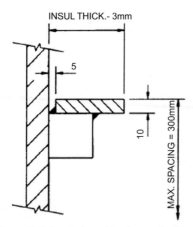

INSUL THICK.- 3mm

5

10

MAX. SPACING = 300mm

Figure 2.6 Insulation clips for vertical vessels. *(Source: CSBP Limited).*

plant or may be threaded for screwing into corresponding nuts, which themselves are welded to the surface to be insulated.

Preferably, the supports shall not penetrate the insulation to a distance greater than about half its thickness unless through–metallic connection between cold and hot surfaces of the insulation can be avoided, or reduced

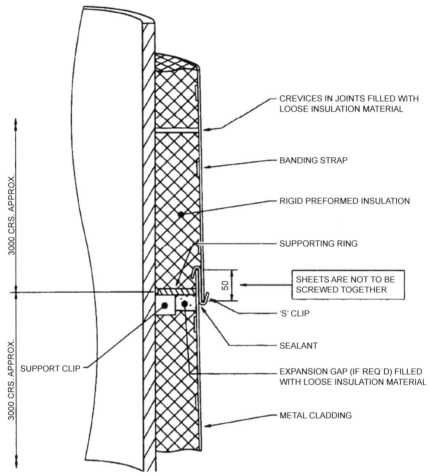

Figure 2.7 Rigid insulation on vertical tanks, vessels, and piping over 400 mm nominal size. (Source: *CSBP Limited*).

to an appropriate minimum. This is of particular importance with insulation over refrigerated plant, or when the external finish over hot insulation is of sheet metal. Sometimes it is possible to interpose a pad of insulating material between the inner welded attachments and the supports of the main insulation system.

The cross-sectional area of each insulation support shall be not greater than that necessary to achieve the required mechanical stability, so that thermal conduction can be reduced to an acceptable minimum. In appropriate cases the support may be nonmetallic, e.g., plastics or wood, and it is possible to take advantage of the fact that many alloy steels have lower thermal conductivity than carbon steel at corresponding temperatures.

Vertical vessels, insulated with block insulation, require a support at the bottom, and, if the vessel is over a certain height, intermediate supports to prevent the insulation from sliding downward. These supports may be angles, plates, rods, or other projections welded or bolted into position around the vessel.

The head insulation on horizontal vessels must also be supported. In most instances, the support for the head insulation is the same as that for the support of the cylindrical insulation, with slots or holes punched for the attachment of straps. The straps then pass from the cylindrical support on one side over the insulation to the support on the other side, and when drawn tight secure the insulation in position.

Depending upon the method of installation various types of insulation supports are required. In summary, insulation supports are:

- Structural angles
- Formed strapping
- Metal plates and rectangular bars
- Metal rods
- Road mesh
- Pins: metal and plastic (welded or adhered to surface)
- Blank nuts
- Rectangular studs (punched, split, etc.)
- Grooved studs
- "S" clips
- Grip nails

Accessories for the above are bolts, nuts, rivets, welding attachments, or adhesives. All these are used to transfer the weight of the insulation to the vessels, pipe, equipment, walls, or structural steel. They are structural members, regardless of whether they support the load in compression or tension.

As a general rule, supports for finishing materials, apart from those for the insulation proper, are not required except when a cleading of metal sheet is to be used over flexible fibrous material. Where such supports have to be used, they may consist of pads or short lengths of compression-resisting insulation, applied directly to the surface to be insulated or mounted over the ends of studs or angle cleats that themselves are attached to the surface.

Alternatively if the insulating material is not of sufficient resistance to compression, particularly over horizontal surfaces, the supports may be in the form of metal rings or "stools" spaced away from the surface on "prongs" to restrict the total area of contact with the insulated surface, so reducing the rate of thermal transmission to an acceptable minimum.

Figure 2.6 illustrates insulation clips for vertical vessels and Figure 2.7 shows rigid insulation on vertical tanks, vessels, and piping over 400 mm nominal size.

2.8.3.3. Securement (Tie Wires, Lacing Wire, Bands, Clips, Washers, Nuts, etc.)

Unless the insulating material can be secured directly to the surface to be insulated, e.g., by means of adhesive, it is usually necessary to provide some form of mechanical accessory to secure it to the permanent attachments. Tie wires, lacing wire, washers, clips, or nuts are the most common when dealing with the various types of attachments and supports.

For cylindrical surfaces it may be satisfactory to secure the complete insulation system by means of circumferential bands that can be tensioned over the outer surface. Sometimes wire netting over the insulating material will serve the same purpose provided that the edges can be laced tightly together. Separate securing accessories may not be required if an integral sheet finish is arranged so that an edge overlap can be secured by means of adhesive.

For large flat surfaces it is usual to secure the insulating material by impaling it over the studs or cleat attachments, using lacing wire to hold it in position. Preferably the lacing wire shall be wrapped around the main attachments and crossed for tensioning. With flexible fibrous mattresses enclosed in wire mesh it may be sufficient to tie the outer mesh to the individual attachments.

Lacing wire and tie wire normally will be of galvanized soft iron but one of the soft alloy steels may be required for special chemical resistance or for use at elevated temperatures or cold services. Typical sizes of wire are 1 mm diameter for general use, or 1.5 mm for heavier use. For refrigeration work, use plastic tape instead of metal wires.

Over flat surfaces or large cylindrical vessels it may be desirable to use multistrand wire that can be tensioned between attachments; wire with seven strands each of 0.7 mm diameter, twisted in rope form, is suitable for heavy work. Other soft wires may be useful for special applications.

Securing wires shall be fastened either by simple wrapping around each attachment or by means of nuts, clips, or washers, as appropriate, at each fixed point. Rapid-fixing clips, which depend on spring-point devices for securement, may become loose in service as a result of corrosion of the points.

Insulation on cylindrical surfaces, with an outer finish of fabric or sheet, may be secured by means of outer circumferential bands of metal or plastic strip. For outer covering of aluminum sheet only aluminum or stainless steel bands shall be used. Metal bands for large vessels shall be of adequate width and thickness to provide stability to the insulation system under the required conditions of service. For overall economy on large vessels, stainless steel banding should be used.

Flexible insulating mattresses that have wire-mesh covering integral in their construction and are stabbed through with intersurface ties shall be impaled over the main attachments and further secured by tightly fastening together adjacent edges of the outer covering mesh. The lacing wires for this purpose should be of similar diameter to that used for the wire mesh. If a final covering of aluminum sheet is to be applied, care shall be taken to avoid direct contact between dissimilar metals.

Lacing wire or other securing devices that is likely to be in direct metallic contact with a final clearing of aluminum sheet shall be coated with a suitable plastic material in order to avoid bimetallic corrosion in the locations of contact. Alternatively, it may be convenient to use aluminum or stainless steel securing materials.

2.8.3.4. Reinforcement

The most commonly used reinforcing materials are galvanized wire hexagon netting or one of the various types of expanded metal, but open–mesh woven glass fabric may be suitable for certain applications, e.g., the reinforcement of weatherproofing compounds applied in paste form. The main uses for metallic reinforcement are with plastic compositions, spray-applied fibrous insulation, wet-applied finishing compositions, and wet finishing cements, but they also are of value for retaining dry fibrous insulation and various types of preformed materials. The size of mesh for wire hexagon netting is normally 25 mm with wire of 1 mm diameter.

Expanded metal will normally vary from 12 mm to 50 mm across the short dimension of the mesh, with thickness of metal varying between 0.5 mm and 1.6 mm, but heavier material may be used, e.g., for reinforcement of insulating concretes.

Galvanized steel reinforcing mesh and securing devices shall not be subjected to temperatures in excess of 65°C under conditions of possible high humidity. Suitable heat-resisting alloy shall be employed for all service temperatures in excess of 400°C. For intermediate temperatures, carbon steel may be used, but this shall be coated with bitumen or paint for protection against corrosion during storage and prior to application on site.

Where substantial mechanical strength is required, e.g., for resistance to compression, it may be desirable to use square-pattern reinforcing mesh, which may vary according to the requirements from strands of 2 mm diameter at 40 mm spacing to strands of 6 mm diameter at 100 mm or 150 mm spacing. Sometimes the component wires may be welded together at the crossing points.

It shall be noted that material of square mesh pattern is likely to distort on expansion at elevated temperatures if it is secured rigidly to attachments on the insulated surface. If metal reinforcement is to be located over the outer layer of insulating or finishing material and is likely to be in direct contact with cleading of a dissimilar metal, precautions shall be taken to avoid electrolytic corrosion action, e.g., by the use of a suitable coating on the inner face of the cleading, or on the reinforcing metal.

Where mechanical strength is required, e.g., for puncture resistance, it is desirable to embed open-weave glass cloth or cotton scrim between layers of weatherproof or vapor-proof mastics.

2.9. HOT INSULATION APPLICATION

This concerns application for systems operating in the temperature range +5°C to +650°C.

Where the surface to be insulated is at a temperature below the dew point of the surrounding air, a vapor barrier should be provided on the exposed (warm) surface.

The piping and equipment shall be insulated after the pressure tests and other necessary tests and inspections have been completed. To facilitate regular inspection of welds and bolted joints, removable portions of insulating and finishing materials shall be provided in the appropriate locations.

All materials shall be subject to inspection and approval by the company to ensure that all materials meet this specification. Insulation application shall also be subject to inspection and any material that has been improperly installed or excessively damaged shall be removed and replaced properly with undamaged material.

All work shall be executed in a neat and workmanlike fashion in strict accordance with these specifications and as called for on drawings covering the work to be done. No changes or deviations will be permitted without advance written approval by the company.

Every precaution shall be taken to see that each day's work is weatherproofed before leaving it for the night. Where this is impractical, a fillet of weatherproof mastic must be placed over the exposed ends of insulation. No insulating work of any type may be performed in rainy weather or when atmospheric condensation is occurring. In the event of doubt regarding the prevailing dew point, the decision will be made by the owner.

2.9.1. Hot Pipework

Piping shall be insulated when coded on mechanical and utility flow diagrams, pipeline lists, and piping and spool drawings. In the event of discrepancies, the flow diagrams shall govern. Insulation classification that may appear on flow diagrams and drawings are defined as following:

- **Ih:** Insulation for heat control for operating temperatures above 21°C
- **Is:** Insulation for personnel protection
- **St:** To be steam traced and insulated
- **Stt:** To be steam traced with insulation cement and insulated
- **Et:** To be electric traced and insulated
- **Ett:** To be electric traced with insulation cement and insulated
- **Sts:** To be steam traced with spacers and insulated
- **Ias:** Insulation for cycling or dual temperature service where the temperature fluctuates from 15°C to 320°C

It is preferable that the insulation of pipework should be carried out with preformed materials where the temperature limitations permit. Where the pipe diameter is too large for molded pipe sections to be used, the pipe should be covered as far as practicable by building up with radiused and beveled lags.

The junction between removable and permanent insulation shall be made readily discernible, e.g., by painting the end or by laying a suitable textile fabric over the end of the permanent insulation.

If the insulation thickness requirement exceeds 75 mm, the insulation shall be applied in multiple layers with a maximum of 75 mm per layer.

Hangers and supports shall be insulated from the pipe surface in a suitable manner, and where applicable should comply with the requirements of BS 3974. Recognized methods using direct-contact support or insulation rings are shown in Figures 2.8, 2.9, 2.10, 2.11, 2.12, 2.13, and 2.14.

Pipe sizes larger than 760 mm shall be insulated and finished as described for vessel insulation.

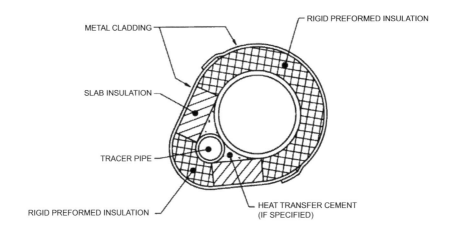

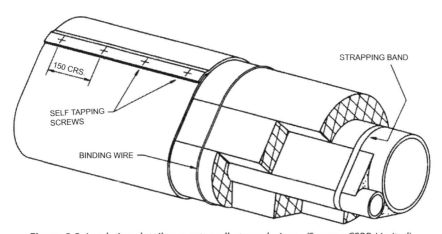

Figure 2.8 Insulation details on externally traced pipes. *(Source: CSBP Limited).*

2.9.2. Application of Insulation to Pipework Heated by External Tracing Pipe

Piping that is steam traced shall be covered with oversize pipe insulation, including the tracer lines. Valves, flanges, unions, and tracer line loops shall not be insulated, unless specified on the piping drawings. Flexible insulation may also be used to insulate traced pipework. If this is used, wire netting or metallic tape or foil shall be applied first to preserve an air space inside the insulation cover.

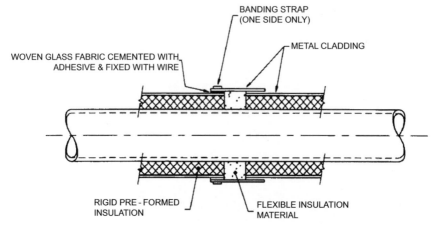

Figure 2.9 Expansion joint in rigid insulation on piping. *(Source: CSBP Limited).*

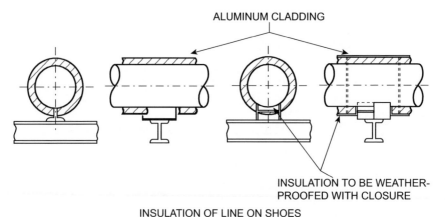

INSULATION OF LINE ON SHOES

Figure 2.10 Pipe shoe for lines 75 mm and larger.

The open ends of the air gap in the region of flanges in the main pipe shall be sealed with a disc of insulating material and finish shall be applied as for the adjacent insulation.

The effectiveness of heat transfer may be increased by:

- Maintaining a hot air space inside the insulation cover to increase the heat transfer area;
- Maintaining direct contact between the tracer pipe and the main pipe by wiring or strapping at intervals;
- Using heat–conducting cement to increase the contact area between the tracer and heated pipe.

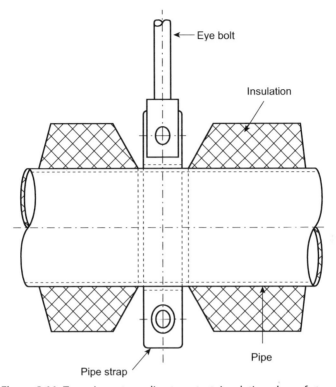

Figure 2.11 Two-piece strap, direct contact, insulation clear of strap.

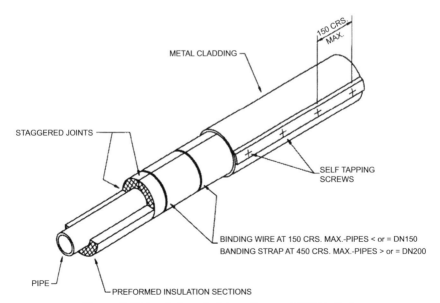

Figure 2.12 Insulation of piping. *(Source: CSBP Limited).*

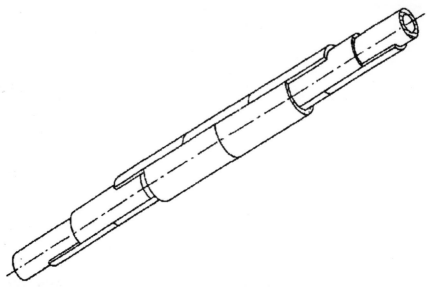

Figure 2.13 Typical method of staggering insulation sections on a straight pipe. (Note: Binding wire or bands on each layer of insulation spaced at intervals of approximately 250 mm.)

When corrosive action between fluid and the main pipe is liable to occur at local hot spots, direct contact between the tracer and main pipe shall be prevented by fitting spacers of low conductivity between the tracer and heated pipe. This may, however, decrease the effectiveness of heat transfer.

The tracer pipe itself shall be looped out and jointed (for example, by a compression fitting) near the main pipe flanges. The exposed length of tracer shall be insulated.

Leak testing shall be satisfactorily completed before the application of thermal insulation.

Electrical tracing should always be applied and insulated in accordance with the manufacturer's instructions. Care shall be taken to avoid mechanical damage, and to protect the tape from water or chemical spillage. The makers shall be consulted before electrical tracing is specified for use in flameproof areas. Preferably, the insulation should be of preformed sections of appropriate inner diameter to fit over the tracing cable on the pipe. Where live electric tracing cables are buried in the insulation a warning notice shall be placed on the outside.

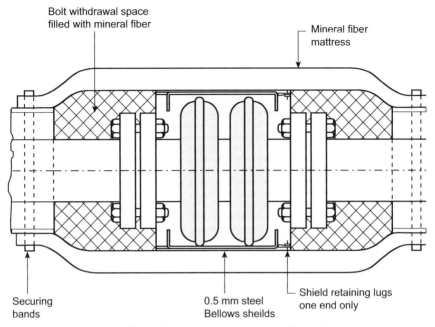

Bolt withdrawal space
filled with mineral fiber

Mineral fiber
mattress

Securing
bands

0.5 mm steel
Bellows sheilds

Shield retaining lugs
one end only

Figure 2.14 Typical method of insulating expansion bellows. (Note: Steel cover to overlap bellows by 40 mm at each end. Cuts to be made in the cover and lugs bent over to locate the cover. The purpose of the shield is to permit free movement of the bellows.)

2.9.3. Vertical Piping

For vertical runs of piping, or near vertical (inclined 45° or more) piping, it is important to prevent downward displacement of the insulating material by the use of appropriate supports, which may be in the form of metal rings, part rings, or studs. These supports shall be located at intervals of not more than 5 meters and, in any case, there shall be a support immediately above each expansion break in the insulation and also above all flanges in vertical lines located as to allow covering flange bolts.

2.9.4. Application
• **Preformed materials**
    Preformed pipe sections shall be fitted closely with all joints butted. Also, cracks and voids in the pipe and any unavoidable gaps in circumferential or longitudinal joints shall be filled with compatible insulating materials such as insulating cement.

Single-layer sectional pipe insulation shall be applied with longitudinal joints staggered and shall be secured with bands or wire at approximately 250 mm spacing and not nearer than 50 mm to the end of the section. An additional layer or layers of sectional pipe insulation also shall be applied with all joints staggered and it (they) shall have an inner layer secured by at least two wires per section with the outer layer secured by bands or wires at approximately 250 mm spacing. After tightening, the ends shall be pressed on to the insulating material. The choice of material for the bands or wires and its corrosion protection shall be based on the environmental conditions.

When sections are held in position and covered by a fabric, this shall be secured by stitching or by the use of an adhesive. The edges of the fabric, if stitched, shall overlap by at least 25 mm; if adhesive tape is used to cover the joints it should preferably be wound with an overlap of at least 25%. Alternatively, with a fabric or sheet outer finish, the whole section may be secured by circumferential bands.

Sections that are split down one side only shall be sprung on to the piping, and secured. Certain types of pipe sections can be secured by corrosion-resisting staples at the joints; these staples shall be not further apart than 100 mm.

To prevent "through joints" where temperatures exceed 260°C, with single layer insulation apply an expansion filler material between the joints of the insulation. This filler should be strips of 2.5 cm thick, $\frac{3}{4}$ lb (0.34 kg) density, long textile fiber resilient glass blanket material. The width of the strips should be the thickness of pipe insulation plus 1.3 cm. Filler is applied and squeezed to a minimum thickness less than 0.16 cm by the pipe insulation during installation process. The projecting portion of the filler is flattened by covering.

The need to dismantle pipework with a minimum disturbance of insulation shall be borne in mind, and permanent insulation shall end sufficiently far from flanges and fittings to enable bolts to be withdrawn. Therefore, unless otherwise specified, insulation shall be stopped at flange or union connections. Clearance shall be stud length plus 25 mm. At each stop, the insulation shall be weatherproofed.

- **Flexible materials**

  Fabric-covered mattresses shall be made from a suitable flexible medium and filled with a suitable filling that contains a minimum of dust or foreign matter. The hem of the fabric cover shall be folded twice before sewing.

The inner faces of mattresses in contact with surfaces above 400°C shall be of glass cloth, ceramic fiber cloth, stainless steel foil, or aluminosilicate paper. The edges of mattresses shall overlap adjoining insulation and be bound with wire. Care shall be taken that air spaces are kept to a minimum and that there are free passages from hot surface to atmosphere. The filling material shall be prevented from packing down by quilting as necessary.

Strip and rope material shall be wrapped spirally around the surface, successive layers being applied to opposite hand. The ends of this material are to be firmly secured and all tie wires buried. Care should be taken that flexible materials are not unduly compressed. Where insulation is fitted in two layers these layers shall be staggered.

Flexible insulating blankets or mats need to be secured by means of circumferential bands of metal or plastic strip, with the exception that the use of circumferential tie wires of 1.0 mm to 1.6 mm diameter is permissible when the ultimate finish is to be of sheet metal. If the finish is to be aluminum sheet, the securing bands have to be of compatible metal.

On vertical pipes of nominal size 100 mm and larger, welded or clamped support rings shall be applied at the upper end and under each flange.

Additional rings may be required at approximately 4 m intervals. For vertical and near-vertical piping, it is important to prevent downward displacement of flexible insulating materials; whereas support from below is suitable for many preformed materials, flexible insulating materials shall be suspended from above, with support rings.

After application, the blankets shall have the required insulation thickness. At the ends of the blankets clamped distance rings shall preferably be attached to the pipe at intervals of approximately 1 m to support the metal jacketing and to prevent compression of the insulation; see Figure 2.15.

2.9.5. Weatherproofing and Finishing

The straight portion of insulated lines shall have the basic insulation covered with aluminum jacket, with all joints lapped 50 mm and arranged to shed water. The jacket shall be secured with bands installed on 230 mm centers. Galvanized steel also may be used where mechanical resistance is important. The metallic sheet may be either flat or corrugated.

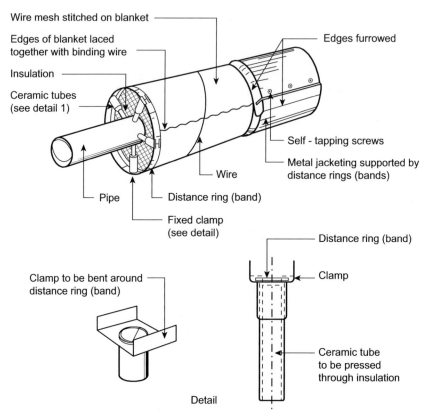

Figure 2.15 Insulation with blankets on piping.

2.9.6. Hot Vessels

Vessels shall be insulated when indicated on the mechanical and utility flow diagram. The thickness shall be as shown on the vessel drawings. In the event of discrepancies, the vessel drawing shall govern.

The shell and head of a vessel shall be insulated with rigid-type or blanket-type insulation. The blocks shall be either curved, flat, or beveled to fit the shell contours, with all joints tightly butted and secured with bands. Bands shall be machine stretched and sealed to prevent slackening.

Vessel skirts, saddle supports, and bottom heads on vessels under skirts shall not normally be insulated unless the bottom is heat traced or insulation is specified by the owner.

The top heads shall be insulated upwards successively from their uppermost insulation support ring to the top. Small diameter vessels of 760 mm outside diameter and under shall be insulated and finished as described for piping insulation.

If the insulation thickness requirement exceeds 75 mm, the insulation shall be applied in not less than two layers.

The need to dismantle associated pipework and inspection covers shall be anticipated, and permanent insulation ended sufficiently far from flanges and fittings to enable bolts to be withdrawn. To facilitate regular inspection of welds and bolted joints, removable portions of insulating and finishing materials shall be provided in the appropriate locations. The junction between removable and permanent insulation shall be made readily discernible, e.g., by painting the end of the permanent insulation or laying a suitable textile fabric over the end.

Inspection covers shall be insulated separately and particular care shall be taken that the insulation value is not less than that provided on the main body of the vessel.

2.9.7. Method of Application

2.9.7.1. Preformed Materials

It may be necessary to cut preformed materials to fit irregular contours. Alternatively, suitable material may be used to provide a regular foundation (see Figures 2.16 and 2.17). All cut faces shall be clean and care shall be taken to butt adjacent edges. In multilayer applications, joints shall be staggered (see Figure 2.18).

Block or preformed insulation on vertical vessels shall be supported by horizontal insulation rings on 3700 mm centers. These rings also form a break in the insulation to allow for wall expansion. The gaps shall be pressure-filled with rock wool. Blocks shall be secured in place with bands of proper size and materials that have been specified in relevant regulations.

Bands shall be spaced on 300 mm centers. Where double layer insulation is used, the inner layer shall be secured by suitable wire with approximately 450 mm spacing. Where necessary, allowance shall be made for expansion by use of stainless steel expander band. After insulation application, gaps and voids, if any, shall be filled with plastic composition of the same materials as that of the insulation.

Insulation on horizontal vessels shall be applied in stagger joint arrangement with all edges securely laced together with 1.5 mm wire or hog rings on 300 mm centers. Nozzle projection through insulation shall be adequately insulated and secured with girdling rings.

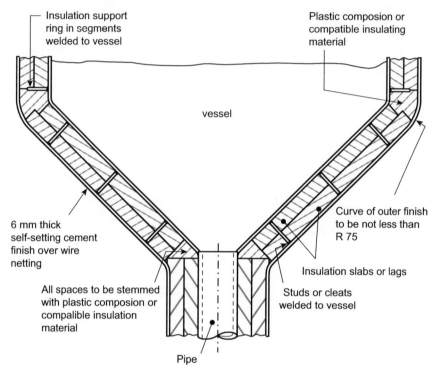

Figure 2.16 Typical method for insulating vessels with conical bottoms.

Exposed vessel heads shall be insulated with blocks secured by bands with floating rings. Band spacing shall be 300 mm maximum at the tangent line. All voids to be filled with plastic composition and the whole to be covered with a 10 mm layer (when dry) of finishing cement

Unexposed vessel heads if insulated shall have the blocks impaled on 3 mm diameter wire pins and speed clips. Over the blocks apply a 10 mm thick layer (when dry) of finishing cement.

2.9.7.2. Blanket Insulation

Insulation on vertical vessels shall be supported by horizontal insulation rings with 2000 mm approximate spacing and shall be applied in a stagger joint arrangement. The top and bottom edges shall be securely laced to the support ring with 1.5 mm diameter wire or hog rings on 300 mm centers. The insulation shall then be secured in place with bands spaced on 300 mm

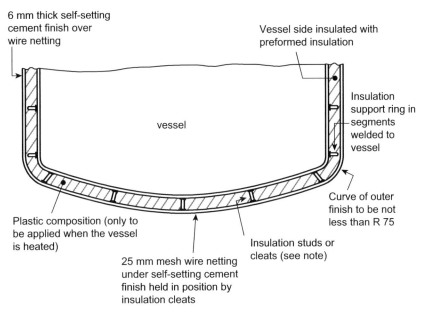

6 mm thick self-setting
cement finish over
wire netting

Vessel side insulated with
preformed insulation

vessel

Insulation
support ring in
segments
welded to
vessel

Curve of outer
finish to be not
less than R 75

Plastic composition (only to
be applied when the vessel
is heated)

Insulation studs or
cleats (see note)

25 mm mesh wire netting
under self-setting cement
finish held in position by
insulation cleats

Figure 2.17 Typical method for the insulation of dished ends of vessels. (Note: Studs or cleats are to be welded to the bottom of the vessel at approximately 300 mm intervals. After applying an initial coating of plastic insulation, sections cut to suitable lengths may be used to build up insulation to correct thickness.)

centers. Nozzle projections through insulation shall be adequately insulated and secured with girdling rings.

Insulation on horizontal vessels shall be applied in a stagger joint arrangement with all edges securely laced together with 1.5 mm wire or hog rings applied on 300 mm centers. Nozzle projections through insulation shall be adequately insulated and secured with girdling rings.

Exposed vessel leads shall have blanket insulation so that all sections are snug, tight, and laced as on the straight sides and secured by a 10 mm to 12 mm round steel rod floating ring in the center and bands on 300 mm maximum centers at the tangent line. Over the blanket, apply a 10 mm thick layer, when dry, of finishing cement.

Unexposed heads on vessels shall have blanket insulation so that all the sections are snug, tight, and laced as on the straight sides and secured by a 10 mm to 12 mm round steel rod floating ring in the center, and bands on 300 mm maximum centers at the tangent line. Over the blanket, apply a 10 mm thick layer, when dry, of finishing cement.

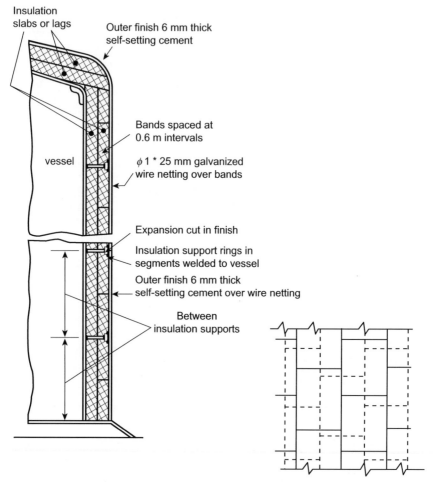

Insulation slabs or lags

Outer finish 6 mm thick self-setting cement

vessel

Bands spaced at 0.6 m intervals

ϕ 1 * 25 mm galvanized wire netting over bands

Expansion cut in finish

Insulation support rings in segments welded to vessel

Outer finish 6 mm thick self-setting cement over wire netting

Between insulation supports

Figure 2.18 Typical method for insulating roof and sides of vessel with internally reinforced roof. (Note: All vertical joints in adjacent layers should be staggered and the horizontal joints in any column of sections in the outer layer should be staggered with respect to those in the adjacent columns in the same layer. Where possible, horizontal joints in the wire netting should not coincide with continuous joints in the sections.)

2.9.7.3. Weatherproofing

Vessel shells shall be finished with the metal jacketing applied directly over block or blanket insulation. On vertical vessels the corrugation in sheeting shall be vertical and the sheets supported on "S" clips.

The horizontal seams of metal jacketing shall be overlapped 75 mm and the vertical seams shall be overlapped 100 mm. Metal jacketing shall be secured with bands spaced not less than one band on each circumferential seam and one at the middle of each sheet so that spacing between bands does not exceed 60 cm. Band joints will consist of one double-pronged seal. When expansion springs are required for insulation securement, the same number of expansion springs per band shall be used on all banding securing the jacketing. Sheet metal screws shall be applied on longitudinal seams to close the fish mouth. Spacing between screws shall be about 20 cm.

Vertical vessels shall have a rain shield applied at the bottom insulation support ring to prevent the entrance of moisture. Opening in the jacketing for nozzles, etc., shall be provided. These openings shall be properly sealed to prevent ingress of water.

On horizontal vessels, the metal sheets shall be lapped a minimum of 75 mm on the circumferential seam and a minimum of 100 mm on the longitudinal seams and installed so that they shed water. Bands shall be installed on 450 mm centers and shall be machine-stressed and fastened under tension.

All vessel-exposed dished ends shall be covered with prefabricated or field-fabricated segmental cladding (orange peel). The cladding shall be secured with bands and seals and joints shall be lapped 50 mm. Lapped joints shall be secured by means of self-tapping screws or rapid-fix rivets.

Heads and compound-curved sections that are difficult to jacket with metal economically shall be covered with weather-barrier coating and reinforcing cloth.

All cutouts in metal jacketing shall be weatherproofed by flashing with aluminum sheet or with mastic.

For vessel diameters under 90 cm, finishing shall be as for pipe. See Figure 2.19 for details of weatherproofing and finishing.

2.9.8. Exchangers

The method of insulation of heat exchanger shells, heads, covers, and nozzles shall be as specified for vessels.

2.9.9. Tanks

Tanks and tank roofs shall be insulated when indicated on tank drawings. Tank insulation may be preformed block insulation, blanket insulation, or sprayed rigid polyurethane foam.

2.9.9.1. Method of Application

• **Preformed materials**

Hand railings on insulated tank roofs shall be installed in accordance with Figure 2.20.

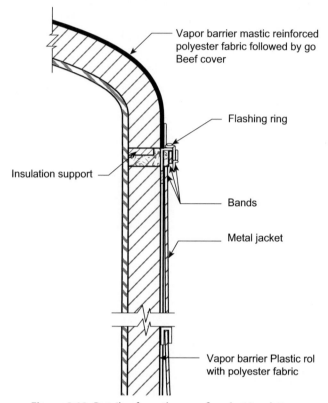

Vapor barrier mastic reinforced
polyester fabric followed by go
Beef cover

Flashing ring

Insulation support

Bands

Metal jacket

Vapor barrier Plastic rol
with polyester fabric

Figure 2.19 Details of weatherproofing, hot insulation.

A rainwater shield and ring cap shall be installed to prevent water ingress into the insulation. The insulation shall be protected with metal jacketing arranged with 50 mm overlaps (with rivets and screws) and supported on "S" clips and the whole jacketing shall be secured with stainless steel bands.

• **Blanket insulation**

Blanket insulation on tank shells shall be applied with a stagger joint arrangement with all seams securely laced together with #16 gauge wire (1.5 mm) on 300 mm centers. In addition the blanket shall be impaled on welding pins and secured with speed clips. Weld pins shall be installed vertically on 3600 mm centers and on 600 mm centers circumferentially. After speed clips have been installed, trim the excess pin close to the speed clip. The insulation shall then be secured in place with 19 mm × 0.51 mm galvanized steel bands spaced on 300 mm centers and fastened

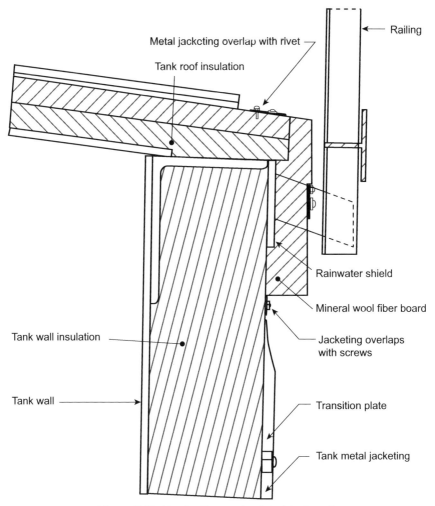

Figure 2.20 Hand railing on insulated tank roof.

to the 19 mm pencil rods installed vertically at approximately 9000 mm spacing around the tank.

- **Sprayed rigid polyurethane foam (PUF)**

Sprayed PUF is intended for use on tank shells with operating temperatures above ambient and up to 90°C. It shall be applied only when overspray cannot cause damage to the adjacent objects, or when adequate protection against overspray is feasible.

Substantial protection from the weather is essential and in very cold weather some heating may be necessary. The temperature of the liquid foam components shall be between 15°C and 25°C at the time of application. The temperature of the surface to be insulated shall be between 20°C and 40°C during spraying. The insulation shall not be sprayed during rain or periods of high winds.

The insulation should be sprayed on a primed or painted surface after removal of any dirt, grease, loose paint, or chalk. Rusted surfaces shall be blast-cleaned and painted. Mill scale and moisture shall be absent.

The insulation shall end at a sufficient height above the tank base to prevent contact between groundwater and insulation, e.g., after rainfall.

The PUF shall be sprayed in layers with each layer being a maximum of 12 mm thick until the specified thickness is obtained. The average finished thickness of sprayed PUF for the total tank surface shall be between 5 mm and 10 mm above the specified thickness. The thickness shall be measured using an electrical magnetic instrument.

The minimum insulation thickness of PUF applied shall be 25 mm.

Provisions shall be made to enable the removal of the insulation from parts that shall remain accessible or remain free of contamination, e.g., staircases, railings, manholes, gauge glasses, or other accessories shall temporarily be covered with plastic foil.

Junctions between permanent and removable insulation shall be properly sealed against ingress of moisture. The insulation shall be made smooth and properly sealed, including the seams where supports, nozzles, stair steps, etc., protrude through the insulation.

PUF used on tank roofs and shells shall be protected against ultraviolet radiation and weather conditions by applying an elastomeric coating of acrylate polymers.

Sprayed PUF without metal sheet covering shall not be used on tank roofs, since otherwise static electricity may occur by walking over the PUF, and furthermore damage to the PUF will cause severe corrosion of the roof plates. However, this prohibition does not include special proprietary tank roof insulation systems containing PUF that is free from these effects.

Failure of the foam to stick to the metal may not be evident immediately and is not readily detected by eye because of the uneven finish of the surface of the foam. It may be detected by tapping the foam, which will give a distinctive hollow sound if it is not stuck to the metal. Special safety precautions have to be observed when spraying polyurethane foams and a fresh-air mask has to be worn.

- **Finishing and weatherproofing**

  Tank shells shall be finished with galvanized steel jacketing applied directly over the insulation. The metal sheet may be corrugated or plain. Corrugations shall run vertically with vertical seams lapped a minimum of 2 corrugations and horizontal seams lapped a minimum of 75 mm and supported on "S" clips. The metal jacketing shall be secured with 19 mm × 0.51 mm stainless steel bands spaced not less than one band on each circumferential lap and one at the middle of each sheet, but not to exceed 600 mm on center.

  Bands shall be fastened to the 19 mm diameter pencil rods installed vertically at approximate 9000 mm spacing around the tank. Voids in the insulation at the pencil rods shall be filled with insulation blanket cut into suitable strips. Over the open joints of the jacketing at the pencil rods apply a batten on corrugated steel of sufficient width to lap 2 full corrugations of the adjacent metal and secure with sheet metal screws at approximately 500 mm of center as required to close seams. All cutouts at the uninsulated nozzles, manways, structural projections, nameplates, code inspection plates, etc., shall be sealed with mastic.

  On tanks and vessels where sufficient changes in circumference would occur as a result of thermal expansion or loading, e.g., asphalt storage, fractionating columns, or large diameter crude storage tanks, insulation and weatherproofing bands shall have breather springs installed as shown in Table 2.2.

- **Irregular surfaces (including machinery)**

  Irregular surfaces whenever required shall be insulated according to the following methods.

  All insulating materials, however fixed, shall be in close contact with the surfaces to which they are applied, unless an air space is required for special reasons. Where the main insulation consists of preformed, or flexible material, all edges or ends shall be closely butted; for multilayer work all joints shall be staggered. Where

Table 2.2 Installed Breather Springs for Various Vessels

Vessel Diameter	Operating Temperature	No. of Springs Required
5" to 8"	Above 79°C (175°F)	One
Over 8" to 15"	All	One
Over 15"	All	One every 15 meters

possible, pipes adjacent to the irregular surfaces should be insulated separately. Care shall be taken not to interrupt the moving member of the machinery when insulating this equipment.

If the insulation is not furnished with irregular surface equipment, especially machinery, it shall be generally covered with insulation blanket securely tied in place of the same thickness as the adjacent pipe insulation. Over the insulation, when dry, apply a 10 mm thick layer of finishing cement. Heat exchanger ends and manhole covers when insulated shall have the required thickness of insulation in block form securely wired together and completely enveloped in flat metal jacketing.

If such machinery, exchangers, etc., are subject to regular maintenance necessitating removal of insulation, consideration shall be given to fabricating the insulation in detachable sections, completely jacketed in cladding that may be easily removed and replaced without damage.

Irregular surfaces where application of molded-type or blanket insulation is impractical shall be insulated to the full specified thickness with insulating cement. Cement shall be applied with a trowel on the surface to be insulated, filling all depressions for their entire depth to eliminate voids of any nature. Care shall be exercised that the thickness of each application will be no greater than that which will set on vertical surfaces without excessive cracking upon subsequent drying. When sufficiently dry, additional applications may be made as required to build cement to the full specified thickness. Where the specified thickness of cement insulation exceeds 3.8 cm, cement shall be reinforced with one layer of 2.5 cm wire netting for each additional 3.8 cm, or a part thereof, uniformly embedded midway between the metal and finished surface.

- **Weatherproofing**

  Where the specified thickness has been applied and secured in place, the outer surface shall receive an adhesive coat of weather-barrier coating brushed on to a minimum thickness. While still tacky, reinforcing cloth shall be stretched taut and thoroughly embedded in the coating.

  Care must be exercised that the weave is not stretched and that cloth is overlapped approximately 3.8 cm to provide strength at the joint equal to that maintained elsewhere. Before the surface becomes dry to touch, a second coat of weather-barrier coating shall be applied by brushing over the reinforcing cloth to a uniform thickness, with a smooth, unbroken surface, and allowing it to dry. Th total thickness of weather-barrier coating and reinforcing cloth, when dry, shall not be less than 0.3 cm. No portion of cloth shall be visible on the finished surface.

Weather-barrier coating shall be carried out 15 cm on to any metal beyond termination of insulation. Application shall be built up in a uniform manner to prevent uneven contraction and any tendency toward surface cracks. Openings around nozzles, manways, etc., shall be made watertight.

All outside corners of insulation shall be rounded, and the weather-barrier coating provided with a double layer of reinforcing cloth.

Care should be taken in applying weather-barrier coating over hot equipment insulation; severe blistering will occur if insulation contains more than 5% moisture. Weather-barrier coating shall not be applied when atmospheric temperature is such that condensate of moisture and ultimate freezing may occur on the finished surface within 24 hours from time of application.

2.10. VALVES, FLANGES, AND FITTINGS

It is preferable that valves and flanges be insulated, but where hidden flange leakage may cause a possible fire or other hazard, e.g., with oil lines, or where repeated access will make it uneconomical, insulation may be omitted. For hot hydrogen duty, a simple sheet-metal shroud should be placed over the flanges to protect them from thermal shock due to changes in atmospheric conditions, while permitting access for safety, etc.

Valves, flanges, and fittings shall be insulated with preformed sections as far as is practicable (see Figure 2.21). It is appreciated that irregular surfaces are difficult to so insulate, but the maximum use of molded sections entails the minimum wastage of material should it be necessary to remove and replace the insulation for inspection purposes. The insulation shall be finished in such a manner that there is free access to instruments. All thermometer pockets, including the boss and welded pad, shall be insulated. For some superheated steam installations it may be desirable to insulate pressure-gauge pipework for a reasonable distance from the tapping point to avoid pressure loss due to cooling.

Some common methods adopted for insulating valves and flanges are described in the following.

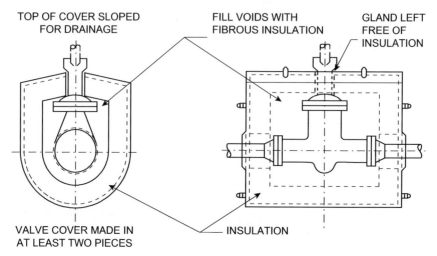

TOP OF COVER SLOPED
FOR DRAINAGE

FILL VOIDS WITH
FIBROUS INSULATION

GLAND LEFT
FREE OF
INSULATION

VALVE COVER MADE IN
AT LEAST TWO PIECES

INSULATION

Figure 2.21 Method of insulating a valve.

- **Flange boxes**

  These are usually made from metal sheet and lined with preformed rigid or flexible insulating material. Direct contact between the metal of the box and the insulated metal surface shall be avoided. A removable type of flange box is shown in Figure 2.22a and b.

- **Mattresses**

  These consist of a glass or silica fiber cloth envelope packed with loosefill insulating material.

- **Plumber's joint**

  This is achieved by the application of plastic composition over flanges or valves, after the insulation on adjacent piping has been appropriately finished. The piping insulation shall terminate at points that allow for easy removal of flange bolts and the chamfered ends of the piping insulation shall be finished and suitably painted. This provides a specific area in which the insulation for the flanged joint or valve may be confined, and also allows for it to be removed periodically without disturbing the piping insulation.

  Weatherproofing of flanges, valves, and fittings are usually secured with encasement in flange boxes or application of weather-barrier coating or metal jacketing over the insulation (see Figure 2.23).

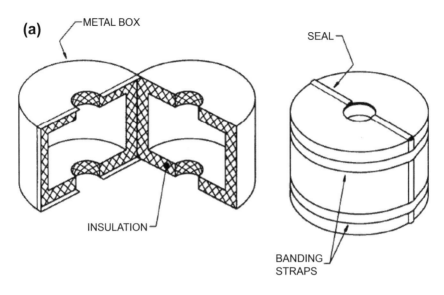

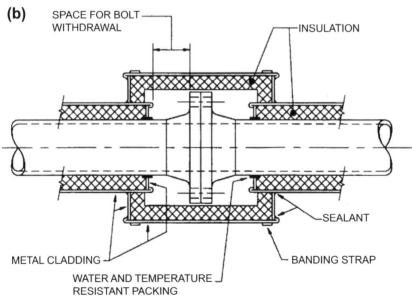

Figure 2.22 (a) Removable flange box; (b) weatherproofing. (Source: CSBP Limited).

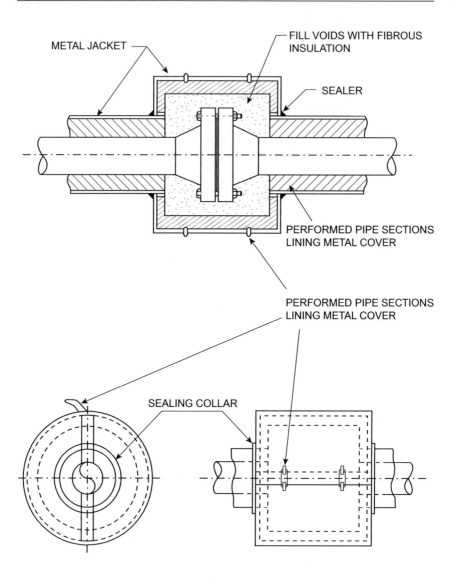

METAL JACKET

FILL VOIDS WITH FIBROUS INSULATION

SEALER

PERFORMED PIPE SECTIONS LINING METAL COVER

PERFORMED PIPE SECTIONS LINING METAL COVER

SEALING COLLAR

COVER TO BE MADE IN TWO OR MORE PIECES

Figure 2.23 Typical methods of insulating pipe flanges.

2.11. COLD INSULATION APPLICATION

This section involves application for systems operating in the temperature range $-100°C$ to $+5°C$. Metal surfaces operating below the dew point shall be given a coat of priming paint, which has to be thoroughly dry before the insulation is applied. The initial protective layer has to be compatible with any adhesive joint sealant and vapor seal used. It shall also be resistant to steam purging temperatures, where applicable.

Cold insulation shall be carried out at ambient temperature. It is sometimes possible to use a liquid with a low freezing point to defrost very small items of an operating plant. (The appropriate safety precautions have to be observed when using a solvent.) If this is done the insulation and vapor seal has to be applied immediately. In general, this practice is not recommended and insulation applied after defrosting shall be regarded as temporary.

The manufacturer's recommendations shall be followed, particularly in respect of:

- The compatibility of the insulation material with adhesives, joint sealants, and vapor sealants
- The maximum temperature of hot-dip adhesives
- The resistance to "cheese knife" cutting if wire ties are used.

Where mechanical support of the insulation is required, wood or plastic skewers, etc., are preferred. Such supports shall be wholly within the insulation.

It is important that the insulation be cut and fitted accurately. It is preferable that molded preformed bends of insulating material be used, as appropriate, to ensure accurate fit without open joints. The practice of filling gaps in joints with plastic composition or rough slivers of insulation is deprecated.

Contraction joints may be required because of the differing rates of thermal movement between the equipment and various types of insulating materials. See Figure 2.24 for some values relating to typical materials in the temperature range −100°C to +20°C.

Where two or more layers of insulating material are required the inner layer shall not be bonded to the vessel or pipework with adhesive, although subsequent layers may be bonded to the appropriate previous layer.

Where the service temperature is very low and where there are appreciably large fluctuations of temperature, and depending on the type of insulating material and the configuration of the system, contraction/expansion joints may be adopted.

Joint-sealer mastic of suitable elasticity shall be applied to the edges of all portions of preformed insulating material, including those of the inner layer.

It is essential that the vapor seal be designed as an integral part of the insulation system and applied as soon as possible so as to keep the insulation dry.

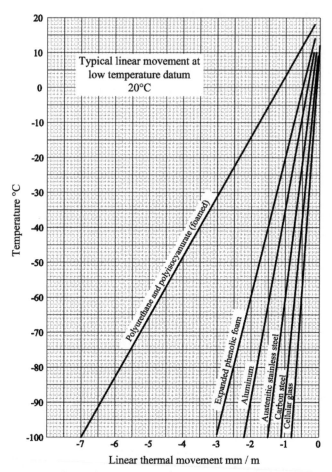

Figure 2.24 Linear thermal movement of various materials between +20°C and −100°C.

The piping and equipment shall be insulated after their pressure tests and other necessary tests and inspections have been completed. To facilitate regular inspection of welds and bottled joints, removable portions of insulating and finishing materials shall be provided in the appropriate locations.

All materials shall be subject to inspection and approval by the owner to ensure that all materials meet this specification. Insulation application shall also be subject to inspection and any material that has been improperly installed or excessively damaged shall be removed and replaced properly with undamaged material. All work shall be executed in a neat and workmanlike fashion in strict accordance with these specifications and as called for on drawings covering the work to be done. No changes or deviations will be permitted without advance written approval by the owner.

Every precaution shall be taken to see that each day's work is weatherproofed before leaving it for the night. Where this is impractical, a fillet of weatherproof mastic must be placed over the exposed ends of insulation. No insulating work of any type may be performed in rainy weather or when atmospheric condensation is occurring. In the event of doubt regarding the prevailing dew point, the decision will be made by the owner.

The insulation classification for cold systems that may appear on flow diagrams and drawings is IC and defined as follows:

IC: To conserve refrigeration, surface condensation, and control fluid temperatures for operating temperatures 21°C and below.

2.11.1. Cold Pipework

Piping shall be insulated when coded on mechanical and utility flow diagrams, pipeline lists, and piping and spool drawings. It is generally good practice to insulate piping with preformed material, although in-situ foam or sprayed foam may be used, especially on larger sizes. It is essential that cellular glass sections not secured to the pipe by adhesives shall have a bore coating of nonsetting compound to act as an anti-abrasive lining; this bore coating has to be allowed to dry before the section is fitted.

When preformed insulating materials are used, two-layer construction is preferred for surfaces operating below -18°C; all joints have to be staggered. Bends shall be insulated with the same type of material cut lobster-back fashion and secured as for straight piping, with one tie per segment, although, alternatively, fully molded or prefabricated bends may be used if these are available.

All attachments fastened to the pipe that protrude through the insulation shall, where possible, be insulated to a distance of four times the insulation thickness and sealed.

Contraction joints shall be provided on long straight runs of piping at approximately 7 m intervals. Joints shall be packed with glass fiber or polyurethane and adequately sealed.

All insulation shall be continuous at supports. A mild steel cradle preformed to the outside diameter of insulation shall be provided at each support. This cradle shall be of sufficient length to prevent undue compression of the insulation due to the weight of the insulated line. Lines larger than 600 mm diameter shall be insulated and finished as described for

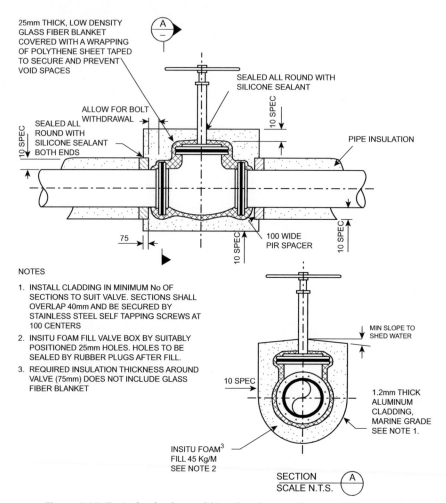

25mm THICK, LOW DENSITY GLASS FIBER BLANKET COVERED WITH A WRAPPING OF POLYTHENE SHEET TAPED TO SECURE AND PREVENT VOID SPACES

A

SEALED ALL ROUND WITH SILICONE SEALANT

ALLOW FOR BOLT WITHDRAWAL

SEALED ALL ROUND WITH SILICONE SEALANT BOTH ENDS

10 SPEC

10 SPEC

PIPE INSULATION

75

100 WIDE PIR SPACER

10 SPEC

10 SPEC

NOTES

1. INSTALL CLADDING IN MINIMUM No OF SECTIONS TO SUIT VALVE. SECTIONS SHALL OVERLAP 40mm AND BE SECURED BY STAINLESS STEEL SELF TAPPING SCREWS AT 100 CENTERS

2. INSITU FOAM FILL VALVE BOX BY SUITABLY POSITIONED 25mm HOLES. HOLES TO BE SEALED BY RUBBER PLUGS AFTER FILL.

3. REQUIRED INSULATION THICKNESS AROUND VALVE (75mm) DOES NOT INCLUDE GLASS FIBER BLANKET

MIN SLOPE TO SHED WATER

10 SPEC

1.2mm THICK ALUMINUM CLADDING, MARINE GRADE SEE NOTE 1.

INSITU FOAM[3] FILL 45 Kg/M SEE NOTE 2

SECTION
SCALE N.T.S.   A

Figure 2.25 Typical valve box cold insulated pipes. *(Source: CSBP Limited).*

vessels. If the required thickness exceeds 75 mm, the insulation shall be applied in not less than two layers.

Expansion joints shall be insulated in such a manner as to allow freedom of movement of the bellows. Applications, sealing, and weatherproofing should be in accordance with the requirements of this book. Figure 2.25 shows typical valve box cold insulated pipes.

2.11.1.1. Application

For single-layer insulation, the ends and butt edges of the preformed insulation shall be buttered with joint sealer and tightly butted so that all voids are eliminated. The insulation shall then be secured with PVC tape bands on 250 mm centers back-taped at least 50 mm from the butt joint.

For multiple-layer insulation all joints shall be buttered with joint sealer as above. In addition each successive layer of insulation shall have a brush coat of nonsetting compound applied to the bore face of the block to be applied. The layers shall be banded in place with PVC tape at approximately 300 mm centers and located 25 mm from the butt joint.

Preformed insulation shall be snugly fitted on the pipe with all joints firmly butted together. Any broken or rounded corners of the insulation shall be cut off and squared so that all voids are eliminated. Where multiple-layer insulation is used, the outer layer shall be placed in such a manner that all joints are staggered.

2.11.2. Weatherproofing

The entire outer surface of the insulation shall receive a tack coat of vapor barrier mastic, 1.5 liters per square meter with a minimum thickness of 1 mm unless specified otherwise, and glass cloth embedded into the surface while still wet, avoiding all wrinkles, pockets, etc., and overlapping the glass cloth a minimum of 50 mm.

A finish coat of vapor barrier mastic shall then be applied to the whole surface 3.0 liters per square meter with a minimum finished and dried thickness of 3 mm unless specified otherwise. Care shall be taken to ensure that all glass cloth and bands are completely covered. The surface shall then be smoothed off with a suitable solvent if metal cladding is not to be applied.

The vapor barrier mastic and glass cloth shall extend at least 150 mm beyond the insulation at all metal projections to ensure a good seal. Heavy fillets of mastic shall be applied to all corners and crevices where water is likely to collect.

Vapor barrier mastic shall also be used as flashing at all possible sources of moisture penetration such as intersections of insulation, nozzles, building walls, valve bonnets, tees, and other protrusions through the surface coating.

Jacketing, where recommended, shall be aluminum sheeting with all joints lapped 50 mm and arranged to shed water. The jacket shall be secured with bands installed at 450 mm centers.

2.11.3. Vessels

Insulation requirements shall be as indicated on the mechanical and utility flow diagrams. Vessel and exchanger flanges, manhole covers, and appurtenance shall be insulated preferably with preformed material. All attachments to the vessel or exchanger such as skirts, supports, ladder and platform clips, etc., shall be covered with insulation for a distance of four times the basic insulation thickness, with the vapor barrier continuing and sealing to the metal. Insulation shall be installed around manholes, exchanger channels, and shell covers so as to allow removal and reuse without damage to the insulation or to adjacent insulation.

Insulation shall be pipe covering for vessels and exchangers smaller than 600 mm outside diameter. Insulation may be blocks, beveled lags, or curved segments for vessels and exchangers 600 mm outside diameter and larger. All such pieces of insulation shall be beveled or shaped to fit closely to the contour of the equipment (or of the inner layer of multilayer insulation). Vessel and heat exchanger heads and transition covers shall have aluminum covers of orange peel design.

Vessels and tubular equipment operating at temperatures -40°C and lower shall have the insulation applied in two layers with joints staggered (see Figure 2.26).

Heads on vessels and tubular equipment shall be insulated with blocks. The butt edges of all segments shall be buttered with an approximately 2 mm thick coating of joint sealer and secured with bands attached to a floating ring in the center and to a band installed at the tangent point of the vessel head. Band spacing shall be 300 mm maximum at the circumference of the vessel.

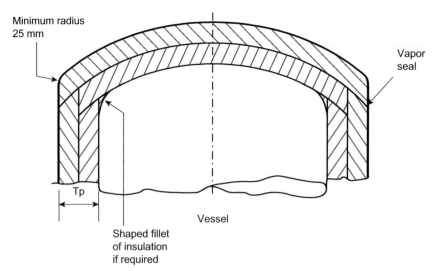

Figure 2.26 Typical method of insulating dished ends of cold vessels.

In addition to the use of adhesive, where it is recommended, each layer shall be banded on. Large slabs will need at least two ties per slab. Wooden skewers or thin clips may be used to hold second and subsequent layers to the first.

Where banding is impracticable, slabs may be impaled on studs that partially penetrate the thickness of insulation. Where differential movement between the vessel and insulating material makes necessary the use of expansion/contraction joints, the positions of these shall be marked off before erection of the insulation begins. Suitable flexible vapor tight cover strips then have to be provided.

Before the erection of the main body of insulation, all protruding pipe stubs, fittings, manhole necks, etc., shall be insulated. The insulation shall extend down any vessel skirt or legs or cradle for a distance not less than four times the thickness of the insulation on the vessel, measured from the surface of the insulation.

2.11.4. Vapor Barrier and Weatherproofing

The entire outer surface of the insulation shall receive a tack coat of vapor barrier mastic, 1.5 liters per square meter with a minimum thickness of 1 mm unless specified otherwise, and glass cloth embedded into the surface while still wet, avoiding all wrinkles, pockets, etc., and overlapping the glass cloth a minimum of 50 mm. A finish coat of vapor barrier mastic shall then be applied to the whole surface, 3.0 liters per square meter, with a minimum finished and dried thickness of 3 mm unless specified otherwise. Care shall be taken to ensure that all glass cloth and bands are completely covered. The surface shall then be smoothed off with a suitable solvent if metal cladding is not to be applied.

At all metal flashing protrusions, insulation terminals, corners, and crevices, a coat of foam sealant shall be applied prior to the mastic and glass cloth that shall extend 150 mm beyond the insulation to ensure a good seal. Heavy fillets of vapor barrier mastic shall be applied to all corners and crevices where water is likely to collect.

Vapor barrier mastic shall also be used at all possible sources of moisture penetration, such as intersections of insulation, nozzles, building walls, and other protrusions through the surface coatings.

Vessel and tubular equipment heads and transitions shall be finished as follows.

A tack coat of mastic shall be applied over the insulation, and glass cloth shall be embedded and pulled down over the vessel shell insulation for a distance of at least 150 mm under the metal jacket, and secured with bands. The fabric shall lap at least 75 mm at all joints and be wrinkle-free. A 1.2 mm thick coat, when dry, of mastic shall then be applied.

Vessels and tubular equipment shells shall be finished with metal jacketing, where specified, applied directly over the vapor barrier coating. On vertical vessels and tubular equipment, the bottom course of metal shall be supported on the bottom insulation support ring. All other horizontal seams of the metal sheets shall be lapped 75 mm circumferentially, supported on "S" clips, with a minimum of 100 mm lap or two corrugations on each vertical seam. Metal coverings shall be secured with bands spaced not less than one band on each circumferential lap and one at the middle of each sheet, but not to exceed 500 mm on center.

No metal screws shall be used. On horizontal vessels, tubular equipment, and fin-tube exchangers, the metal sheets shall be lapped a minimum of 75 mm on the circumferential seam and a minimum of 100 mm or two corrugations on the longitudinal seams. Bands shall be installed on 460 mm centers and shall be machine-stressed and fastened under tension. On vertical and horizontal vessels, the junction between the mastic head and metal jacket shall be sealed with a mastic fillet.

2.12. VALVES, FLANGES, FITTINGS, AND IRREGULAR SURFACES

2.12.1. Insulation Application

All valves, flanges, and fittings on insulated cold piping shall be insulated with preformed or mitered insulation of the same thickness as the adjacent piping insulation, and secured with bands and joint sealer in the same manner as that specified for straight piping. Care shall be taken to insure tight joints between fittings and straight pipe insulation.

Pumps, compressors, or other irregularly shaped equipment shall be enclosed in block insulation with voids filled with glass fiber. Butt edges of all segments shall be buttered with an approximately 2 mm coating of joint sealer and shall be securely banded or tied in place.

2.12.2. Weatherproofing

Fittings and flanges shall have a tack coat of mastic applied directly over the insulant, followed by a layer of glass fabric installed wrinkle-free. Over the glass fabric a 1.2 mm thick layer, when dry, of mastic weather coat shall be troweled to a smooth finish. This finish shall extend approximately 50 mm under the adjacent pipe weatherproofing jacket. Vertical joints between the mastic and pipe jacket shall be sealed to prevent entrance of moisture.

Pumps, compressors, or other irregular shapes shall have a tack coat of mastic applied to the outer surface and glass cloth shall be embedded into the mastic. The fabric shall lap at least 75 mm at all joints and shall be wrinkle free. A 1.2 mm thick coat, when dry, of mastic shall then be applied.

2.13. COLD SPHERES

Insulation requirements shall be as indicated on the mechanical flow diagrams. The thickness of insulation shall be as shown on the sphere drawings. Flanges, manhole covers, and all appurtenances shall be insulated. All attachments to the sphere such as ladder and platform clips shall be covered with insulation for a distance of four times the basic insulation thickness.

Insulation shall extend over and down support legs a distance of not less than 900 mm from the lowest juncture of the leg with the sphere, and shall be supported by angle or plate welded around the sphere support leg. Insulation shall be installed around manholes and shell covers so as to allow removal and reuse without damage to the insulation or to adjacent insulation. Insulation shall be single-layer construction and shall be properly vapor sealed.

2.13.1. Insulation Application

Prior to application of insulation, all metal surfaces shall be cleaned and primed. The primer shall be fully cured before application of the insulation. The sphere shall have a bar or plate support of such size as to support 13 mm the thickness of the insulation welded around the shell at the equator.

Insulation shall be blocks, beveled lags, or curved segments beveled or shaped to fit closely to the contour of the equipment with all joints butted snugly together. A full coating of latex hydraulic cement adhesive shall be applied to the erection face of the insulant and a sealer to one side and two adjacent edges of each block. Joints shall be tightly butted and shall be staggered where possible. Latex cement shall be equal to Benjamin Foster Flexfas adhesive 82-10.

On the bottom half of the sphere, the insulation shall be banded in addition to being adhered. Bands shall be fastened to a 13 mm × 1.80 m maximum carbon steel floating ring at the bottom of the sphere and to the equator support on 900 mm centers maximum.

2.13.2. Weatherproofing

Over the insulation a tack coat of mastic shall be applied. While still tacky, glass cloth shall be laid smooth and thoroughly embedded into the mastic. The fabric shall lap at least 75 mm at all joints. A 1.2 mm thick coat, when dry, of mastic shall then be applied over the entire surface.

To ensure seal at the junction of the insulation and the equator support, the insulation shall be cut in such a manner as to provide a flush surface over the ring, and a close fitting around the ring. Whenever needed, over the vapor barrier/weather coat mastic, a galvanized iron jacket may be fitted.

2.14. PREFABRICATED UNDERGROUND PIPE SYSTEMS

The prefabricated underground systems have all the same insulation requirements as insulation above grade does. The systems shall fulfill the thermal, physical, and economic requirements of the other insulation systems. Likewise, these requirements make it necessary to have a number of different types of underground systems.

These systems shall have watertight joint closure of the insulation and jacket in the field with a special jacket fitting and sealing technique. Another form of watertight jacketing is application of reinforced mastic jackets, which are installed after the insulation is installed.

Suitable expansion–contraction joints or loops for process pipe and jacket shall be provided. In Figure 2.27a an underground system using prefabricated cellular glass insulation is illustrated.

In this system the pipe insulation, fitting insulation, and expansion loop (or joints) shall be installed after the pipe is in position in the trench. The pipe shall be held up by temporary supports, which are removed as the insulation is installed. The inner surface of the insulation shall be protected from abrasion by anti-abrasion coating to prevent pipe movement from wearing away the insulation.

Special loop insulation, to take care of the change in configuration of pipe, shall be provided. The insulation surface shall be protected with a sealed laminate of glass fabric, aluminum bonded with bituminous, high-molecular-weight polymers.

(a)

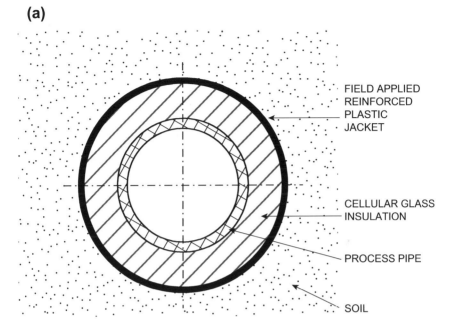

FIELD APPLIED
REINFORCED
PLASTIC
JACKET

CELLULAR GLASS
INSULATION

PROCESS PIPE

SOIL

(b)

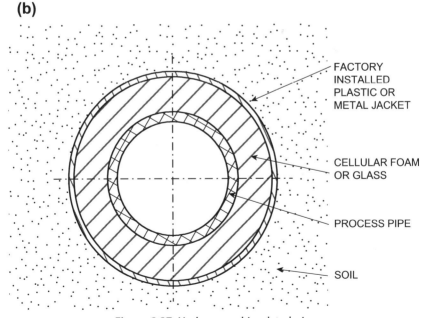

FACTORY
INSTALLED
PLASTIC OR
METAL JACKET

CELLULAR FOAM
OR GLASS

PROCESS PIPE

SOIL

Figure 2.27 Underground insulated pipe.

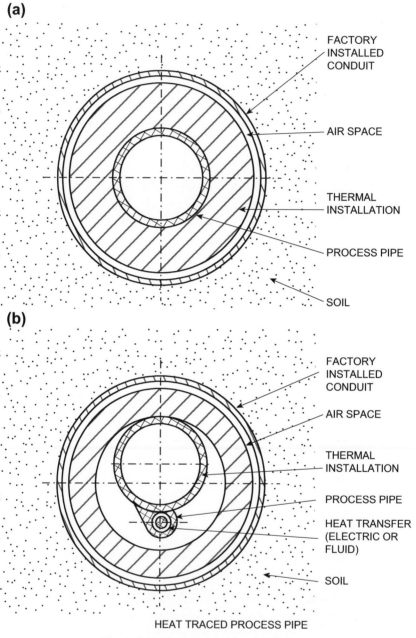

(a)

FACTORY INSTALLED CONDUIT

AIR SPACE

THERMAL INSTALLATION

PROCESS PIPE

SOIL

(b)

FACTORY INSTALLED CONDUIT

AIR SPACE

THERMAL INSTALLATION

PROCESS PIPE

HEAT TRANSFER (ELECTRIC OR FLUID)

SOIL

HEAT TRACED PROCESS PIPE

Figure 2.28 Underground insulated pipe.

Care shall be taken that this protective jacket is watertight before earth is backfilled over the insulated pipe.

In Figure 2.27b a factory prefabricated insulated pipe system where the insulation fills the void between the pipe and outer jacket or conduit is shown.

In most instances, the insulation used in these systems is cellular organic foam or glass.

Fittings and connections are preformed for the particular systems and must be designed to fulfill the piping requirements. Figure 2.28a shows a conduit and the insulation is installed to provide an air space between it and the conduit. The insulation may be cellular glass, glass fiber, mineral fiber, calcium silicate, or the equivalent.

Figure 2.28b shows a pipe in a conduit system where heat is supplied by an external electric or fluid tracer. For detailed information on design and application of a thermally insulated underground piping system, reference is made to BS 4508 Part 1 through 4 and BS CP 3009.

2.15. INSPECTION AND MAINTENANCE OF EXISTING INSULATION SYSTEMS

The consequences of failing insulation systems are very often detected only in an advanced state and are reflected in high repair and maintenance costs. Routine maintenance practice shall be extended with a system of scheduled inspections, preventive maintenance, and a long-term maintenance program.

2.15.1. Inspection

- **Purpose of inspection**

  The purpose is to detect shortcomings at the earliest possible time in order to prevent uncontrolled deterioration of the insulation system and consequential underlagging corrosion.

- **Items to be inspected**

  Of prime importance are the integrity of the jacketing system and the thickness of the insulating material. The inspection shall be aimed at detecting shortcomings/defects of these items. At locations with sagged insulation (e.g., pipelines or horizontal vessels) or at transitions from vertical to horizontal pipes, corrosion may be expected. The condition of hot water/steam lines is crucial as minor leaks in these lines will promote underlagging corrosion in the main product lines.

- **Inspection techniques**

  Visual inspection is still the most widely used method of inspection for insulation systems and on pipeline surfaces or equipment for corrosion checks. However, the frequent removal of insulation for visual inspection of underlagging corrosion is impractical. Inspection methods using conventional gamma radiography, flash radiography (x-ray), or infrared scanning are used in these inspections. Flash radiography has proven to be a quick and effective method of assessing the external condition of piping under insulation.

- **Inspection frequency**

  The optimal frequency of inspections depends on a number of factors such as plant size, previous maintenance programs, and the type of insulation. For hot insulation, inspection shall be carried out once a year.

- **Inspection program**

  The plant shall be divided into a number of areas or zones. In each area an inspection route must be determined such that all major equipment and pipelines can be inspected. All the items to be inspected shall be listed on an inspection list. During the inspection all items shall be checked off. Damage and failures shall be reported.

  In this way, no equipment or pipelines will be overlooked and the condition of the insulation at that moment will be recorded. Based on such records a plan of action and a budget can be made for preventive and programmed maintenance.

- **Inspection survey**

  Causes of shortcomings or damage to the insulation system should be investigated and rectified. The renewing of damaged insulation without determining the actual cause is an incorrect approach towards preventive maintenance.

2.15.2. Maintenance

- **Preventive maintenance**

  After an inspection survey has been completed the reported damage and remarks shall be translated into a plan of action for remedial and preventive maintenance. The recommendations for preventive maintenance refer to situations or structures that need to be modified to prevent future or repeated damage to insulation or the underlying surfaces.

Technical shortcomings in design shall be rectified, for example:
- Repositioning of supports and brackets to eliminate water ingress;
- The installation of rainwater shields.

Damage caused by personnel or equipment can be prevented by:
- Installation of a walkway over insulated pipes in a pipe track;
- Rerouting of foot traffic by putting up hand railings;
- Avoidance of fire drills on or near insulated tanks or equipment;
- Instruction and monitoring of third parties, such as painters, cleaners, and scaffolders.

Damaged or saturated insulation shall be discarded and the insulated metal surfaces cleaned, preferably grit-blasted, and painted before installing the new insulating material. A program of preventive maintenance will eliminate more expensive repairs later on.

- **Programmed maintenance**

Based on the results of inspection surveys, the scope of long-term insulation maintenance can be determined and priorities can be set. In order to systematically control the upgrading of existing insulations in a plant, it is recommended to divide the various units into manageable areas, indicated on a plot plan, and to carry out the work area by area. Simultaneously, maintenance painting in the same area shall be scheduled. The progress of work can then be properly recorded and costs for scaffolding will decrease substantially as compared to when pipelines are followed or when work is carried out randomly throughout the plant.

2.16. SURFACE PREPARATION FOR THERMAL INSULATION INSTALLATION

This section gives the minimum requirements for surface preparation of substrates prior to thermal insulation installation to protect against corrosion, both for initial construction and/or maintenance.

The standard includes the minimum requirements for surface preparation of ferrous metals, nonferrous metals, and nonmetallic surfaces (e.g., masonry materials and wood). Applicable methods of surface preparation such as degreasing, pickling, manual cleaning, flame cleaning, and blasting are discussed in this section. Recommendations are made regarding the selection of appropriate method(s) of surface preparation with reference to the substrate and the coating (including metallic coating and electroplating) to be applied.

Surface preparation is (are) the method(s) of treating the surface of substrate prior to application of coating (painting, coating, and lining, etc.). Typical contaminants that shall be removed during surface preparation are moisture, oil, grease, corrosion products, dirt, and mill scale.

The surface prepared must achieve a level of cleanliness and roughness suitable for the proposed coating and permit good adhesion of the thermal insulation expenditure on the preparation work shall be in a reasonable proportion to the purpose and to the nature of the coating. The personnel carrying out surface preparation work must have the expertise and technical know-how to enable them to carry out the work in a technically satisfactory and operationally reliable manner.

The surfaces must be accessible and adequately illuminated. The relevant accident prevention regulations and safety provisions must be observed.

All surface preparation work must be properly quality-controlled and inspected. Each subsequent coating may only be applied when the surface to be coated has been prepared in accordance with the principles of the applicable construction standard.

The handling of parts or assemblies, after cleaning, shall be kept to a minimum. When handling is necessary, clean gloves or similar protection shall be used. Canvas, PVC, or leather are suitable materials for gloves.

Cleaned surfaces shall be insulated as soon after cleaning as is practical and before detrimental corrosion or recontamination occurs.

There is no single, universal method of cleaning by which all surfaces can be prepared for the application of protective insulating, and the cleaning method for any given type of article must be carefully selected and properly carried out.

2.16.1. Selection of Cleaning Method(s)

The cleaning method(s) shall be selected with reference to the following considerations. The choice between blast cleaning, acid pickling, flame cleaning, and manual cleaning is partly determined by the nature of the coating to be applied. It should be appreciated, however, that insulation applied to a properly prepared (e.g., blast cleaned) surface will always last longer than similar coating applied to flame cleaned or manually cleaned surfaces.

2.16.2. Initial Condition of Surface (Rust Grade)

The initial condition of surfaces for preparation, which, among other factors, determines the choice and mode of execution of the preparation measures and the relevant reference sample to be used, must be known.

The rust level in accordance to SIS 05 59 00 or ISO 8501-1 can be classified as follows:

A = Steel surface covered with firmly adhesive scale and largely free of rust

B = Steel surface with beginning of spalling of scale and beginning of rust attack.

C = Steel surface from which scale has been rusted away or can be scraped off but which exhibits only a few rust pits visible to the eye.

D = Steel surface from which the scale has been rusted away that is exhibiting numerous visible rust pits.

2.16.3. New Construction (Uncoated Surfaces)

The grade of steel, as well special treatments or methods that have an effect on the preparation, e.g., use of cold rolling or deep drawing methods, should be investigated to determine proper treatment of new uncoated surfaces.

2.16.4. Maintenance (Coated Surface)

The following information should be known to conduct proper maintenance of coated surfaces:

• Rust level of coated surfaces according to DIN 53210 and ASTM D 610;
• Type of coating (e.g., type of binder and pigment, metal coating), approximate coat thickness and date when carried out;
• Extent of blistering according to DIN 53209, and ASTM D 714;
• Additional information, e.g., on adhesion, cracking, chemicals and other contaminants, and also other significant phenomena.

2.16.5. Cleanliness of the Surfaces

• **Removal of contaminants/coats of material different from the metal**

This includes removal of dirt, dust, soot, ash, concrete, coal slag, sand, moisture, water, acids, alkalis, soap, salts, encrustations, growths, fluxes, oily and greasy contaminants, earlier coatings and cementations that are loose, rusted under, or unusable as an adhesion surface, and corrosion products of metallic coatings. Unless otherwise specified the best cleaning method for the job shall be selected.

• **Removal of coats of material related to the metal (scale and rust removal)**

Removal of firming adherent scale is only possible with the following methods of removing rust:

- Blasting;
- Flame cleaning;
- Pickling;
- Manual cleaning.

With each of these methods only specific surface condition can be produced and particular levels of cleanliness achieved correspondingly, the appearance of the prepared surface is dependent not only on the level of cleanliness but also on the method of removing rust used.

- **Standard level of cleanliness**

Standard level of cleanliness for prepared steel surfaces listed in Tables 2.3 and 2.4 shall apply. Unless otherwise specified by the company, the level of cleanliness of uncoated surfaces shall be in accordance with manual cleaning and the provisions of Table 2.3 for flame cleaning and pickling.

- **Blasting**

During blasting, parts that are not to be worked, parts already coated, and the environment should be protected against the blasting abrasives thrown out. With chemically contaminated surfaces, prewashing may be advisable, and with coarse coats of rust on plates, preliminary derusting with impact tools may be advisable.

The standard levels of cleanliness for blasted surfaces are described as follows:

- **Sa 1: Light blast cleaning**

When viewed without magnification, the surface shall be free from visible oil, grease, and dirt, and from poorly adhering mill scale rust, paint, coatings, and foreign matter (see Table 2.3). The reference photographs are BSa1, CSa1, and DSa1.

- **Sa 2: Thorough blast cleaning**

When viewed without magnification, the surface shall be free from visible oil, grease, and dirt, and from most of the mill scale, rust, paint, coatings, and foreign matter. Any residual contamination shall be firmly adhering (see Note 1).

(**Note 1:** For previously painted surfaces that have been prepared for renewed painting, only photographs with rust grade designations D or C (for example DSa2½ or CSa2½) may be used for the visual assessment.)

Staining shall be limited to no more than 33% of 1300 square mm of

Table 2.3 Standard Levels of Cleanliness for Prepared Steel Surfaces

Standard Level of Cleanliness	Rust Removal Method	Initial Condition of Steel Surface	Reference Sample Photographs	Essential Characteristics of the Prepared Steel Surface	Remarks
Sa 1	Blasting	B C D	B Sa 1 C Sa 2 D Sa 1	Only loose scale, rust, and loose coatings are removed.	These standard levels of cleanliness apply to blasting a) of uncoated steel surface, and b) of coated steel surfaces if the coatings are also removed sufficiently to obtain the required level of cleanliness.
Sa 2		B C D	B Sa 2 C Sa 2 D Sa 2	Virtually all scale, rust, and coatings are removed, i.e., only as much firmly adhesive scale, rust, and coating residue remain (no continuous coats) as to correspond with the overall impression of the reference sample photographs.	
Sa 2½		A B C D	A Sa 2½ B Sa 2½ C Sa 2½ D Sa 2½	Scale, rust, and coatings are removed to the extent that the residues on the steel surface remain merely as faint shading as a result of coloring of the pores.	

(Continued)

Table 2.3 Standard Levels of Cleanliness for Prepared Steel Surfaces—cont'd

Standard Level of Cleanliness	Rust Removal Method	Initial Condition of Steel Surface	Reference Sample Photographs	Essential Characteristics of the Prepared Steel Surface	Remarks
Sa 3		A B C D	A Sa 3 B Sa 3 C Sa 3 D Sa 3	Scale, rust, and coatings are completely removed (when observed without magnification).	
St 2	Manual or mechanical rust removal	B C D	B St 2 C St 2 D St 2	Loose coatings and loose scale are removed; rust is removed to the extent that the steel surface after subsequent cleaning exhibits a faint luster from the metal.	
St 3		B C D	B St 3 C St 3 D St 3	Loose coatings and loose scale are removed; rust is removed to the extent that the steel surface after subsequent cleaning exhibits a distinct luster from the metal. In special cases removal of firmly adhesive coatings, e.g., by grinding, scraping, or	Normally requires machining

Fl	Flame cleaning	A	A Fl	by using pickling agents is possible. If required, this should be additionally agreed.	Subsequent thorough mechanical brushing is always necessary.
		B	B Fl	Coatings, scale, and rust are removed to the extent that residues on the steel surface remain merely as shades in various hues.	
		C	C Fl		
		D	D Fl		
Be	Pickling	A		Coating, residues, scale, and rust are completely removed.	Coatings must be removed in a suitable manner before pickling.
		B			
		C			
		D			

Table 2.4 Comparison of Standard Levels of Cleanliness According to SIS 055900 with Other Rust Removal Levels or Quality Classes

Standard Level of Cleanliness	Rust Removal Method	SIS 055900	DIN 18364	BS 4232 (Only for Blasting)	SSPC-VIS
Sa 1		Sa 1	–	–	Brush off SP 7
Sa 2		Sa 2	Rust removal level 2	Third quality	Commercial SP 6
Sa 2½		Sa 2½	Rust removal level 3	Second quality	Near white SP 10
Sa 3		Sa 3		First quality	White metal SP 5
St 2		St 2	Essentially less than rust removal level 1	–	Band tool cleaning SP 2
St 3		St 3	Less than rust removal level 1	–	Power tool cleaning SP 3
Fl	Flame cleaning	–	Rust removal level 2	–	Flame tool cleaning SP 4
Be	Pickling	–	Rust removal level 3	(CP 3012)	Pickling SP 8

surface area and may consist of light shadows, slight streaks, or minor discolorations caused by stains of rust, stains of mill scale, or stains of previously applied paint. Slight residues of rust and paint may also be left in the bottoms of pits if the original surface is pitted.

- **Sa 2½: Very thorough blast cleaning**
  When viewed without magnification, the surface shall be free from visible oil, grease, and dirt, and from mill scale, rust, paint, coatings, and foreign matter. Any remaining traces of contamination shall show only as slight stains in the form of spots or stripes.

- **Sa 3: White blast cleaning**
  This is blast cleaning to visually clean steel when viewed without magnification; the surface shall be free from visible oil, grease, and dirt, and shall be free from mill scale, rust, paint, coatings, and foreign matter. It shall have a uniform metallic color.

- **Manual rust removal**
  The standard levels of cleanliness St 2 and St 3 are used for hand and power tool surface cleaning.

  - **St1:** This grade is not included as it would correspond to a surface unsuitable for painting.

  - **St 2:** Thorough hand and power tool cleaning:
    - When viewed without magnification, the surface shall be free from visible oil, grease, and dirt, and from poorly adhering mill scale, rust, paint, coatings, and foreign matter. This grade is only accepted for spot cleaning.

  - **St 3:** Very thorough power tool cleaning
    - This is the same as for St 2 but the surface shall be treated much more thoroughly to give a metallic sheen arising from the metallic substrate.

- **Thermal rust removal, flame cleaning**
  For this type of cleaning, the standard level of cleanliness is Fl according to Table 2.3.

  - **Fl:** When viewed without magnification, the surface shall be free from mill scale, rust, paint, coatings, and ISO foreign matter. Any remaining residue shall show only as a discoloration of the surface (shades of different colors).

- **Chemical rust removal, pickling**
  For this type of cleaning, the standard level of cleanliness is Be according to Table 2.3.

- **Be:** When viewed without magnification the surface shall be free from coating residues, scale, and rust (see Table 2.4).

A comparison of standard levels of cleanliness according to SIS 055900 with other rust removal levels is shown in Table 2.4.

2.16.6. Rust Converters, Rust Stabilizers, and Penetrating Agents

This refers to the use of so-called rust converters, rust stabilizers, and similar means for chemically converting the corrosion products of the iron into stable iron compounds. This also applies to penetrating agents intended to inhibit rust. The use of these materials is not permitted in this section.

2.16.7. Influence of Environmental Conditions on the Cleaning and Cleanliness of Surfaces

Storing of unprotected steel in urban, industrial, or marine atmospheres, including the practice of "rusting off" scale by weathering, shall be avoided, because otherwise a higher level of cleanliness may be necessary owing to the deposit of corrosive substances that will occur. Steel shall as far as possible be prepared and protected in state A or A to B (see Table 2.3).

No surfaces shall be prepared for coating during rain or other precipitation. If the possibility of condensation is to be absolutely excluded, the temperature of the surface to be worked must with certainty remain above the dew point of the surrounding air. If the work still has to be carried on, even under conditions deviating from those required, provision shall be made at the planning stage for special measures, e.g., covering, enclosing in a tent, warming the surfaces or drying the air.

Preparatory works in the region of plants subject to explosion or fire risk require special measures (e.g., low-spark or flame-free methods).

2.16.8. Testing of the Cleanliness of the Prepared Surfaces

The surfaces shall be tested after subsequent cleaning. For all the standard levels of cleanliness according to Table 2.4 visual testing is in practice sufficient. A check shall be made as to

- whether the prepared surfaces exhibit the essential characteristics stated for the appropriate level of cleanliness in Table 2.3 and shown by way of example in the corresponding reference sample photograph, or
- whether there is adequate conformity to any agreed reference surfaces.

The reference sample photographs shall in particular show the interrelationships between the levels of cleanliness. The comparison is made without magnification (magnifying glass). For standard level of cleanliness Bc (see Tables 2.3 and 2.4), by nature of the method, reference samples are unnecessary. If there are doubts as to freedom from oil or grease of the surface, this shall be checked.

2.16.9. Degree of Roughness (Surface Profile)

Blast cleaning produces a roughened surface and the profile size is important. The surface roughness (average peak-to-valley height) achieved for each quality of surface finish depends mainly upon the type and grade of abrasive used. Unless otherwise specified by the company, the amplitude of surface roughness of steel work shall be within 0.1 mm to 0.03 mm for painting, coating, and lining. Table 2.5 gives the range of maximum and average maximum profile heights of some abrasive to be expected under normal good operation conditions (wheel and nozzle). If excessively high air pressure or wheel speed is used, the profile may be significantly higher.

- **Methods of measurement**

  The methods described herein are some of the suitable methods for measuring surface roughness. The company shall decide the method(s) of measurement to be used.

  - **Sectioning**

    A metallurgical section is prepared and the surface profile measured under a suitable microscope using a micrometer eyepiece.

  - **Grinding**

    The thickness of the blast-cleaned specimen is measured with a flat-ended micrometer. The surface is then ground until the bottoms of only the deepest pits are just visible. A further thickness measurement is then taken.

  - **Direct measurement by microscope**

    The blast-cleaned specimen, or a replica, is viewed through a suitable microscope, first focusing on the peak and then focusing on the lowest adjacent trough, noting the necessary adjustment of focus.

  - **Profile tracing**

    A blast-cleaned specimen is traversed with a diamond or sapphire stylus and the displacement of the stylus as it passes over peaks and

Table 2.5 Typical Maximum Profiles Produced by Some Commercial Abrasive Media

Abrasive	Typical Profile Maximum (mm)	Height (mm) Avg. Maximum
Steel Abrasives		
Shot S 230	0.074	0.056
Shot S 280	0.089	0.064
Shot S 330	0.096	0.071
Shot S 390	0.117	0.089
Grit G 50	0.056	0.04
Grit G 40	0.086	0.061
Grit G 25	0.117	0.078
Grit G 14	0.165	0.13
Mineral Abrasives		
Flint shot (medium–fine)	0.089	0.068
Silica sand (Medium)	0.10	0.074
Boiler slag (Medium)	0.117	0.078
Boiler slag (Coarse)	0.152	0.094
Heavy Mineral Sand (medium–fine)	0.086	0.066

Profile heights shown for steel abrasives were produced with conditioned abrasives of stabilized operating mixes in recirculating abrasive blast cleaning machines. Profile heights produced by new abrasives will be appreciably higher. Cast steel shot: Hardness 40 to 50 Rockwell C. Cast steel grit: Hardness 55 to 60 Rockwell C.

troughs is recorded. For instruments and the procedures for the measurement see ISO 3274 and ISO 4288.

- **Comparator disc**

    The comparator disc (Keane–Tator) is a field instrument to determine anchor pattern depth of the blasted surface. The comparator disc is composed of five sections, each with a different anchor pattern depth. To use this instrument, place the disc on the blasted surface and visually select the reference section most closely approaching the roughness.

2.16.10. Temporary Protection of Prepared Surfaces Against Corrosion and/or Contamination

Temporary protection is necessary if the proposed coating (primer or total coating) cannot be applied to the prepared surface before its level of cleanliness has changed (e.g., by formation of initial rust). The same applies to parts on which the coating is not to be applied. It is normal to use wash primers and shop primers as well as phosphatizing.

2.16.11. Preparation of Surfaces Protected by Temporary Coats or by Only Part of the Purposed Coats until Subsequent Coating Is Applied

Before further coating, all contaminants as well as any corrosion and weathering products that have been produced in the meantime shall be removed in an appropriate manner described in this standard. Assembly joints and damaged areas of the primer coating shall be again derusted.

- **Preparation of joint areas (welded, riveted, and bolted joints)**

    The best method of removing residues of welding electrodes and welding or riveting scale, is blasting or manual cleaning. The method shall be approved by the company.

- **Preparation of surfaces of shop primers, base coating, or top coatings for further coating**

    It may be necessary to apply a little solvent to an existing coating or, for example in the case of two-component coatings that have cured, to roughen them with sandpaper or steel wool or lightly blast them and then remove the dust so that the following coating adheres satisfactorily. Surfaces of existing coatings (especially primers rich in zinc) must not be burnished or smeared by mechanical brushing or similar methods so that later coatings no longer adhere satisfactorily. If an existing shop primer or base coating is not in a condition suitable to provide a base for further coatings or is not compatible with such further coating, it shall be removed. The methods of surface preparation shall be approved by the company.

- **Preparation of hot-dip galvanized or hot-dip aluminum–coated surfaces for other coatings**

    Defective areas in the production of the metal coating:

    Defective areas in the metal coating must be prepared and repaired in such a manner that the corrosion-protecting action of the coating is restored and that the adhesion and protective action of further coating are not impaired.

    Defective areas are repaired by building up with solder or sprayed zinc or shall be prepared for subsequent coating in the same way as hot-dip galvanized or hot-dip aluminum coated surfaces, or sprayed zinc-coated and sprayed-aluminum surfaces. Coated surfaces shall not, however, be cleaned with alkaline detergents.

- *Preparation of unweathered hot-dip galvanized or hot-dip aluminum coated surfaces:*

    During subsequent treatment, transport, or assembly, contamination may occur, for example by grease, oil, marking, or coding inks. These shall be removed, for example, by brushing off and rinsing with special detergent or solvents.

- *Preparation of weathered hot-dip galvanized or hot-dip aluminum coated surfaces:*

   According to the period of weathering and the site where the surfaces are located, apart from surface contaminations, various corrosion products of the coating metal or steel that become soluble or poorly adherent contaminants shall be removed, according to the extent and nature of the deposits:

   - In the case of hot-dip galvanized surfaces, e.g., oxidic compounds and various salts, by dry brushing (brushes with plastics bristles) or washing with water with a detergent added, or by water or steam cleaning.
   - In the case of hot-dip aluminum-coated surfaces by brushing or washing with suitable solvents, cold detergents, or emulsion cleaner. If necessary, it may also be advisable to clean by water or steam cleaning; steam cleaning with the addition of weak phosphoric acid cleaner may also be used.
   - For mechanical preparation of severely attacked surfaces, wire brushes, scrapers, emery discs, and blasting methods are suitable.

2.16.12. Procedure Qualification

The preparation procedure specification shall incorporate the full details of the following, but is not limited to these issues:

- Cleaning of components and method of cleaning including oil and grease;
- Cleaning medium and technique;
- Blast cleaning finish, surface profile, and surface cleaning, in the case of blast cleaning;
- Dust and abrasive removal;
- Drying time and temperature;
- Recleaning technique.

2.17. DEGREASING

Degreasing is used for complete removal of oil, grease, dirt, and swarf from the surfaces that are to be protected by painting, coating, and lining. The five following main cleaning methods, which in turn may consist of different processes, are generally used for surface cleaning of substrates:

- Hot solvent cleaning;
- Cold solvent cleaning;

Table 2.6 Guide to Selection of Degreasing Methods

Degreasing Method	Scope	Submethod	Cleaning Material	Variation of Cleaning Process	Condition of Surface after Treatment	Cleaning Apparatus	Special Points	Safety Precautions
Hot solvent cleaning	All materials and all type of surfaces	Not water rinsable	Trichloroethylen to BS 580 1, 1, 1, trichloroethane to BS 4487 methylene chloride perchloroethylene to BS 1593 triclorotrifluorethane	1– Vapor immersion 2– Liquid immersion 3– Jetting 4– Ultrasonic cleaning	Dry	Specially designed apparatus essential. Continuous or batch operation	Will not remove soaps, sweat, or chemical residues. Not suitable for painted articles, or parts containing rubber and certain other nonmetallic materials. Ultrasonic cleaning specially suitable for removing fine solid particles.	Adequate shop ventialation, ard correct operation of plant to avoid excessive inhalation of narcotic vapor. No smoking
		Water rinsable	Emulsifiable blenc of cresybic acid (BS 524) and o-dichlorobenzene (BS 2944 grade B)	1– Immersion (bath temp 60°C)				

(Continued)

Table 2.6 Guide to Selection of Degreasing Methods—cont'd

Degreasing Method	Scope	Submethod	Cleaning Material	Variation of Cleaning Process	Condition of Surface after Treatment	Cleaning Apparatus	Special Points	Safety Precautions
Cold solvent cleaning	All metals and all types of surfaces and shapes not heavily contaminated	Not water rinsable	Trichloroethylen to BS 580 1, 1, 1- trichloroethane to BS 4487 perchloroethylene to BS 1593 trichlorotrifluoro-ethane with spirit to BS 245 white spirit/ solvent naphthas to BS 479 aromatic solvents coal tar solvents	1- Immersion 2- Brushing 3- Wiping 4- Spraying 5- Ultrasonic system	Dry or from petroleum solvents slightly oily	Tanks or apparatus of special design. Continuous or batch operation	Complete grease removal not certain. Soaps, sweat, or chemical residues not removed	Adequate shop ventilation, no smoking, strict provision against risk if petroleum solvents are used
		Water rinsable	Dichloromethane-based mixture to BS 1994 trichloroethylene-based mixture	1- Immersion (liquid) 2- Brushing				
Emulsifiable solvent cleaning	All metal parts of accessible shape, not heavily contaminated	–	Hydrocarbon, e.g., white spirit with an emulsifying agent	1-Immersion (liquid) 2-Brushing	Wet and possibly slightly oily	Tanks or apparatus of special design. Continuous or batch operation	Will remove sweat and certain chemical residues. Suitable for painted surfaces.	Precautions against contact of concentrated material with skin. Fire precautions if petroleum solvents are used
				Spraying	Wet and possibly slightly oily	Steam or water jet apparatus	Suitable for large assemblies which cannot be dismantled.	

Method	Application	Type	Composition	Process	State	Apparatus	Remarks
Aqueous alkaline and detergent cleaning	Parts without a highly finished surface and of accessible shape. Strong alkalis for ferrous metals, milder alkalis for nonferrous metals and general purposes cleaning, organic detergent alone (as solution or emulsion) for tin plate and light duty cleaning.	Hot alkaline	Mixture of sodium hydroxide and sodium metasilicate pentahydrate	1- Immersion 2- Electrolytic cleaning 3- Ultrasonic cleaning	Wet	Tanks or apparatus of special design. Continouos or batch operation	Will remove sweat and certain chemical residues. Milder alkalis suitable for certain types of painted surfaces. Not suitable for composite items containing rubber, leather, fabric, or wood. Ultrasonic cleaning specially suitable for removing fine solid particles. Suitable for large assemblies which cannot be dismantled.
		Mild alkaline	Mixture of metasilicate pentahydrate and sodium carbonate				Strong alkalis require protective measures: eyeshields and rubber gloves to prevent possible damage to eyes and skin during handling.
Steam cleaning		Detergent	Detergent	Ultrasonic immersion	Wet and possibly slightly oily	Ultrasonic apparatus	
		–	Steam alone or mixture of steam with detergent or alkaline material	Steam cleaning	Wet	Steam injector apparatus	

- Emulsifiable solvent cleaning;
- Aqueous alkaline and detergent cleaning;
- Steam cleaning.

The choice of method(s) depends upon the material of the substrate, and the type, shape, and condition of the surface that is to be cleaned.

Table 2.6 gives a guide to selection of degreasing methods. The specification shall specify the degreasing method(s) as appropriate.

CHAPTER THREE

Material Selection for Thermal Insulation

Contents

Alireza Bahadori, Thermal Insulation Handbook for the Oil, Gas, and Petrochemical Industries
© 2014 Elsevier Inc.
http://dx.doi.org/10.1016/B978-0-12-800010-6.00003-4

In this chapter, the emphasis is on comparison of the thermal insulation efficiency of the different frequently used insulation materials based on their thermal properties. The insulators are made from different materials, each with a different thermal conductivity, absorptivity, diffusivity, and resistivity.

Depending on how large or small the value of these thermal properties, a particular insulation material may be more efficient in terms of thermal insulation than another.

Based on this premise, we will in this work investigate the thermal properties and thermal insulation efficiency of the most frequently used insulation materials and develop some guides based on their thermal properties that can be used to predict and compare the suitability of the insulation materials in terms of thermal response.

This chapter consists of nine sections as follows:
1. Mineral Fiber Preformed Pipe Thermal Insulation
2. Mineral Fiber Block and Board Thermal Insulation
3. Mineral Fiber Blanket and Blanket-Type Pipe Thermal Insulation.
4. Calcium Silicate Preformed Block and Pipe Thermal Insulation
5. Cellular Glass/Foam Glass Thermal Insulation
6. Corkboard and Cork Pipe Thermal Insulation
7. Spray-Applied Rigid Cellular Polyurethane (PUR) and Poly-isocyanurate (PIR) Thermal Insulation
8. Preformed Rigid-Cellular Polyurethane (PUR) and Polyisocyanurate (PIR) Thermal Insulation
9. Miscellaneous Materials for Use with Thermal Insulation

Sections 3.1 through 3.5 are categorized as hot thermal insulation whereas Sections 3.6 through 3.8 are classified as cold thermal insulation.

Note: Cellular glass/foam glass hot thermal insulation may also be used as cold thermal insulation.

3.1. MINERAL FIBER PREFORMED PIPE THERMAL INSULATION

The specification in this section covers the minimum requirements for composition, sizes, dimensions, physical properties, inspection, packaging, and marking of mineral wool preformed pipe insulation for use on pipe surfaces up to 600°C. If insulation is to be used at higher temperatures, the actual temperature limits shall be as agreed upon between the manufacturer and the company.

For satisfactory performance, properly installed protective vapor barriers shall be used in low-temperature applications to prevent movement of moisture through or around the insulation towards the colder surface.

Figure 3.1 shows a typical mineral wool insulation. It is used for hot applications and comes in different formats:
- Preformed high-density resin: bonded molded pipe sections for pipes;
- Rigid boards and wire mat for tanks and boilers.

Mineral wool is a noncombustible insulation and is produced from spun slag. Densities range from 60kg/m³–160kg/m³. The thermal conductivity is 0.035W/mk.

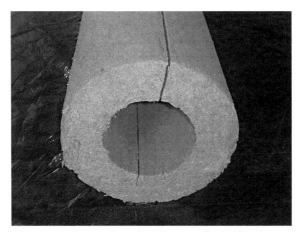

Figure 3.1 Mineral wool thermal insulation for pipes.

3.1.1. Materials and Manufacture

- **Composition**

    The mineral fiber preformed thermal insulation for pipes shall be manufactured from mineral fiber substances such as rock or slag from a molten state into fibrous form and shall be made rigid with organic binder. The binder shall be phenolic or the manufacturer's standard.

- **Jacket (facing)**

    The pipe insulation may be jacketed or plain; this will be specified by the purchaser. If the purchaser chooses jacketed insulation, the type of jacket will be aluminum foil or kraft paper according to the purchaser's specification.

3.1.2. General Properties

The mineral fiber preformed pipe insulation shall have the following general properties.

The insulation shall have suitable resistance to sunlight in order to minimize shrinkage and have as low a water vapor permeability as possible. It shall be noncombustible and shall be resistant to vermin and fungus, be free from objectionable odors, and shall not react with any process chemicals present.

The insulation shall be asbestos-free and shall not be a health hazard in case of contact or during work. The insulation shall have only traces of corrosive materials, e.g., water–soluble chlorides, which attack stainless steels, or high–alkalinity materials, which attack aluminum jacketing.

3.1.3. Chemical Composition

The chemical composition for the major constituents of mineral fiber preformed pipe insulating materials shall be as follows when tested by spectrometric methods:

- SiO_2 = 30–45 weight %
- Al_2O_3 = 8–15 weight %
- TiO_2 = 2–4 weight %
- Fe_2O_3 = 2.5 max. weight %
- CaO = 30–35 weight %
- MgO = 6–12 weight %
- Na_2O = 0–1 weight %
- K_2O = 0–1 weight %
- P_2O_5 = 0–1 weight %
- Water-soluble chloride: approximately 6 mg/kg when tested in accordance with ASTM C871

If the water-soluble chloride content of the insulation material exceeds the amount specified in the chemical composition above, sufficient amounts of sodium silicate (Na_2SiO_3) inhibitor has to be added to the insulation material. An acceptable proportion of sodium plus silicate ions to chloride ions as found by leaching from the insulation when tested in accordance with ASTM Test Method C871 shall be in accordance with Figure 3.2.

With reference to Figure 3.2, the minimum allowable value of sodium plus silicate will be 50 ppm when tested in accordance with ASTM Test Method C692, providing leachable (water-soluble) chloride is higher than 10 mg per kg of insulation material.

The pH of leach water shall be measured in accordance with ASTM Test Methods C871 and shall be greater than 7.0 but not greater than 11.7 at 25°C.

3.1.4. Physical Properties

The physical properties of mineral fiber preformed pipe insulation shall be as follows:

- **Linear shrinkage**

    Linear shrinkage for length shall be a maximum of 2% after soaking heat and shall be tested in accordance with ASTM Test Method C356 at the manufacturer's temperature limit.

- **Density**

    The minimum density shall be agreed upon between the purchaser and manufacturer. As a reference, the value 48 Kg/m³ for a temperature

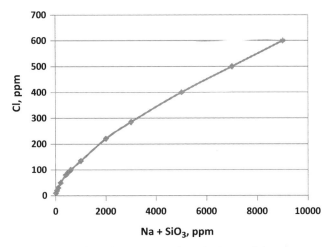

Figure 3.2 Acceptability of insulation material on the basis of the plot points of the Cl and $(Na+SiO_3)$ analysis.

up to 450°C and 96 Kg/m^3 for a temperature up to 650°C should be used in accordance with ASTM Test Method C302.
- **Thermal conductivity**
  The maximum thermal conductivity value shall be 0.065 W/mK at 200°C and 0.073 W/mK up to 450°C in accordance with ASTM Test Method C335.

3.1.5. Water Vapor Sorption

Water vapor sorption shall be a minimum 5% by weight in accordance with ASTM Test Method C1104.

3.1.6. Shapes, Sizes, and Dimensions

The mineral fiber preformed pipe insulation shall be supplied as hollow cylinders, split lengthwise on one or both sides of the cylindrical axis (depending on the company's request) and shall be furnished in sections or segments in a length of 1 meter unless otherwise agreed upon between the company and manufacturer.

The minimum and maximum thickness shall be specified by the company but in no case shall be less than 13 mm. Thicknesses greater than 150 mm may be furnished in multiple layers. The dimensional standard of the inner and outer diameter of insulation for nominal pipe size shall be in accordance with ASTM C585.

Table 3.1 Dimensional Tolerance Values

Dimension	Tolerance
Length	±3 mm
Thickness	±3% (The minimum and maximum thickness shall be specified by the company but in no case shall be less than 13 mm)

3.1.7. Dimensional Tolerances

The average measured length and thickness of any individual section or segment shall not differ from the specified dimensions by more than the data in Table 3.1.

When installed on the pipe of the specified size, sections shall fit snugly and shall have tight longitudinal and circumferential joints.

3.1.8. Workmanship

The insulation shall not have visible defects that will adversely affect its installation or service quality.

3.1.9. Sampling

The insulation shall be sampled for the purpose of tests in accordance with the criteria in ASTM C390.

3.1.10. Inspection and Rejection

- **Inspection**

    The manufacturer and/or supplier shall be responsible for carrying out the tests and inspections required by this standard specification, using his own or other reliable facilities, and he shall maintain complete records of all such tests and inspections. Such records shall be available for review by the purchaser. The manufacturer and/or supplier shall furnish to the purchaser a certificate of inspection stating that each lot has been sampled, tested, and inspected in accordance with this standard specification and has been found to meet the requirements specified.

    The supplier shall afford the purchaser's inspector all reasonable facilities necessary to satisfy that the material is being produced and furnished in accordance with this standard specification. Such inspections in no way relieve the manufacturer and/or supplier of his responsibilities under the term of this standard specification.

The purchaser reserves the right to perform any inspections set forth in this standard specification where such inspections are deemed necessary to assure that supplies and services conform to the prescribed requirements.

The purchaser's inspector shall have access to the material subject to inspection for the purpose of witnessing the selection of the samples, the preparation of the test pieces, and the performance of the test(s). For such tests, the inspector shall have the right to indicate the pieces from which the samples will be taken in accordance with the provisions of this standard specification.

- **Rejection**

  If the inspection of the sample shows failure to conform to the requirements of this chapter's specification, a second sample from the same lot shall be tested and the results of this retest averaged with the result of the original test.

  Upon retest, failure to conform to this standard specification shall constitute grounds for rejection.

  In case of rejection, the manufacturer or supplier shall have the right to reinspect the rejected lot and resubmit the lot for inspection after removal of that portion of the lot not conforming to the specified requirement.

3.1.11. Packaging and Marking

- **Packaging**

  Unless otherwise specified by the purchaser, the mineral fiber preformed pipe insulation shall be packaged in a manufacturer's standard commercial container approved by the purchaser.

  Overseas consignments shall be packed in double-corrugated cartons incorporating weatherproof paper because of the greater handling involved and the possibiltiy of exposure to wet conditions. If the size of sections exceeds the practical size for cartons, wooden or strong mesh crates may be used.

  Mineral fiber preformed pipe insulation materials shall not be unpacked at the site until they are required for use.

  Stacking of packed mineral fiber preformed pipe insulation during transportation shall be in accordance with the manufacturer's recommendation.

- **Marking**

    Unless otherwise agreed upon between the purchaser and manufacturer and/or supplier, containers shall be marked as follows:
- Purchase order number
- Name and type of material
- Thickness of insulation and pipe size in case of pipe insulation
- Quantity of material in the container
- Date of manufacture
- Supplier's name
- Origin of manufacture
- Destination

3.1.12. Storage

The insulation shall be stored in a dry atmosphere under cover and inspected at intervals not exceeding three months. Cartons shall be stored end up and be stacked no more than three high.

3.2. MINERAL FIBER BLOCK AND BOARD THERMAL INSULATION

The specification in this section covers the minimum requirements for the chemical composition, dimensions, physical properties, inspection, packaging, and marking of rigid and semi-rigid mineral fiber block and board thermal insulation for use on surfaces up to 540°C. If the insulation is to be used at higher temperatures, the actual temperature limits shall be as agreed upon between the manufacturer and company.

For satisfactory performance, properly installed protective vapor barriers must be used in low-temperature application to prevent movement of moisture through or around the insulation towards the colder surface.

3.2.1. Material and Manufacture

The mineral fiber preformed block and board thermal insulation shall be manufactured from mineral substances such as rock or slag processed from a molten state into fibrous form. The boards shall not contain nonfibrous pieces of these materials that have any dimensions exceeding 10 mm. Mineral fiber preformed block and board insulation shall be rigid or semi-rigid material composed of mineral fibers with or without binder.

3.2.2. General Properties

The mineral fiber preformed block and board thermal insulation shall have the following general properties.

It shall be noncombustible and shall be resistant to vermin and fungus, be free from objectionable odors, and shall not react with any process chemicals present.

The insulation shall be asbestos-free and shall not be a health hazard in case of contact or during work.

The insulation shall have only traces of corrosive materials, e.g., water-soluble chlorides, which attack stainless steels, or high-alkalinity materials, which attack aluminum jacketing.

3.2.3. Chemical Composition

The optimum chemical composition for the major constituents of mineral fiber preformed block and board thermal insulation shall be as follows, when tested by spectrometric methods:

- SiO_2 = 30–45 weight %
- Al_2O_3 = 8–15 weight %
- TiO_2 = 2–4 weight %
- Fe_2O_3 = 2.5 max. weight %
- CaO = 30–35 weight %
- MgO = 6–12 weight %
- Na_2O = 0–1 weight %
- K_2O = 0–1 weight %
- P_2O_5 = 0–1 weight %
- Water-soluble chloride: approximately 6 mg/kg when tested in accordance with ASTM C-871

If the water-soluble chloride content of the insulation material exceeds the amount specified in the chemical composition above, a sufficient amount of sodium silicate (Na_2SiO_3) inhibitor shall be added to the insulation material. An acceptable proportion of sodium plus silicate ions to chloride ions as found by leaching from the insulation when tested in accordance with ASTM Test Method C871 shall be in accordance with Figure 3.1.

With reference to Figure 3.1, the minimum allowable value of sodium plus silicate will be 50 ppm when tested in accordance with ASTM Test Method C692, providing leachable (water-soluble) chloride is higher than 10 mg per kg of insulation material.

The pH of leach water shall be measured in accordance with ASTM Test Method C871, and shall be greater than 7.0 but not greater than 11.7 at 25°C

3.2.4. Physical Properties

The physical properties of mineral fiber preformed block and board insulation shall be as follows:

- **Linear shrinkage**

    Linear shrinkage for length shall be a maximum of 2% after soaking heat and tested in accordance with ASTM Test Method C356 at the manufacturer's temperature limit.

- **Thermal conductivity**

    The maximum thermal conductivity value shall be 0.065 W/mK at 200°C when tested in accordance with ASTM Test Method C177.

- **Temperature of use**

    When tested in accordance with ASTM Test Method C411 at the intended use temperature, insulation for use above ambient shall show no physical changes that adversely affect its service qualities.

- **Bulk density**

    The density of the block and board mineral fiber thermal insulation shall be a maximum of 300 kg/m^3 and variation from the specified density shall not exceed ±15%.

- **Moisture absorption (water vapor)**

    When tested in accordance with ASTM C553 insulation for use below ambient temperature, the insulation shall gain no more than 1% volume.

- **Recovery after compression**

    When tested in accordance with BS 3958: Part 5, the recovery after compression shall be not less than 95% of the original thickness.

3.2.5. Size of Board and Slab

The length and width of board and slab shall be as per the manufacturer's standard dimension unless otherwise agreed upon between the company and manufacturer. The thickness of board and slab shall be specified by the purchaser. However the thickness shall be within the range of 25 to 100 mm in 13 mm increments.

Figure 3.3 shows some typical mineral wool board insulation.

3.2.6. Dimensional Tolerances

The average measured length, width, and thickness of the board and slab thermal insulation shall not differ from the dimensions specified by more than that stated in Table 3.2.

Figure 3.3 Mineral wool board insulation.

Table 3.2 Dimensional Tolerances for the Board and Slab
Thermal Insulation

Dimension	Tolerance in mm
Length	±12
Width	±6
Thickness	+6, -3

3.2.7. Workmanship

The insulation shall not have visible defects that will adversely affect its service qualities.

3.2.8. Sampling

The insulation shall be sampled for the purpose of tests in accordance with ASTM Test Method C390.

3.2.9. Inspection and Rejection

• **Inspection**

   The manufacturer and/or supplier shall be responsible for carrying out all the tests and inspections required by this standard specification, using his own or other reliable facilities, and he shall maintain complete records of all such tests and inspections. Such records shall be available for review by the purchaser. The manufacturer and/or supplier shall furnish to the purchaser a certificate of inspection stating that each lot has been sampled, tested, and inspected in accordance with this standard specification and has been found to meet the requirements specified.

The supplier shall afford the purchaser's inspector all reasonable facilities necessary to satisfy that the material is being produced and furnished in accordance with this standard specification. Such inspections in no way relieve the manufacturer and/or supplier of his responsibilities under the terms of this standard specification.

The purchaser reserves the right to perform any inspections set forth in this standard specification where such inspections are deemed necessary to assure that supplies and services conform to the prescribed requirements.

The purchaser's inspector shall have access to the material subject to inspection for the purpose of witnessing the selection of the samples, the preparation of the test pieces, and the performance of the test(s). For such tests, the inspector shall have the right to indicate the pieces from which the samples will be taken in accordance with the provisions of this standard specification.

- **Rejection**

If the inspection of the sample shows failure to conform to the requirements of this standard specification, a second sample from the same lot shall be tested and the results of this retest averaged with the result of the original test. Upon retest, failure to conform to this specification shall constitute grounds for rejection.

In case of rejection, the manufacturer or supplier shall have the right to reinspect the rejected lot and resubmit the lot for inspection after removal of that portion of the lot not conforming to the specified requirement.

3.2.10. Packaging and Marking

- **Packaging**

Unless otherwise specified by the purchaser, the mineral fiber block and board thermal insulation shall be packaged in a manufacturer's standard commercial container approved by the purchaser.

Overseas consignments shall be packed in double-corrugated cartons incorporating weatherproof paper because of the greater handling involved and the possibility of exposure to wet conditions. If the size of sections exceeds the practical size for cartons, wooden or strong mesh crates may be used.

Mineral fiber block and board insulation materials shall not be unpacked at the site until they are required for use. Stacking of packed mineral fiber block and board insulation during transportation shall be in accordance with the manufacturer's recommendation.

3.2.11. Storage

The insulation shall be stored in a dry atmosphere under cover and inspected at intervals not exceeding three months. Cartons shall be stored end up and be stacked no more than three high.

3.3. MINERAL FIBER BLANKET AND BLANKET-TYPE PIPE THERMAL INSULATION

This insulation is made from basalt rock and slag. This combination results in a noncombustible product with a melting point of approximately 2150°F (1177°C), which gives it excellent fire resistance properties. This mineral wool is a water repellent yet vapor permeable material.

Mineral wool blanket insulation is flexible, lightweight, water repellent, fire resistant, and sound absorbent. It is suitable for use in many industries, including petrochemical, power-generating plants, boilers, furnaces, towers, ovens, and drying equipment. The product is noncombustible and can be specified with confidence where fire performance is of concern.

In summary, some benefits are:

- Flexibility
- Noncombustibility
- Low moisture sorption
- Fire resistance
- Excellent thermal resistance
- Does not rot or sustain vermin
- Does not promote growth of fungi or mildew
- Made from natural and recycled materials

This section covers minimum requirements for chemical composition, physical properties, dimensions, inspection, packaging, and marking of mineral blanket and blanket-type pipe insulation for use on heated surfaces operating at temperatures up to and including 650°C. For specific applications the actual temperature limit shall be agreed upon between the manufacturer and the purchaser. The insulation shall be faced on one or both sides with a flexible metal mesh.

3.3.1. Material and Manufacture

- **Composition**

   The mineral fiber blanket and blanket-type pipe thermal insulation shall be manufactured from mineral substances such as rock or slag processed from a molten state into fibrous form. The blanket shall not contain nonfibrous pieces of these materials that have any dimensions exceeding 10 mm.

- **Facing**

    The blanket shall be faced on one or both sides with woven wire mesh or expanded metal, and held together by wire or twine extending from one face to the other, spaced not more than 120 mm apart. The ties shall not become detached from the facing when pressure is applied over the surface of the blanket. When both sides are to be faced, units may have the same or different types on the two sides.

    The standard types of metal mesh used as facings are as follows:

- Woven wire mesh, mild steel no. 20 to 22 gauge (approximately 0.33 to 0.73 mm) having hexagonal-shaped opening, galvanized after weaving;
- Expanded metal lath (copper bearing not galvanized mild steel) having diamond-shaped openings; 6-10 mm short way of mesh (SWM), 0.45 mm thick strand, 1.22 kg/m^2 mass.

    Other types or compositions of facings may be specified.

3.3.2. General Properties

The mineral fiber blanket thermal insulation shall have the following general properties.

The insulation shall have suitable resistance to sunlight and low water permeability.

It shall be noncombustible and shall be resistant to vermin and fungus, be free from objectionable odors, and shall not react with any process chemicals present. The insulation shall be asbestos-free and shall not be a health hazard in case of contact or during work. The insulation shall have only traces of corrosive materials, e.g., water-soluble chlorides that attack stainless steels or high-alkalinity materials that attack aluminum jacketing.

3.3.3. Chemical Composition

The chemical composition for the major constituents of mineral fiber blanket thermal insulation may be as follows:

- SiO_2 = 30–45 weight %
- Al_2O_3 = 8–15 weight %
- TiO_2 = 2–4 weight %
- Fe_2O_3 = 2.5 max. weight %
- CaO = 30–35 weight %

- MgO = 6–12 weight %
- Na_2O = 0–1 weight %
- K_2O = 0–1 weight %
- P_2O_5 = 0–1 weight %
- Water-soluble chloride: approximately 6 mg/kg when tested in accordance with ASTM C871

If the water-soluble chloride content of the insulation material exceeds the amount specified in the chemical composition above, a sufficient amount of sodium silicate (Na_2SiO_3) inhibitor shall be added to the insulation material. An acceptable proportion of sodium plus silicate ions to chloride ions as found by leaching from the insulation when tested in accordance with ASTM Test Method C871 shall be in accordance with Figure 3.1.

With reference to Figure 3.1, the minimum allowable value of sodium plus silicate will be 50 ppm when tested in accordance with ASTM Test Method C692, providing the leachable (water-soluble) chloride is higher than 10 mg per kg of insulation material. The pH of leach water shall be measured in accordance with ASTM Test Methods C871 and shall be greater than 7.0 but not greater than 11.7 at 25°C.

3.3.4. Physical Properties

The physical properties of mineral fiber blanket and blanket-type pipe insulation shall be as follows:

- **Thermal conductivity**
    The maximum thermal conductivity shall be 0.065 W/mK at 200°C when tested in accordance with ASTM Test Method C177.
- **Density**
    The density of thermal insulation shall be as agreed upon between the manufacturer and purchaser.
- **Handleability**
    Each piece of metal-mesh-covered insulation shall be sufficiently coherent to permit transportation and installation as a unit.
- **Moisture content**
    The moisture content of the materials shall not exceed 5% by mass when conditioned at high humidity in accordance with BS 2972.
- **Compressibility and resilience**
    When tested, the thickness of the specimen while under pressure shall be not greater than the nominal thickness plus 3 mm. After removal of

the pressure, the thickness shall be not less than the nominal thickness minus 3 mm.

Note: The objects of this test are to ensure that a mattress supplied at greater than nominal thickness is sufficiently compressible to be fitted at its nominal thickness and that a mattress is sufficiently resilient to recover its nominal thickness after being subjected to compression.

- **Limiting temperature and thickness**

   The manufacturer shall state the maximum limiting temperature and limiting thickness at that temperature.

   When a sample is heated in accordance with BS 2972 at the stated maximum limiting temperature of use, the material shall maintain its general form and not suffer visible deterioration of the fibrous structure.

- **Linear shrinkage**

   Linear shrinkage for length shall be a maximum of 4% at the maximum used temperature.

   Note: The color changes are not relevant.

3.3.5. Dimensions

The standard sizes of metal–mesh blanket insulation may be as specified herein or as agreed upon between the manufacturer and purchaser:

- Length: 1.25 and 2.5 m
- Width: 0.6 m
- Thickness: 25 to 150 mm in 13 mm increments

   Note: Thicknesses over 75 mm may be composed of two or more blankets plied together.

3.3.6. Dimensional Tolerances

The average measured length, width, and thickness shall differ from the manufacturer's standard dimensions.

3.3.7. Workmanship

The insulation shall not have visible defects that will adversely affect the service quality.

3.3.8. Sampling

The insulation shall be sampled for the purpose of tests in accordance with ASTM C390 or BS 2972 as specified in the relevant test.

3.3.9. Inspection and Rejection

• **Inspection**

The manufacturer and/or supplier shall be responsible for carrying out all the tests and inspections required by this standard specification, using his own or other reliable facilities, and he shall maintain complete records of all such tests and inspections. Such records shall be available for review by the purchaser. The manufacturer and/or supplier shall furnish to the purchaser a certificate of inspection stating that each lot has been sampled, tested, and inspected in accordance with this standard specification and has been found to meet the requirements specified.

The supplier shall afford the purchaser's inspector all reasonable facilities necessary to satisfy that the material is being produced and furnished in accordance with this standard specification. Such inspections in no way relieve the manufacturer and/or supplier of his responsibilities under the terms of this standard specification.

The purchaser reserves the right to perform any inspections set forth in this standard specification where such inspections are deemed necessary to assure that supplies and services conform to the prescribed requirements.

The purchaser's inspector shall have access to the material subject to inspection for the purpose of witnessing the selection of the samples, the preparation of the test pieces, and the performance of the test(s). For such tests, the inspector shall have the right to indicate the pieces from which the samples will be taken in accordance with the provisions of this standard specification.

• **Rejection**

If the inspection of the sample shows failure to conform to the requirements of this standard specification, a second sample from the same lot shall be tested and the results of this retest averaged with the result of the original test.

Upon retest, failure to conform to this standard specification shall constitute grounds for rejection. In case of rejection, the manufacturer or supplier shall have the right to reinspect the rejected lot and resubmit the lot for inspection after removal of that portion of the lot not conforming to the specified requirement.

3.3.10. Packaging and Marking

• **Packaging**

Unless otherwise specified by the purchaser, the mineral fiber blanket and blanket-type pipe thermal insulation shall be packaged in the manufacturer's standard commercial container approved by the purchaser.

Overseas consignments shall be packed in double-corrugated cartons incorporating weatherproof paper because of the greater handling involved and the possibility of exposure to wet conditions. If the size of sections exceeds the practical size for cartons, wooden or strong mesh crates may be used.

Mineral fiber blanket insulation materials shall not be unpacked at the site until they are required for use. Stacking of packed mineral fiber blanket insulation during transportation shall be in accordance with the manufacturer's recommendation.

3.3.11. Storage

The insulation shall be stored in a dry atmosphere under cover and inspected at intervals not exceeding three months. Cartons shall be stored end up and be stacked no more than three high.

3.4. CALCIUM SILICATE PREFORMED BLOCK AND PIPE THERMAL INSULATION

Nonasbestos calcium silicate insulation board and pipe insulation features light weight, low thermal conductivity, and high temperature and chemical resistance.

Calcium silicate is a rigid, high density material used for high-temperature applications ranging from 250°F (121°C)–1000°F (540°C). The relation between temperature and thermal conductivity is indicated in Figure 3.4.

The specification in this section covers the minimum requirements for composition, dimensions, physical properties, inspection, packaging, and marking of calcium silicate preformed block and pipe thermal insulation for use on surfaces with a temperature up to 920°C. For specific applications, the actual temperature limit shall be agreed upon between the manufacturer and the purchaser.

Calcium silicate block and pipe high-temperature insulation is manufactured using an industry preferred filter press method that provides accurate dimensional tolerances and superior compressive and flexural strength. These attributes facilitate installation and provide exceptional

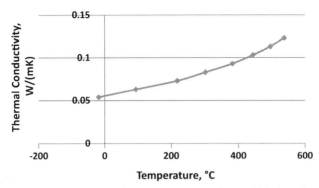

Figure 3.4 Thermal conductivity of calcium silicate preformed block and pipe thermal insulation.

resistance to mechanical abuse. Calcium silicate insulation is asbestos-free and meets or exceeds all of the requirements of ASTM C533, Type I.

Calcium silicate insulation is recommended for use on equipment and piping operating at temperatures from ambient to 250°F (121°C) indoors and protected outdoors up to 1200°F (649°C). It is ideally suited for industrial use in areas such as the petrochemical and power-generating industries where energy conservation, process control, personal protection, and fire protection are prerequisites.

3.4.1. General Properties

The insulation shall have suitable resistance to sunlight in order to minimize shrinkage. It shall be noncombustible and shall be resistant to vermin and fungus, be free from objectionable odors, and shall not react with any process chemicals present.

The insulation shall be asbestos-free and shall not be a health hazard in case of contact or during work. The insulation shall have only traces of corrosive materials, e.g., water-soluble chlorides that attack stainless steels or high-alkalinity materials that attack aluminum jacketing.

3.4.2. Composition

Calcium silicate thermal insulation shall be composed predominately of reacted hydrous calcium silicate and usually incorporates a fibrous reinforcement. Asbestos shall not be used as a component in the manufacture of the material.

3.4.3. Physical Properties

*   **Thermal conductivity**

    The thermal conductivity shall not exceed 0.079 W/mK for type I and 0.039 W/mK for type II at 204°C when tested in accordance with the following ASTM Test Methods:

    *   **Block insulation:** Test Method C177 or C518 using 38 ± 13 mm thick specimen. Curved block shall be trimmed to provide plane parallel surfaces.
    *   **Pipe insulation:** Test Method C335 using 38 ± 13 mm thick specimen of pipe insulation as supplied for fit to 76 mm nominal steel pipe.
*   **Bulk density**

    The maximum bulk density of the dry material shall be 240 kg/m^3 for type I and 352 Kg/m for type II when tested in accordance with ASTM Test Methods C303 for block insulation and C302 for pipe insulation.
*   **Flexural strength**

    The flexural strength shall not be less than 250 kN/m^2 when tested in accordance with ASTM Test Methods C203 for block insulation and C446 for pipe insulation.
*   **Compressive strength**

    The reduction in thickness shall not exceed 5% under a compressive load of 500 kN/m^2 when tested in accordance with ASTM Recommended Practice C165.
*   **Weight loss by tumbling**

    The maximum loss in weight when tested in accordance with ASTM Test Method C421 shall be as shown in Table 3.3.
*   **Linear shrinkage**

    Linear shrinkage after heat soaking shall be a maximum of 2% when tested according to ASTM Test Method C356 at the insulation temperature limit listed in the specification or the manufacturer's temperature limit, whichever is higher.
*   **Hot surface performance**

    Hot surface performance shall be tested in accordance with ASTM Test Method C411. The temperature shall be as listed in the specification or the manufacturer's temperature limit.

Table 3.3 Weight Loss by Tumbling

Duration	Weight Loss %
After first 10 minutes	20
After second 10 minutes	40

- **Warpage**

    The maximum warpage shall be 6 mm.
- **Cracking**

    No cracks shall completely penetrate through the insulation thickness, but minor surface cracks on hot faces are acceptable.
- **Surface burning characteristic**

    The Test Method shall be ASTM E84 and the maximum flame spread index and smoke density index for calcium silicate block and pipe thermal insulation shall be zero.

3.4.4. Standard Shape, Size, and Dimension

Calcium silicate preformed thermal insulation shall be supplied in the form of flat block, curved block, grooved block, and pipe section with the following dimensions:

Note: All the suppliers provide the full range of standard shapes and sizes listed. Conversely other shapes and sizes may be available. The supplier shall be consulted for details of the range offered.

- **Flat block**

    Flat block shall be furnished in a length of 450 to 1000 mm, a width of 150 to 1000 mm, and in thicknesses from 25 to 150 mm in 13 mm increments. The thicknesses greater than 75 mm may be furnished in two or more layers when specified by the purchaser.
- **Curved block**

    Curved block shall be furnished in a length of 1000 mm, a width of approximately 150 mm or 300 mm, a thickness of 38 to 100 mm in 13 mm increments, and curved to inside radii of over 419 mm. Individual dimensions shall conform to those specified by the manufacturer.
- **Grooved block**

    Grooved block shall be furnished in a length of 1000 mm, a width of 305 mm or 458 mm, and in thickness from 25 to 150 mm in 13 mm increments. The size and spacing of grooves shall be as specified by the manufacturer. Long edges of grooved block may be furnished beveled as specified by the manufacturer.
- **Pipe section**

    Calcium silicate pipe insulation shall be supplied either as hollow cylindrical shapes split in half lengthwise (in a plane including the cylindrical axis) or as curved segments. The pipe insulation shall be furnished in sections or segments in a length of 1000 mm to fit standard size pipe and tubing, and in nominal thickness from 25 to 150 mm, in 13 mm increments. Thicknesses greater than 75 mm may be furnished in

Table 3.4 Dimensional Tolerance Values

	Block	Pipe
Length	±3 mm	±3 mm
Width	±3 mm	–
Thickness	±3 mm	±3 mm

two or more layers, as specified by the purchaser. Individual dimensions shall conform to those specified by the manufacturer.

3.4.5. Dimensional Tolerances

The average measured length, width, and thickness shall not differ from the manufacturer's standard dimensions by more than that shown in Table 3.4.

• **Pipe insulation**

The following additional dimensional tolerances apply only to calcium silicate pipe insulation supplied as half sections.

• **Fit and closure**

When fitted to the appropriate size pipe by banding on 230 mm centers, the longitudinal seams on both sides of the pipe insulation shall close to within 1.6 mm along the entire length of the section.

• **Concentricity**

The inner bore of the pipe insulation shall be concentric with the outer cylindrical surface. The deviation from concentricity shall not exceed 3 mm or 5% of the wall thickness, whichever is greater.

• **Half-section balance**

The plane formed by the slit between the half sections shall include the cylindrical axis. Deviation of the slit plane from the cylindrical axis over a 914 mm length shall not exceed 3 mm.

• **Grooved block**

The following additional requirements apply only to calcium silicate block insulation containing grooves that is intended for installation over curved surfaces 508 mm in diameter or larger.

• **Fit and closure**

When fitted to the curved surface, the grooves shall close to 3 mm or less through the depth of the groove. The exposed surface crack shall not open more than 3 mm. The insulation will pivot at the ungrooved area at the bottom of the groove with the groove closing and the exposed ungrooved surface cracking. The deeper the groove is in the insulation, the smaller the crack on the exposed surface will be.

3.4.6. Workmanship

Since some requirements for this material are not easily defined by a numerical value, the insulation shall not have visible defects that will adversely affect its service qualities.

3.4.7. Sampling

The insulation shall be sampled in accordance with ASTM C390. Specific provisions for sampling shall be agreed upon between the purchaser and the supplier as part of the purchase contract.

3.4.8. Inspection and Rejection

• **Inspection**

The manufacturer and/or supplier shall be responsible for carrying out all the tests and inspections required by this standard specification, using his own or other reliable facilities, and he shall maintain complete records of all such tests and inspections. Such records shall be available for review by the purchaser.

The manufacturer and/or supplier shall furnish to the purchaser a certificate of inspection stating that each lot has been sampled, tested, and inspected in accordance with this standard specification and has been found to meet the requirements specified. The supplier shall afford the purchaser's inspector all reasonable facilities necessary to satisfy that the material is being produced and furnished in accordance with this standard specification. Such inspections in no way relieve the manufacturer and/or supplier of his responsibilities under the terms of this standard specification.

The purchaser reserves the right to perform any inspections set forth in this standard specification where such inspections are deemed necessary to assure that supplies and services conform to the prescribed requirements.

The purchaser's inspector shall have access to the material subject to inspection for the purpose of witnessing the selection of the samples, the preparation of the test pieces, and the performance of the test(s). For such tests, the inspector shall have the right to indicate the pieces from which the samples will be taken in accordance with the provisions of this standard specification.

- **Rejection**

    If the inspection of the sample shows failure to conform to the requirements of this standard specification, a second sample from the same lot shall be tested and the results of this retest averaged with the result of the original test.

    Upon retest, failure to conform to this standard specification shall constitute grounds for rejection.

    In case of rejection, the manufacturer or supplier shall have the right to reinspect the rejected lot and resubmit the lot for inspection after removal of that portion of the lot not conforming to the specified requirement.

3.4.9. Packaging and Marking

- **Packaging**

    Unless otherwise specified by the purchaser, the calcium silicate preformed block and pipe thermal insulation shall be packaged in the manufacturer's standard commercial container approved by the purchaser.

    Overseas consignments shall be packed in double-corrugated cartons incorporating weatherproof paper because of the greater handling involved and the possibility of exposure to wet conditions. If the size of sections exceeds the practical size for cartons, wooden or strong mesh crates may be used.

    The insulation materials shall not be unpacked at the site until they are required for use.

    Stacking of packed calcium silicate insulation during transportation shall be in accordance with the manufacturer's recommendation.

- **Marking**

    Unless otherwise agreed upon between the purchaser and manufacturer and/or supplier, containers shall be marked as follows:

- Purchase order no.
- Name and type of material
- Sizes of thermal insulation
- Quantity of material in the container
- Date of manufacture
- Supplier's name
- Origin of manufacture
- Destination

3.4.10. Storage

The insulation shall be stored in a dry atmosphere under cover and inspected at intervals not exceeding three months. Cartons shall be stored end up and be stacked no more than three high.

3.5. CELLULAR GLASS/FOAM GLASS THERMAL INSULATION

Foam glass thermal insulation (Figure 3.5) is a lightweight, rigid insulating material composed of millions of completely sealed glass cells, each an insulating space. This all–glass, closed–cell structure provides an unmatched combination of physical properties ideal for piping and equipment above-ground, as well as underground, indoors, or outdoors, at operating temperatures from −450°F to +900°F (−268°C to +482°C):

- Resistant to water in both liquid and vapor forms
- Noncorrosive
- Noncombustible/nonabsorbent of combustible liquids
- Resistant to most industrial reagents
- Dimensionally stable under a variety of temperature and humidity conditions
- Superior compressive strength

Figure 3.5 Sample cellular glass/foam glass thermal insulation. *(Source: Pittsburgh Corning Corporation).*

- Resistant to vermin, microbes, and mold

  Some additional benefits of this type of insulation (in particular, FOAMGLAS®) are detailed in the following.

  FOAMGLAS insulation's diversity of properties results in an equally unmatched combination of benefits, proven over decades of in-the-field performance:
- Constant, long-term energy efficiency provides low, predictable energy costs.
- Enhanced process control allows improved, consistent product quality.
- Minimal maintenance/repair/replacement of insulation *or* facility infrastructure reduces life cycle costs.
- Fire resistance protects the insulated equipment, and helps minimize subsequent plant shutdown time.
- Virtual elimination of the potential for auto–ignition from absorbed combustible liquids or fire from condensed low-temperature gases.
- It has proven durability for underground and exterior installations.
- Manufacturing of FOAMGLAS insulation puts no stress on the atmosphere's ozone layer … while its long-term thermal efficiencies reduce energy demand and the effects of burning fossil fuels on the environment.

The specification in this section covers the minimum requirements for the composition, physical properties, dimensions, inspection, packaging, and marking of cellular glass thermal insulation intended for use on surfaces operating at temperature between −268 and 427°C.

3.5.1. Classification

The cellular glass insulation may be furnished in the following types and classes as required by the company:
- Type I: Flat Block
- Type II: Pipe and Tubing Insulation
  - Class 1: Regular (uncovered)
  - Class 2: Jacketed
- Type III: Special Shapes
- Type IV: Board

3.5.2. Material and Manufacture

The material shall consist of a glass composition that has been foamed or cellulated under molten conditions, annealed, and set to form a rigid incombustible material with hermetically sealed cells. The material shall be trimmed into blocks of standard dimensions or commercial sizes.

The pipe insulation, board, and special shapes such as curved sidewall segments and head segments shall be fabricated from standard blocks.

3.5.3. Physical Properties

The physical properties of cellular glass insulation shall be as follows. Since all cellular glass is produced initially in block form and is only cut to form pipe, curved, or segmental insulation without additional treatment, the physical testing of cellular glass shall be done on block and dry specimens.

- **Density**

    The density of insulation varies depending on the other physical properties and can be between 112 to 152 kg/m^3 when tested in accordance with ASTM Test Method C303.

- **Thermal conductivity**

    The thermal conductivity of cellular glass thermal insulation shall not be greater than that shown in Table 3.5 when tested in accordance with ASTM Test Method C177.

- **Compressive strength**

    Compressive strength shall not be less than 490 kPa when tested in accordance with ASTM Test Method C165.

- **Water vapor transmission**

    Water vapor transmission of cellular glass thermal insulation shall be zero according to ASTM Test Method E96.

- **Flexural strength**

    The minimum flexural strength of the insulation shall be 414 kPa when tested in accordance with ASTM Test Method C203.

- **Hot surface performance**

    The hot surface performance of the insulation, when tested in accordance with ASTM Test Method C411 at a maximum test temperature of 427°C and maximum rate of heating at 111°C/hr, shall

Table 3.5 Thermal Conductivity of Cellular Glass Thermal Insulation

Mean Temperature (°C)	Thermal Conductivity (W/mK)
+20	0.045
0	0.042
−20	0.038

indicate maximum warpage of 3 mm and no open cracks shall completely penetrate through the insulation thickness.

- **Fire resistance**

    The insulation as per BS 476 shall be noncombustible.

- **Modulus of elasticity**

    The modulus of elasticity of insulation when tested in accordance with ASTM Test Method C623 shall not be less than 980 MPa.

3.5.4. Standard Sizes and Dimensions

Blocks shall be rectangular sections and shall be true to form and dimension, with the corners square and the sides and edges parallel. The typical sizes and thicknesses are shown in Table 3.6. The required sizes and thicknesses shall be as specified by the company:

- **Type II: Pipe and tubing insulation**

    Cellular glass pipe and tubing insulation shall be fabricated in either 457 mm or 610 mm nominal length and/or as will be specified by the company. Cellular glass pipe insulation for 152 mm (NPS 6) and smaller pipe shall be fabricated to a minimum thickness of 25 mm. Pipe insulation for larger than 152 mm (NPS 6) shall be made to a minimum 38 mm thickness. Sizes shall conform to ASTM Practice C585.

    Cellular glass pipe insulation for pipe sizes up to 250 mm (NPS 10) shall be supplied in half sections. For pipe sizes with diameter exceeding 250 mm (NPS 10) and for equipment up to 7000 mm diameter, cellular glass insulation shall be supplied in radiused and beveled segments having

Table 3.6 Typical Sizes and Thicknesses

Length (mm)	Width (mm)	Thickness (mm)
457	305	38.1
		44.5
		50.8
		63.5
		76.2
		102
		127
610	457	50.8
		63.5
		76.2
		102
		127

a width on the outside radius of between 140 and 160 mm for diameters up to 1000 mm, and a width on the outside radius between 210 and 435 mm for diameters from 1000 mm up to 7000 mm.

For equipment in excess of 7000 mm diameter, cellular glass insulation shall be supplied in flat blocks that are a minimum of 300 mm wide. For equipment heads and spherical tanks, special factory radiused and beveled segments to suit the curvature of heads/tanks requiring not more than one field cut for proper closure shall be used.

All circumferential and longitudinal joints in pipe sections and radiused and beveled segments and slabs shall be of the butt type.

- **Type III: Special shapes**
  The dimensions of special shapes shall be specified by the company.
- **Type IV: Board**
  The standard available dimensions of board shall be 610 mm wide by 1219 mm long by 38 mm or 51 mm thick.

3.5.5. Dimensional Tolerances

The average measured length, width, and thickness for Types I, II, and IV shall not differ from the supplied dimensions by more than the tolerances listed in Table 3.7.

Manufacturing tolerances for the bore diameter and wall thickness of cellular glass pipe insulation are as given in ASTM Practice C585.

The following additional tolerances apply only to cellular glass pipe and tubing insulation applied in half sections.

- **Fit and closure**
  When fitted to appropriate size pipe by banding on 230 mm centers, the longitudinal joints on both sides of the pipe insulation shall close to within 1.6 mm along the entire length of the section.
- **Concentricity**
  The inner bore of the pipe insulation shall be concentric with the outer cylindrical surface. Deviation from concentricity shall not exceed 3.2 mm or 5% of the wall thickness, whichever is greater.

Table 3.7 Dimensional Tolerances

Dimension (mm)	Block	Pipe	Board
Length	±1.6	±3.2	±3.2
Width	±1.6	–	±1.6
Thickness	±1.6	±13.2	±1.6

- **Half-section balance**

  The plane formed by the slit between half sections shall include the cylindrical axis. Deviation of the split plane from the cylindrical axis over a 610 mm length shall not exceed 3.2 mm.

  Dimensional tolerances for Type III shall be decided upon between the manufacturer and the company.

3.5.6. Workmanship

Since some requirements for this material are not easily specified by numerical value, the insulation shall have no visible defects that will adversely affect its service qualities.

3.5.7. Sampling

The insulation shall be sampled for the purpose of testing in accordance with ASTM Criteria C390.

3.5.8. Inspection and Rejection

- **Inspection**

  The manufacturer and/or supplier shall be responsible for carrying out all the tests and inspections required by this standard specification, using his own or other reliable facilities, and he shall maintain complete records of all such tests and inspections. Such records shall be available for review by the purchaser. The manufacturer and/or supplier shall furnish to the purchaser a certificate of inspection stating that each lot has been sampled, tested, and inspected in accordance with this standard specification and has been found to meet the requirements specified.

  The supplier shall afford the purchaser's inspector all reasonable facilities necessary to satisfy that the material is being produced and furnished in accordance with this standard specification. Such inspections in no way relieve the manufacturer and/or supplier of his responsibilities under the terms of this standard specification.

  The purchaser reserves the right to perform any inspections set forth in this standard specification where such inspections are deemed necessary to assure that supplies and services conform to the prescribed requirements.

  The purchaser's inspector shall have access to the material subject to inspection for the purpose of witnessing the selection of the samples, the preparation of the test pieces, and the performance of the test(s). For such tests, the inspector shall have the right to indicate the pieces from which the samples will be taken in accordance with the provisions of this standard specification.

- **Rejection**

    If the inspection of the sample shows failure to conform to the requirements of this standard specification, a second sample from the same lot shall be tested and the results of this retest averaged with the result of the original test.

    Upon retest, failure to conform to this standard specification shall constitute grounds for rejection.

    In case of rejection, the manufacturer or supplier shall have the right to reinspect the rejected lot and resubmit the lot for inspection after removal of that portion of the lot not conforming to the specified requirement.

3.5.9. Packaging and Marking

- **Packaging**

    Unless otherwise specified by the purchaser, the cellular glass thermal insulation shall be packaged in the manufacturer's standard commercial container approved by the purchaser.

    Overseas consignments shall be packed in double-corrugated cartons because of the greater handling involved. If the size of sections exceeds the practical size for cartons, wooden or strong mesh crates may be used.

    The insulation materials shall not be unpacked at the site until they are required for use.

    Stacking of packed cellular glass insulation during transportation shall be in accordance with the manufacturer's recommendation.

- **Marking**

    Unless otherwise agreed upon between the purchaser and manufacturer and/or supplier, containers shall be marked as follows:

- Purchase order number
- Name and type of material
- Class of material (if required)
- Sizes of thermal insulation
- Shape of insulation in case of special shapes
- Quantity of material in the container
- Date of manufacture
- Supplier's name
- Origin of manufacture

Table 3.8 Amendments to Glass-Fiber Thermal Insulation

Form of Insulation	Temperature Limit (°C)	Density (max. kg/m³)
Blanket	500	128
Pipe section	450	192
Board/slab	450	192

3.5.10. Storage

The insulation shall be stored in a dry atmosphere under cover and inspected at intervals not exceeding three months. Cartons shall be stored end up and be stacked no more than three high.

3.5.11. Special Considerations for Hot Thermal Insulation

- **Glass fiber thermal insulation**

    Glass fiber thermal insulation shall have the properties of mineral fiber thermal insulation as specified in pervious parts with the amendments that are shown in Table 3.8.

- **Ceramic fiber blanket thermal insulation**

    Ceramic fiber thermal insulation shall be refractory oxides consisting primarily of alumina and silica, with small amounts of impurities permitted, processed from a molten state into fibrous form without binder.

    The insulation material shall be in accordance with ASTM C892, Grade 6, with the following amendments:

- The insulation shall be suitable for use up to 1150°C.
- The minimum density shall be 96 kg/m³ with a maximum thermal conductivity of 0.09 W/mK at 450°C.
- ASTM C892–High-Temperature Fiber Blanket Thermal Insulation.

3.6. CORKBOARD AND CORK PIPE THERMAL INSULATION

Along with a clearly superior ductility and the ability to withstand tension that may cause cracking, mortars based on natural hydraulic lime (NHL) and cork aggregates also have a strong resistance to water vapor, taking advantage of the full potential of the cork insulation panels (Figure 3.6). Cork is a renewable raw material that is 100% natural and has acoustic and anti-vibration properties when used as a thermal insulation material – its

Figure 3.6 Corkboard and cork pipe thermal insulation. *(Source: Modern Enviro, Co.).*

addition in mortars offers the guarantee of increased durability, without the loss of resistance and adhesion characteristics.

Some advantages of this insulation are the following:
- Excellent thermal protection
- High thermal inertia
- Reduction of energy costs
- Superior environmental performance
- 100% natural and renewable insulation
- Recyclable system
- Promotes the cork-oak tree forest, responsible for the annual capture of 5 million tons of CO_2, and for the biodiversity of 180 different species in Portugal alone
- Breathable and healthy system
- High-dimensional stability of the insulation panel
- Low maintenance cost
- Material durability
- Excellent acoustic and anti-vibration protection
- Does not release toxic fumes in case of fire

The specification in this section covers the minimum requirements for material, physical properties, dimensions, tolerances and inspection of baked cork thermal insulation in the form of board and pipe for use on surfaces operating at temperatures below approximately 80°C. For specific applications, the temperature limit shall be agreed upon between the manufacturer and the purchaser.

3.6.1. Materials and Manufacture

- **Composition**

    Corkboard and cork pipe insulation shall be composed of compressed and baked granulated cork, without added binder.

- **Corkboard**

    Corkboard shall be reasonably smooth without undue voids. The edges shall be firm, the corners square, and the sides and ends parallel. When specified, one side of the board shall be sanded to a smooth finish.

- **Cork pipe insulation**

    Cork pipe insulation is supplied in various thicknesses, grouped into four classes according to the operating temperature of the pipe on which it is to be applied. Table 3.9 shows the different classes.

Note: The wall thicknesses of cork pipe insulation are selected not only to reduce heat gain or loss but also to prevent condensation on the outer surface of the insulation under particular conditions of ambient temperature and humidity. Detailed information as to the thickness, heat gain, and condensation prevention is available from the manufacturer, or may be calculated.

Cork pipe insulation shall be supplied either as hollow cylindrical shapes split in half lengthwise or as beveled lagging. Cork pipe insulation is usually factory finished on the outside with an asphalt mastic coating. When stipulated by the purchaser, it shall be supplied without the coating. For further information see ASTM Guide C647 and Practice C755. The purchaser shall specify the class and shape of insulation as required by the job.

3.6.2. Physical Properties

The insulation shall have the following minimum physical properties.

The average thermal conductivity at a mean temperature of 297 K shall be a maximum of 0.042 W/mK for corkboard when tested in accordance with ASTM Test Method C177 and Test Method C518 and

Table 3.9 Cork Pipe Insulation Classes

Class	Pipe Temperature Range
Light-duty thickness	Above 2°C
Medium-duty thickness	2 to −18°C
Heavy-duty thickness	−18 to −32°C
Special thickness	Below −32°C

0.048 W/mK for cork pipe when tested in accordance with ASTM Test Method C335.

The density of corkboard insulation shall be 96 to 128 kg/m^3 when tested in accordance with ASTM Test Method C303 and shall be 112 to 224 kg/m^3 for cork pipe insulation when tested in accordance with ASTM Test Method C302.

The flexural strength average shall be a minimum of 103 kPa for corkboard insulation when tested in accordance with ASTM Test Method C203 with the following modification: the preferred specimen size shall be 305 mm long by 76 mm wide and 51 mm thick, and the test conditions shall conform to those listed in the following note. For thicknesses less than 51 mm, the span shall be five times the thickness. Use a testing speed of 152 ± 25 mm/min.

The deflection average shall be a minimum of 6.4 mm and shall be determined at the same time as the flexural strength test. Measure the total deflection at midspan when the specimen reaches the maximum load.

Note: Tests for density, flexural strength, and deflection shall normally be made under prevailing atmospheric conditions. In the case of dispute, tests shall then be run on test specimens conditioned until weight equilibrium is obtained at 23 ± 1°C and 50 ± 2% relative humidity.

3.6.3. Standard Sizes and Dimensions

• **Corkboard**

The standard sizes of corkboard are shown in Table 3.10. Other sizes shall be available on request.

• **Cork pipe insulation**

Cork pipe insulation shall be made to fit nominal pipe sizes from 6.4 to 915 mm inclusive and tubing sizes from 6.4 to 152 mm inclusive. All sizes shall be furnished in sections 915 mm long. Each section shall be true to shape and roundness and shall be a neat fit on an average pipe or tube of the size for which the section was designed. The edges and ends shall be square. The purchaser shall specify the site and dimensions of the insulation.

Table 3.10 Standard Sizes of Corkboard

Width (mm)	Length (mm)	Thickness (mm)
305	915	25, 38, 51, 76, 102
610	915	51, 76, 102
915	915	51, 76, 102

3.6.4. Dimensional Tolerances

- **Corkboard**

    Tolerances for length, width, and thickness of corkboard shall be as follows:
    - Width (mm): ±1.6
    - Length (mm): ±3.2
    - Thickness: +0, −1.6

        When one side of the board is sanded, the thickness tolerance shall be +0, −3.2 mm.

- **Cork pipe insulation**

    The tolerance for length and thickness of cork pipe insulation shall be as follows:
    - Length (mm): 915 ± 2.4
    - Thickness (mm): ±1.6

3.6.5. Workmanship

The insulation shall not have visible defects that will adversely affect its performance in service. For cork insulation the moisture-proof coating, when finished, shall be continuous and free of holes.

3.6.6. Sampling

The insulation shall be sampled for the purpose of the tests in accordance with ASTM Criteria C390.

3.6.7. Inspection and Rejection

- **Inspection**

    The manufacturer and/or supplier shall be responsible for carrying out all the tests and inspections required by this standard specification, using his own or other reliable facilities, and he shall maintain complete records of all such tests and inspections. Such records shall be available for review by the purchaser. The manufacturer and/or supplier shall furnish to the purchaser a certificate of inspection stating that each lot has been sampled, tested, and inspected in accordance with this standard specification and has been found to meet the requirements specified.

    The supplier shall afford the purchaser's inspector all reasonable facilities necessary to satisfy that the material is being produced and furnished in accordance with this standard specification. Such inspections in no way relieve the manufacturer and/or supplier of his responsibilities under the term of this standard specification.

The purchaser reserves the right to perform any inspections set forth in this standard specification where such inspections are deemed necessary to assure that supplies and services conform to the prescribed requirements.

The purchaser's inspector shall have access to the material subject to inspection for the purpose of witnessing the selection of the samples, the preparation of the test pieces, and the performance of the test(s). For such tests, the inspector shall have the right to indicate the pieces from which the samples will be taken in accordance with the provisions of this standard specification.

- **Rejection**

  If the inspection of the sample shows failure to conform to the requirements of this standard specification, a second sample from the same lot shall be tested and the results of this retest averaged with the result of the original test.

  Upon retest, failure to conform to this standard specification shall constitute grounds for rejection. In case of rejection, the manufacturer or supplier shall have the right to reinspect the rejected lot and resubmit the lot for inspection after removal of that portion of the lot not conforming to the specified requirement.

3.6.8. Packaging and Marking

- **Packaging**

  Unless otherwise specified or agreed upon between the manufacturer and purchaser, corkboard and cork pipe insulation shall be packaged in the manufacturer's standard commercial container.

  Overseas consignments shall be packed in double-corrugated cartons incorporating weatherproof paper because of the greater handling involved and the possibility of exposure to wet conditions. If the size of sections exceeds the practical size for cartons, wooden or strong mesh crates may be used.

  Corkboard and cork pipe insulation materials shall not be unpacked at the site until they are required for use. Stacking of insulation during transportation shall be in accordance with the manufacturer's recommendation.

3.6.9. Storage

The insulation shall be stored in accordance with the manufacturer's recommendations.

3.7. SPRAY-APPLIED RIGID-CELLULAR POLYURETHANE (PUR) AND POLYISOCYANURATE (PIR) THERMAL INSULATION

The specification in this section covers the minimum requirements for types, composition, physical properties, inspection, packaging, and marking of spray-applied rigid-cellular polyurethane and polyisocyanurate intended for use as thermal insulation for service temperatures between –100°C to +85°C or as agreed upon between the manufacturer and the purchaser.

3.7.1. Designation

For the purpose of this specification, the materials are divided into four types as follows:

- **Type 1:** Polyurethane foams (PUR) suitable for general use
- **Type 2:** Polyurethane foams suitable for use where there is a requirement for greater resistance to compressive forces
- **Type 3:** Polyisocyanurate foams (PIR) suitable for general use
- **Type 4:** Polyisocyanurate foams suitable for use where there is a requirement for greater resistance to compressive forces

     The designation shall consist of a three-component code comprising the following items, in the order presented:

2. Foam type

3. Thermal conductivity selected in accordance with Table 3.11

3.7.2. Composition

The material shall be of rigid polyurethane or polyisocyanurate foam as specified by the purchaser.

     Note: No requirement for odor is included as its assessment is largely subjective. However it is recommended that the material should be free from objectionable odor.

3.7.3. Manufacture

Spray-applied rigid-cellular polyurethane thermal insulation is produced by the catalyzed chemical reaction of polyisocyanates with polyhydroxyl compounds, with the addition of other compounds such as stabilizers and blowing agents.

Table 3.11 Thermal Conductivity of the Foam

Thermal Conductivity (W/mK)
0.015
0.016
0.017
0.018
0.019
0.02
0.021
0.022
0.023
0.024
0.025
0.026
0.027
0.028
0.029
0.03
0.031
0.032

Note: These values are 30-day values for quality-control purposes. For corresponding long-term design values, the manufacturer's advice should be sought.

Spray-applied rigid-cellular polyisocyanurate thermal insulation is produced by the catalyzed polymerization of polyisocyanates, usually in the presence of polyhydroxyl compounds, with the addition of other compounds such as stabilizers and blowing agents.

The materials shall be capable of being mixed and applied using commercial polyurethane spray equipment. In most cases the thermal insulation is formed directly on the surface to be insulated. The foam injection process shall be compatible with the atmospheric site conditions and a minimum temperature of 15°C shall be maintained for foam components.

3.7.4. Physical Properties

The products are to be delivered on site in two components ready for use with the following foam properties.

The insulation shall be self-extinguishing. Density after molding shall be $45 \pm 5\%$ kg/m^3 in accordance with ASTM D1622. The free rise density shall be 28 ± 2 kg/m^3 to ensure good compaction of the foam and good homogeneity due to high compression rate.

Thermal conductivity at 20°C mean temperature per ASTM C177 measured on 25 mm thick foam, cut on both sides and aged at 21°C for 180 days shall be no greater than 0.023 W/mK. The thermal conductivity for freshly blown foam shall be no greater than 0.020 W/mK. The minimum percentage of closed cells shall be 90% when tested in accordance with ASTM D2856. The leachable halides content shall not exceed 30 ppm.

The maximum water vapor permeability shall be 4.4 ng/(Pa.s.m) for Types 1 and 3 and 3.6 ng/(Pa.s.m) for Types 2 and 4 when tested in accordance with ASTM E96 at 24°C.

The maximum water absorption shall be 3% volume when tested in accordance with ASTM D2842. The linear coefficient of thermal contraction shall be $(50{-}100) \times 10^{-6}$ m/m°C according to ASTM D696.

The response to thermal and humid aging shall be a maximum of 12% volume change for Types 1 and 3 and a maximum of 6% volume change for Types 2 and 4 when tested in accordance with ASTM Test Method D2126. Expose 305 by 25 mm specimens to 70 ± 2°C and 97 ± 3% relative humidity for 168 ± 2h. Measure after $24 \pm \frac{1}{2}$h and 168 ± 2h.

Surface burning characteristics shall be determined in accordance with ASTM Test Method E84 at the end use thickness and the results shall be reported.

Tensile strength shall be the following when tested in accordance with ASTM D1623:
- At room temperature = 500–700 kPa
- At −196°C = 600–800 kPa

    Tensile modulus shall be the following when tested in accordance with ASTM D1623:
- At room temperature = average 14 MPa
- At −196°C = average 28.5 MPa

    Compressive strength shall be at least 100 kPa for Types 1 and 3 and 210 kPa for Types 2 and 4 at 10% deformation and at 20°C when tested in accordance with ASTM D1621. A different compressive strength may be manufactured when agreed upon between the manufacturer and purchaser.

    The above thermal and structural properties shall be supplemented by the following data:
- Test reports on compressive, tensile, and shear strength and moduli at 21°C and also at −165°C (ASTM C165).

- Thermal conductivity versus temperature curve of the foam from $-165°C$ to $65°C$ with adequate definition as per ASTM C177 or C518 measured on sample cut from freshly blown foam, after initial cure, parallel to foam rise. A minimum of six data points are required.
- Thermal conductivity versus time curve of foam aged for 180 days at $21°C$ and 50% RH measured by ASTM C518 at $24°C$ parallel to foam rise. Adequate points are required to define curve.
- Thermal conductivity versus temperature curve of foam aged for 180 days from $-40°C$ to $65°C$ measured by ASTM C518 parallel to foam rise. A minimum of six data points are required.
- Contraction/expansion coefficients versus temperature curves from $-165°C$ to $21°C$.

3.7.5. PUR Materials for Pipe Supports

For pipe sizes 150 mm and under, the foam shall be 160 kg/m^3 high-density molded polyurethane with a minimum ultimate compressive strength of 2 MPa at $20°C$ and a design stress of 735 kPa. For pipe sizes 200 mm through 600 mm, the foam shall be 224 kg/m^3 high-density molded polyurethane with a minimum ultimate compressive strength of 4 MPa at $20°C$ and a design stress of 1.15 MPa.

For pipe sizes 600 mm and larger, the foam shall be 320 kg/m^3 high-density molded polyurethane with a minimum ultimate compressive strength of 7 MPa at $20°C$ and a design stress of 1.8 MPa.

Thermal conductivity: When tested at a cryogenic temperature of $-160°C$, the maximum apparent thermal conductivity shall be as follows:

- 160 kg/m^3 0.022 W/mK
- 224 kg/m^3 0.025 W/mK
- 320 kg/m^3 0.035 W/mK

The above values shall be substantiated by the vendor by independent laboratory test reports.

3.7.6. Preformed Spacers

Preformed spacers shall be of PUR/PIR with the following minimum properties:

- The spacers shall consist of fully monolithic molded (180°) half pipe sections with a minimum density of 50 kg/m^3, designed to form compartments for the in-situ molding operation.

- The foam shall be self-extinguishing.
- The minimum percentage of closed cells shall be 90% per ASTM D2856.
- The compressive strength shall be a minimum of 240 kPa.
- The thermal conductivity at 20°C mean temperature per ASTM C177, measured on 25 mm thick foam cut on both sides and aged at 21°C for 180 days shall be no greater than 0.025 W/mK.
- The leachable halides content shall not exceed 90 ppm.

3.7.7. Sampling

For purposes of sampling, the lot shall consist of all the polyurethane liquid components purchased at one time.

3.7.8. Unit Sample

The unit sample shall consist of approximately 23 kg of each of the two liquid components as required to prepare the foam test specimens specified in the following section. Samples may be drawn from representative bulk storage or from one or more shipping containers.

Sampling for qualification tests shall be in accordance with statistically sound practice. Qualification tests will be conducted on the physical properties.

Sampling for inspection tests shall be for properties agreed upon between the manufacturer and the purchaser.

3.7.9. Test Specimen Preparation

A finished foam insulation test panel shall be made by spray application consistent with the manufacturer's recommendation including temperature of the liquid components, ambient temperature, temperature and type of the substrate, type and operation of spray equipment, and thickness of foam per pass. Unless otherwise specified and reported, the ambient and substrate temperature shall be 24 ± 3°C. The relative humidity must not exceed 80%. The test panels shall be of a sufficient quantity and size to satisfy test requirements.

Note: About 15m^2 of finished foam should be sufficient. Specific panel sizes and thickness should be selected based on the requirement of the individual tests.

The test panels shall be allowed to cure for at least 72 h at 23 ± 1°C and 50 ± 5% relative humidity prior to cutting or testing for physical properties.

Core specimens, when required, shall be obtained by removing both the external skin and the boundary skin found at the substrate/foam interface. A trim cut on each face to a depth of 3 to 6 mm is generally sufficient. Core specimens may obtain one or more internal skins at spray pass boundaries.

3.7.10. Inspection and Rejection

- **Inspection**

    The manufacturer and/or supplier shall be responsible for carrying out all tests and inspections required by this standard specification, using his own or other reliable facilities, and shall maintain complete records of all such tests and inspections. Such records shall be available for review by the purchaser. The manufacturer and/or supplier shall furnish to the purchaser a certificate of inspection stating that each lot has been sampled, tested, and inspected in accordance with this standard specification and has been found to meet the requirements specified.

    The supplier shall afford the purchaser's inspector all reasonable facilities necessary to satisfy that the material is being produced and furnished in accordance with this standard specification. Such inspection in no way relieves the manufacturer and/or supplier of his responsibilities under the terms of this standard specification.

    The purchaser reserves the right to perform any inspections set forth in this standard specification, where such inspections are deemed necessary to assure that the supplies and services conform to the prescribed requirements.

    The purchaser's inspector shall have access to the material subject to inspection for the purpose of witnessing the selection of the samples, the preparation of the test pieces, and the performance of test(s). For such tests, the inspector shall have the right to indicate the pieces from which the samples will be taken in accordance with the provision of this standard specification.

- **Rejection**

    If the inspection of the sample shows failure to conform to the requirements of this standard specification, a second sample from the same lot shall be tested and the result of this retest averaged with the result of the original test.

    Upon retest, failure to conform to this standard specification shall constitute grounds for rejection.

In case of rejection, the manufacturer or supplier has the right to reinspect the rejected lot and resubmit the lot for inspection after removal of that portion of the lot not conforming to the specified requirement.

3.7.11. Certification

When specified in the purchase order or contract, a manufacturer's certification shall be furnished to the purchaser that the material was manufactured, sampled, tested, and inspected in accordance with this standard specification and has been found to meet the requirements. When specified in the purchase order or contract, a report of the test results shall be furnished.

Upon the request of the purchaser in the contract or order, the certification of an independent third party indicating conformance to the requirements of this standard specification may be considered.

3.7.12. Packaging and Marking

- **Packaging**

    Unless otherwise specified or agreed upon between the manufacturer and the purchaser, the liquid components shall be packaged in the manufacturer's standard commercial containers. Each container shall be blanketed with dry air and nitrogen and sealed.

- **Marking**

    Each container shall be marked with the following information:
- Polyisocyanate (A component) or resin (B component).
- Name of the manufacturer
- Manufacturer's product designation
- Manufacturer's lot number or the date of production, or both
- Net weight of the contents and gross weight of the container and contents
- Instruction for safe handling and recommended storage temperatures
- Mixing instructions
- Listing agency label if available
- Destination

3.7.13. Storage

The storage of the liquid components shall be in accordance with the manufacturer's recommendations.

3.7.14. Health and Safety Precautions

The manufacturer shall provide the purchaser information regarding any hazards and recommended protective measures to be employed in the safe installation and use of the material.

3.8. PREFORMED RIGID-CELLULAR POLYURETHANE (PUR) AND POLYISOCYANURATE (PIR) THERMAL INSULATION

The specification in this section covers minimum requirements for types, composition, physical properties, dimensions, tolerance, inspection, packaging, and marking of preformed rigid-cellular polyurethane and polyisocyanurate material intended for use as thermal insulation for pipe work and equipment for service temperatures between −100 and +140°C.

It applies to slab cut and molded pipe sections and radiused and beveled lags.

For specific applications, the actual temperature limits shall be agreed upon by the manufacturer and purchaser.

The application of a suitable vapor-retardant material may be required in conjunction with the application of this insulant where the service temperatures are to be generally below ambient.

3.8.1. Designation

For the purpose of this specification, the materials are divided into four types as follows:

- **Type 1:** Polyurethane foams suitable for general use
- **Type 2:** Polyurethane foams suitable for use where there is a requirement for greater resistance to compressive forces
- **Type 3:** Polyisocyanurate foams suitable for general use
- **Type 4:** Polyisocyamerate foams suitable for use where there is a requirement for greater resistance to compressive forces

  The designation shall consist of a two-component code comprising the following items, in the order presented.

1. Foam type
2. Thermal conductivity selected in accordance with Table 3.12

   An example of the designation required for a Type 2 foam with a thermal conductivity of 0.02 W/mK is as follows:

Foam type: Type 2

Thermal Conductivity: 0.02 W/mk

Table 3.12 Thermal Conductivity
of the Foam

Thermal Conductivity (W/mK)
0.015
0.016
0.017
0.018
0.019
0.02
0.021
0.022
0.023
0.024
0.025
0.026
0.027
0.028
0.029
0.03
0.031
0.032

Note: These values are 30-day values for
quality-control purposes. For corresponding
long-term design values, the manufacturer's
advice should be sought.

3.8.2. Composition

The material shall consist of rigid polyurethane or rigid polyisocyanurate foam with closed cell structure.

The material shall be suitably formulated to ensure that, when tested by the method described in BS 4375: 1974, a test specimen of 150 mm × 50 mm × 13 mm exposed to a small flame shall show an extent burnt of less than 125 mm for rigid polyurethane foam and less than 25 mm for rigid polyisocyanurate.

Note: Materials indicated by PUR are substantially composed of polyurethane linkage and those indicated by PIR are substantially composed of polyisocyanurate linkage.

3.8.3. Manufacture (PUR)

Preformed rigid cellular polyurethane thermal insulation is produced by the chemical reaction of polyisocyanates with polyhydroxyl compounds, usually in the presence of catalysts, cell stabilizers, and blowing agents.

Preformed rigid cellular polyisocyanurate (PIR) thermal insulation is produced by the polymerization of polymeric polyisocyanates, usually in the presence of polyhydroxyl compounds with the addition of catalysts, cell stabilizers, and blowing agents.

3.8.4. Physical Properties

Preformed PUR/PIR shall have the following minimum properties.

- Thermal conductivity at $20°C$ mean temperature per ASTM C177 measured on 25 mm thick foam, cut on both sides and aged at $21°C$ for 180 days shall be no greater than 0.023 W/mK. The thermal conductivity for freshly blown foam shall be no greater than 0.020 W/mK.
- Minimum percentage of closed cells shall be 90% when tested in accordance with ASTM D2856.
- Maximum water vapor permeance at $38°C$ and 100% relative humidity shall be 4.38×10^{-3} µg/NS when tested in accordance with ASTM E96.
- Linear coefficient of thermal expansion/contraction shall be 70×10^{-6} m/m per $°C$ when tested in accordance with BS 4370.
- The insulation shall be self-extinguishing and have 90% retention of weight in accordance with ASTM D3014.
- The maximum leachable halides content shall be 90 ppm.
- **Structural properties**

  The density and the chemical formulation of foam shall be selected in such a way that the following relation is satisfied:

$$\frac{\sigma(1 - \phi)}{E\alpha(T_2 - T_1)} \geq 1.5$$

where

σ = tensile strength of the foam at $-165°C$ (minimum value of all directions in kPa).

E = tensile modulus of the foam at $-165°C$ (maximum value of all directions in kPa).

α = average linear contraction coefficient of the foam from $-165°C$ up to $+21°C$ (maximum value of all directions).

$T_2 - T_1 = 180°C$; temperature difference between cold surface and surroundings.

$\varphi = 0.4$; Poisson's ratio at $-165°C$, estimated value. Other values may be used if substantiated by experimental data.

The formula above is a safety factor, expressing the ratio of the tensile strength of the material and the tensile stress induced in the material under cryogenic temperatures.

The density shall be a minimum of 45 kg/m^3 when tested in accordance with ASTM D1622.

Compressive strength shall be at least 100 kPa for Types 1 and 3 and 210 kPa for Types 2 and 4 at 10% deformation and at 20°C when tested in accordance with ASTM D1621. Different compressive strengths shall be manufactured if agreed upon between the manufacturer and purchaser.

The above thermal and structural properties shall be supplemented by the following data:

- Test reports on compressive, tensile, and shear strength and moduli at 21°C and also at −165°C (ASTM C165).
- Thermal conductivity versus temperature curve of the foam from −165°C to 65°C with adequate definition as per ASTM C177 or C518 measured on sample cut from freshly blown foam, after initial cure, parallel to foam rise. A minimum of six data points are required.
- Thermal conductivity versus time curve of foam aged for 180 days at 21°C and 50% RH measured by ASTM C518 at 24°C parallel to foam rise. Adequate points are required to define curve.
- Thermal conductivity versus temperature curve of foam aged for 180 days from −40°C to 65°C measured by ASTM C 18 parallel to foam rise. A minimum of six data points are required.
- Contraction/expansion coefficients versus temperature curves from −165°C to 21°C.

3.8.5. Shapes and Sizes

Where supplied in two halves, pipe sections shall be oversized to accommodate contractions in accordance with Table 3.13.

The longitudinally mating faces shall be flat and in the same plane, so that when the two pieces are put together no gap exists between the mating surfaces.

Note: It is common practice for the mating faces while still being flat in the lengthwise direction to have a variable profile in the radial direction. This is acceptable provided that the mating surfaces so created still fit snugly together. In many cases this practice enhances the "snugness" of the fit.

The ends shall be flat and normal to the longitudinal axis of the section.

For all sizes up to and including 508 mm diameter, the PUR/PIR shall be applied in two half sections.

Table 3.13 Contraction Gap Between Pipe/Equipment Surface and Inner Layer of PUF

Minimum Gap (mm)

Operating Temperature (°C)	Operating Temperature Pipe/Equipment Outside Diameter Range (mm)			
	D < 219	219 < D < 406	406 < D < 610	610 < D < 800
−196 < T < −75	2	4	6	7
−75 < T < −40	1	2	3	4

Note: The diameter of preformed PUR sections shall be the contraction gap plus the outside diameter of the pipe/equipment. The maximum contraction gap shall not exceed 1.25 times the minimum gap.

For sizes over 508 mm and equipment between 500 mm and 3600 mm in diameter, the PUR/PIR shall be supplied in radiuses and beveled segments having a width on the outside radius of minimum 450 mm.

For equipment in excess of 3600 mm diameter, the PUR/PIR may be supplied in flat beveled segments.

For equipment heads, the PUR/PIR shall be preferably supplied all in radiused and beveled head segments for diameters between 900 mm and 3600 mm.

For equipment heads up to 900 mm, these shall be supplied in one-piece blocks.

For equipment heads over 3600 mm, flat and beveled blocks shall be used.

The mating beveled edges shall be flat, so that when they are put together to form a cylinder, no gaps exist between abutting lags.

The ends shall be flat and normal to the longitudinal axis of the lag.

3.8.6. Molded Components

All molded items shall be free from grease or other molding release agent that will adversely reduce the adhesion of insulation, mastic, and adhesives.

3.8.7. Color Identification

PIR foam shall be supplied with a pink color. PUR foam shall be supplied in any other color or without added color as required.

3.8.8. Dimensional Tolerances

The dimensions of product supplied shall not deviate from those specified by more than the appropriate tolerances given in either Table 3.14 or Table 3.15. For the slab, the permissible thickness deviations shall be ±1.5 mm.

Table 3.14 Dimensional Tolerances for Pipe Sections and Lags

| Dimensions (mm) | Permissible Deviations | |
	Molded Pipe Sections	Cut Pipe Sections and Lags
Lengths	3	3
Bores less than 150	1.5	1.5
Bores 150 and above	3	1.5
Outside diameters less than 150	1.5	
Outside diameters 150 and above	3	1.5

Note: For single-layer components or the first layer of a multilayer component, the tolerance on the bore is given on the quoted pipe outside diameter. For the second or subsequent layer(s) of multilayer components, it is given on the outside diameter of the mating inner layer.

Other slab parameters shall be as follows:

- **Squareness**

  The thermal insulation boards shall not be out of square by more than 1.5 mm/30 cm of width or length.

- **Straightness**

  Unless otherwise specified, the slab shall be furnished with straight edges that shall not deviate by more than 0.8 mm/30 cm.

- **Flatness**

  The slabs shall not depart from absolute flatness by more than 1.5 mm/30 cm of width or length.

  The trueness, squareness, and flatness shall be determined in accordance with ASTM Practice C550, except that a straight edge of length longer than the dimension being determined shall be used.

Table 3.15 Dimensional Tolerances for Slabs

Lengths or Widths	Permissible Deviations of Lengths or Widths	Maximum Differences in the Length of the Diagonals of Rectangular Slabs
Up to and including 100	±1	3
Over 100 up to and including 1000	±1.5	5
Over 1000 up to and including 2000	±2.5	7

3.8.9. Workmanship

This insulation shall not have visible defects that will adversely affect the service properties. Material shall have a uniform fine cellular structure.

3.8.10. Sampling

The sampling for the qualification tests shall be in accordance with ASTM Criteria C390. Conduct the qualification tests in accordance with the physical properties in Section 3.8.4.

3.8.11. Specimen Preparation

A period of at least 72 h must elapse from the time of manufacture of the rigid cellular PUR or PIR until the cutting of any test specimen.

Unless otherwise specified, the test specimen shall be conditioned at $23 \pm 2°C$ and $50 \pm 5\%$ relative humidity for at least 12 h prior to testing.

3.8.12. Inspection and Rejection

- **Inspection**

    The manufacturer and/or supplier shall be responsible for carrying out all the tests and inspections required by this standard specification, using his own or other reliable facilities, and shall maintain complete records of all such tests and inspections. Such records shall be available for review by the purchaser. The manufacturer and/or supplier shall furnish to the purchaser a certificate of inspection stating that each lot has been sampled, tested, and inspected in accordance with this standard specification and has been found to meet the requirements specified.

    The supplier shall afford the purchaser's inspector all reasonable facilities necessary to satisfy that the material is being produced and furnished in accordance with this standard specification. Such inspection in no way relieves the manufacturer and/or supplier of his responsibilities under the term of this standard specification.

    The purchaser reserves the right to perform any inspections set forth in this standard specification, where such inspections are deemed necessary to assure that supplies and services conform to the prescribed requirements.

    The purchaser's inspector shall have access to the material subject to inspection for the purpose of witnessing the selection of the samples, the preparation of the test pieces, and the performance of test(s). For such tests, the inspector shall have the right to indicate the pieces from which the samples will be taken in accordance with the provisions of this standard specification.

- **Rejection**

    If the inspection of the sample shows failure to conform to the requirement of this standard specification, a second sample from the same lot shall be tested and the results of this retest averaged with the result of the original test.

    Upon retest, failure to conform to this standard specification shall constitute grounds for rejection.

    In case of rejection, the manufacturer or supplier shall have the right to reinspect the rejected lot and resubmit the lot for inspection after removal of that portion of the lot not conforming to the specified requirement.

3.8.13. Packaging and Marking

- **Packaging**

    Unless otherwise specified by the purchaser, the preformed rigid cellular polyurethane and polyisocyanurate foam thermal insulation shall be packaged in the manufacturer's standard commercial container approved by the purchaser.

    Overseas consignments shall be packed in double-corrugated cartons incorporating weatherproof paper because of the greater handling involved and the possibility of exposure to wet conditions. If the size of sections exceeds the practical size for cartons, wooden or strong mesh crates may be used.

    PUR/PIR insulation materials shall not be unpacked at site until they are required for use.

    Stacking of packed PUR/PIR preformed insulation during transportation shall be in accordance with the manufacturer's recommendation.

3.8.14. Storage

The storage of PUR/PIR thermal insulation shall be in accordance with the manufacturer's recommendation.

3.8.15. Health and Safety Hazards

The installation and use of thermal insulating materials may expose individuals to health and safety hazards. The manufacturer shall provide the buyer appropriate information regarding any hazards known to him associated with the designated end use of this product, and shall recommend the protective measures to be employed in the safe installation and use.

3.9. MISCELLANEOUS MATERIALS TO BE USED WITH THERMAL INSULATION

The specification in this section covers the minimum requirements for the properties of filler insulation, vapor barriers, joint sealants, adhesive materials, metallic jacketing, and accessory materials. Unless otherwise specified, the test methods used shall be approved by the purchaser.

3.9.1. Filler Insulation

- **Medium-density rock wool/glass wool board**

  Medium-density rock wool/glass wool board stock for contraction joints:
  - Density: 32 kg/m^3 ($\pm$10%).
  - Thermal conductivity at 25°C mean temperature: 0.035 W/mK or less.
  - The material shall be free from delamination or fiber fallout when tested to BS 2972.

- **Low-density rock wool/glass wool blankets**

  Low-density rock wool/glass wool blankets for packing voids between piping/equipment and insulation:
  - Density: 24 kg/m^3 ($\pm$10%)
  - Thermal conductivity at 25°C mean temperature: 0.034 W/mK or less.
  - The material shall be free from delamination and fiber fallout when tested to BS 2972.

3.9.2. Vapor Barriers

- **Primary metal vapor barrier**

  The vapor barrier shall be corrugated aluminum sheet, Alloy 3003 or 5005 in accordance with ASTM B209. Corrugation shall be 5 mm. The material shall have a factory-applied moisture barrier of a laminate of chemically inert polyethylene and a layer of kraft paper (without adhesives), to ensure the highest adhesion with the foam. The thickness of the vapor barrier shall be at least 0.2 mm minimum.

Table 3.16 The Properties of the Primary Vapor Barrier Mastic

Property	Unit	Value
Temperature resistance	°C	−40 to +120
Water vapor transmission ASTM E96 at 0.75 mm dry film thickness	perms★	0.02
Average nonvolatile percent	vol %	30
Fire resistance		
ASTM E-84/76A flame spread		20
Smoke developed		5
Fuel contribution		10
Spread of flame (BS 476)	Class 1	
Specific gravity	kg/l	1.1−1.3
Consistency	Thixotropic soft paste	
Minimum applied thickness (dry film)	mm	1.2
Minimum coats	%	15

WVT = rate of water vapor transmission, g/hm^2
Δp = vapor pressure difference, mm Hg
★ Perms = permeance =WVT/Δp(ASTM E96)

- **Primary vapor barrier mastic**

  The vapor barrier mastic shall be a heavy-duty, highly durable flexible elastomeric polymer "Hypalon"-based fire-retardant coating with exceptional dry film strength and good puncture resistance and shall have the minimum properties shown in Table 3.16.

Table 3.17 The Properties of the Secondary Vapor Barrier Mastic

Physical Properties	Unit	Value
Polyester layer	μm	≥ 12
Aluminum foil	μm	≥ 25
Elongation: LD	%	≥ 50
Elongation: TD	%	≥ 50
Tensile strength: LD	kg/15 × 100 mm^2	≥ 8
Tensile strength: TD	kg/15 × 100 mm^2	≥ 10

Chemical Properties	Unit	Value
Humidity absorption	%	0.3 max.
Water vapor transmission	perms★	<0.001

WVT = rate of water vapor transmission, g/hm^2
Δp = vapor pressure difference, mm Hg
★ Perms = permeance = WVT/Δp(ASTM E96)

Table 3.18 The Properties of the Vapor Stop Coating/Adhesive

Property	Unit	Value
Specific gravity	kg/l	1.1−1.2
Nonvolatile content	vol %	55−60
Tensile strength (ASTM D412)	kg/cm^2	> 8
Two-component minimum thickness (dry film)	mm	1.2
Service temperature	°C	−196 to +120
Halides	ppm	≤ 10

- **Secondary foil vapor barrier**

   The vapor barrier shall be a three-layer lamination of polyester film/aluminum foil/polyester film, with the minimum properties in Table 3.17.
- **Vapor stop coating/adhesive**

   This shall be a two-part elastomeric material suitable for use in cryogenic conditions with the minimum properties in Table 3.18.
- **Glass fiber–reinforced epoxy (GRE) vapor stop**

   The epoxy resin used shall be Epikote 816 with an aromatic/cycle-aliphatic amine type curing agent. The fabric glass fiber reinforcement shall be of low alkali fibrous glass that is compatible with the epoxy resin. The minimum weight of glass fiber reinforcement shall be approximately 220 g/m^2.

   The epoxy resin and glass fiber shall be capable of producing a reinforced vapor stop over the surface of the PUR insulation; it shall be a self-extinguishing type and shall withstand ambient/subambient conditions and ultraviolet light exposure. The GRE vapor stop shall have the following minimum properties:
- Tensile strength (ASTM D3039):
  - Longitudinal: 15.7 MPa
  - Circumferential: 78.5 MPa
- Maximum water vapor transmission: 0.03 perms*
- Minimum dry film thickness: 5.5 mm: comprising a 5.0 mm reinforced and 0.5 mm unreinforced outer layer.

3.9.3. Joint Sealant/Adhesive Materials

The joint sealant used to seal all cellular glass insulation for dual temperatures and those parts in the insulation system that shall be readily removable shall have the minimum properties mentioned in Table 3.19.

Table 3.19 Joint Sealant/Adhesive Materials

Physical Properties	Unit	Value
Consistency		Soft paste
Color		Gray or white
Average nonvolatile content	vol %	98
Flammability (dry) Low flame Service temperature (Material shall remain soft and tough in service and shall not crack or shrink with thermal cycling)		−70 to +150
Water vapor transmission (at 3 mm dry-film thickness)	perms★	≤ 0.03

WVT = rate of water vapor transmission, g/hm^2
Δp = vapor pressure difference, mm Hg
★ Perms = permeance = WVT/Δp(ASTM E96)

The fabrication adhesive used for bonding together pieces of pre-fabricated PUR/PIR and cellular glass insulation for removable fittings/flange covers and valves shall be a fire resistive two-part component material with the minimum properties in Table 3.20.

The adhesive used to bond insulation and/or supports to piping/equipment in cold insulation shall be a virtually 100% solids-free, three-component cryogenic adhesive, enabling bonding between nonporous surfaces without evaporation problems; it shall have the minimum properties in Table 3.21.

The anti-abrasive coating to be applied as a bore coating to all cellular glass in insulation, and to be applied to aluminum piping and/or piping

Table 3.20 Properties of the Fabrication Adhesive

Property	Unit	Value
Color		Gray
Consistency		Soft paste
Specific gravity	kg/l	1.3−1.8
Flammability: wet		Nonflammable
Flammability: dry		Fire resistive
Service temperature	°C	−70 to +90

Table 3.21 The Adhesive Used to Bond Insulation and/or Supports to Piping/Equipment in Cold Insulation

Property	Unit	Value
Color		Black
Consistency		Thixotropic
Specific gravity	kg/l	1.8
Service temperature range	°C	−196 to +121

Table 3.22 The Anti-Abrasive Coating to Be Applied as a Bore Coating to All Cellular Glass

Property	Unit	Value
Color		Gray/off-white
Consistency		Brushing
Specific gravity	kg/l	1.8
Average nonvolatile content	Vol %	50
Flammability: wet		Nonflammable
Flammability: dry		Fire resistive
Service temperature	°C	−196 to +121

under vibrational influences, shall be an oil-modified urethane compound designed for the specified purpose and having the minimum properties illustrated in Table 3.22.

For service temperatures above 121°C a high-temperature anti-abrasive coating shall be used.

The metal sealant suitable for gun extrusion to a minimum of 4 mm at all overlaps in the metallic jacketing shall be a tough flexible elastomer-based material, comprising polymeric vapor sealant of butyl rubber with permanent flexibility, good adhesion, and the minimum properties shown in Table 3.23.

3.9.4. Metallic Jacketing

Galvanized sheeting shall not be used over insulation on or near (austenitic) stainless steel and/or nickel alloy piping and equipment. Galvanized sheeting is vulnerable to corrosion in coastal and arid areas.

In areas with potential fire hazards, aluminized steel or stainless steel sheeting shall be used.

Table 3.23 Properties of the Flexible Elastomer

Property	Unit	Value
Color		Light gray
Consistency		Soft paste
Specific gravity	kg/l	1.55
Average nonvolatile content	wt %	65−70
Flammability: wet		Nonflammable
Flammability: dry	BS 476, Part 7	Class 1
Service temperature	°C	−70 to +120

- **Jacketing material**

  Metallic jacketing materials shall be one of the following:

  - **Aluminized steel**

    The material shall be aluminized steel Type 2, coating designation T_2-100 in accordance with ASTM A463.

  - **Aluminum sheet**

    The material shall be of a quality meeting ASTM B209M requirements and shall be half hard, Alloy 1060, Temper H14, with a factory-applied polykraft moisture retarder permanently bonded to the inner side.

  - **Stainless steel sheet**

    The material shall be of a quality meeting the requirements of ASTM A167 Type 304L or 316L.

  - **Galvanized steel sheet**

    The material shall be in accordance with ASTM A653 and ASTM A924; hot-dip galvanized to ASTM A525 coating designation G 90.

  - **Coated steel**

    The material shall be mild carbon steel with protective coating

  - **Thickness of sheets**

    The thickness of the sheets shall be as shown in Table 3.24.

3.9.5. Accessory Materials

The metallic fastening materials shall be as following:

- Stainless steel Type 302 or 304 in accordance with ASTM A167
- Galvanized mild steel in accordance with ASTM A653 and ASTM A924
- Aluminum in accordance with ASTM B209, Alloy 1060-H14

Table 3.24 Thickness of the Sheets

Outside Diameter of Insulation (mm)	Aluminum		Aluminized or Galvanized Steel		Stainless Steel	
	Flat	Corrugated	Flat	Corrugated	Flat	Corrugated
Up to 150	0.8	0.6 (5 mm corrugations)	0.6	0.4 (5 mm corrugations)	0.6	0.4 (5 mm corrugations)
150 to 450	0.9	0.7 (5 mm corrugations)	0.8	0.6 (5 mm corrugations)	0.8	0.6 (5 mm corrugations)
Over 450	1.2	1 (5 mm corrugations)	0.8–1	0.6 (5 mm corrugations)	0.8	0.6 (5 mm corrugations)

Removable covers for flanges, valves, etc., shall be made of 1.0 mm sheets, independent of the type of material.

Bands shall be 10 mm to 20 mm wide with a minimum thickness of 0.5 mm. The material shall be Type 304 stainless steel, galvanized mild steel, or aluminum as appropriate.

Expander type bands shall be of a stainless steel type, and capable of remaining in tension during heating and cooling cycles experienced in normal operation.

Breather springs for bands shall be of a stainless steel type and capable of remaining under tension during the heating and cooling cycles experienced in normal operation. Seals for banding shall be of the wing-type construction, Type 302 or 304 stainless steel.

Quick-release toggle fasteners for securement of removable boxes shall be a Type 304 or 302 stainless steel spring shackle lock and shall be sized commensurate with the weight of the box concerned. They shall be spot-welded to the stainless steel bands, which are to be incorporated in the box design.

Rivets where permitted in place of screws for metal fabrication shall be the expanding hard aluminum blind rivet pop type of 4 mm diameter or the stainless steel pop blind-eye type of the following sizes: for pipework and equipment up to 1000 mm they shall be 3.2 mm diameter; for pipework and equipment over 1000 mm they shall be 4.8 mm diameter.

Screws required for metal jacket fabrication shall be slotted pan head self-tapping Type A No. 8 × 12 mm long in accordance with BS 4171, complete with chloroprene rubber washers under the head. The material may be hard aluminum alloy for aluminum jackets and Type 302 or 304 stainless steel for steel jackets.

"S" and "J" clips shall be formed from 20 mm wide, 0.5 mm thick banding of Type 302 or 304 stainless steel. The reinforcing membrane for embedding between the coats of the vapor barrier mastic in preformed PUR/PIR insulation or cellular glass shall be a high-strength resilient synthetic fabric, having a minimum of 20×10 threads per 25 mm and a weight of 33 g/m^2, or an open weave glass cloth, with a weave lock finish and having a minimum of 18×12 threads per 25 mm.

The tape used to secure the inner layers of PUR/PIR to the pipework shall be a pressure-sensitive glass fiber reinforced tape that is a minimum of 19 mm wide.

The vapor barrier cover for contraction joints in the outer layer of insulation shall be a corrugated butyl rubber sheet with 1.2 mm minimum thickness.

The adhesive for the secondary foil vapor barrier shall be an adhesive suitable for adhering the polyester and staying flexible within the service temperature range from $-60°C$ up to ambient.

The glued overlaps of the secondary foil vapor barrier shall be sealed off with a 50 mm wide adhesive tape of similar material (adhesive on one face) capable of sealing the foil within the temperature range from -60°C up to ambient.

The tape to seal all longitudinal/circumferential joints of primary metal vapor barrier shall be a composite-type aluminum foil, reinforced and with a self-bonding layer made of butyl rubber on the back face. It shall have a minimum width of 75 mm to allow for sufficient cover of the joints. The expanded metal for reinforcement shall have a 6 mm to 10 mm mesh, bitumen-coated or galvanized mild steel or Type 302 or 304 stainless steel.

Wire mesh poultry netting shall have 25 mm hexagon openings of 0.9 mm minimum diameter wire, carbon steel galvanized after weaving. Wire to secure the insulation shall be 1.5 mm (16 gauge) diameter soft annealed galvanized mild steel wire for hot insulation systems and stainless steel Type 302 or 304 for cold insulation systems.

Support rings for metal sheeting shall be made of:
- 1.5×25 mm carbon steel strip for an outside diameter of insulation up to 125 mm
- 3×25 mm carbon steel strip for an outside diameter of insulation above 125 mm up to 760 mm
- 5×30 mm carbon steel strip for an outside diameter of insulation above 760 mm

Mineral fiber hydraulic-setting thermal insulating and finishing cement shall be asbestos-free and shall be in accordance with ASTM C449. When cement is to be used in contact with austenitic stainless steel, the product shall conform to the requirement of ASTM C 795.

Glass fiber cloth, 4×4 strand/cm open mesh, shall be used as reinforcement or as a bonding agent for coating and mastic.

3.9.6. Inspection and Rejection

- **Inspection**

    The manufacturer and/or supplier shall be responsible for carrying out all the tests and inspections required by this standard specification, using his own or other reliable facilities, and he shall maintain complete records of all such tests and inspections. Such records shall be available for review by the purchaser. The manufacturer and/or supplier shall furnish to the purchaser a certificate of inspection stating that each lot has been sampled, tested, and inspected in accordance with this standard specification and has been found to meet the requirements specified.

The supplier shall afford the purchaser's inspector all reasonable facilities necessary to satisfy that the material is being produced and furnished in accordance with this standard specification. Such inspections in no way relieve the manufacturer and/or supplier of his responsibilities under the terms of this standard specification.

The purchaser reserves the right to perform any inspections set forth in this standard specification where such inspections are deemed necessary to assure that supplies and services conform to the prescribed requirements.

The purchaser's inspector shall have access to the material subject to inspection for the purpose of witnessing the selection of the samples, the preparation of the test pieces, and the performance of the test(s). For such tests, the inspector shall have the right to indicate the pieces from which the samples will be taken in accordance with the provisions of this standard specification.

- **Rejection**

   If the inspection of the sample shows a failure to conform to the requirements of this standard specification, a second sample from the same lot shall be tested and the results of this retest averaged with the result of the original test.

   Upon retest, failure to conform to this standard specification shall constitute grounds for rejection.

   In case of rejection, the manufacturer or supplier shall have the right to reinspect the rejected lot and resubmit the lot for inspection after removal of that portion of the lot not conforming to the specified requirement.

CHAPTER FOUR

Cryogenic Insulation Systems for LNG Industries

Contents

Liquefied natural gas (LNG) is expected to play a progressively important part in the global energy markets. The market for LNG is growing bigger than any other market for energy resources. As demand for LNG increases, an efficient and safe way to ship and store LNG fuel systems should be offered. Since LNG is normally carried by ship at $-163°C$, the functional

Alireza Bahadori, Thermal Insulation Handbook for the Oil, Gas, and Petrochemical Industries
© 2014 Elsevier Inc.
http://dx.doi.org/10.1016/B978-0-12-800010-6.00004-6

requirements of LNG shipping include cryogenic reliability due to thermal cyclic stresses and high thermal insulation performance for safe and efficient transportation of LNG.

In order to increase the insulation performance as well as structural safety, cryogenic vacuum pipe with multilayer insulation has been recently developed.

In order to guarantee the LNG cold temperature of $-160°C$, high-quality insulation installation in accordance with strict specifications is essential. Cryogenic insulation restricts the inflow of atmospheric heat into the pipe or process equipment, keeping the liquid cold and allowing it to retain its form. In a gas plant, cryogenic insulation is a critical path activity, and in the majority of locations where insulation is installed in an LNG plant, access is limited and restricted.

Apart from the challenge of designing a system of structural integrity at cryogenic temperatures, the main problem with most cold insulation systems is ingress and accumulation of humidity over time. Regardless of the combination of insulation materials, vapor barriers, and jacketing systems, there will be joints and link-up regions, and hence the potential for moisture ingress.

The vacuum caused by cooling, and the inevitable vapor pressure directed from the warm to cold side, can force moisture into and through the system, reducing insulation efficiency and potentially causing frost wedging of insulation and corrosion of the tank or pipe structure.

Insulation systems consisting of polyisocyanurate (PIR) rigid foam or cellular glass insulation, applied in multiple layers, in combination with high-performance vapor barriers, are used in order to produce the necessary degree of insulation. The increasing use of cryogenic fluids in natural gas processing and transport requires much higher thermal insulation efficiency than the usual industrial insulations.

Mechanical and weather protection for the installed insulation is achieved by the application of specified cladding materials, which include stainless steel, aluminized steel, or UV-cured glass-reinforced plastic (GRP).

The PIR rigid foam used for pipe, equipment, and storage tank insulation is produced specifically for use in LNG insulation systems, and has to have the required mechanical properties in order to accommodate the thermal stresses involved, in addition to possessing the necessary thermal properties.

Cellular glass insulation is also used for cold insulation of pipe work, equipment, storage tanks, and ships. Cellular glass is a lightweight material consisting of millions of packed glass cells. Each glass cell provides long-term performance, and together they can insulate the material under all possible conditions. These materials are used in both the on-site insulation application to pipe work and equipment and the off-site pre-insulation of pipe work and equipment and are applied in strict accordance with LNG industry standards.

4.1. POLYURETHANE

The industrial and commercial success of polyurethane occurred during the post-war period (1945–1955), with the first relevant applications in the field of elastomer and protective coatings.

The chemical reaction of polyurethanes belongs to the family of thermoset polymers; it is a polymeric chain obtained by polyaddition reaction between a polyisocyanate and a polyol, forming a sequence of urethane bonds (-NH-(CO)-O-). This reaction is exothermic and has no byproducts.

4.1.1. Benefits and Advantages of Polyurethane Insulation

- **Low thermal conductivity**

    Heat is a form of energy, always moving from a higher to lower temperature. The low thermal conductivity rating of rigid polyurethane foam, one of the lowest values among commonly used insulating materials, allows efficient retention of heat flow.

- **Strength**

    The good balance between the weight, mechanical strength, and insulation properties of polyurethane foam (CORAFOAM®) demonstrates its versatility as an insulating material. These qualities allow it to be used in applications that require insulation with a combination of load-bearing, impact-resistant, and weight- and space-saving properties, as well as ease of installation and maintenance.

    This polyurethane foam provides a very favorable ratio of physical-mechanical properties versus density; further enhancement of the overall properties is achieved when bonded with facing materials such as metal or plasterboard.

- **Lightness**

    Polyurethane rigid foams are cellular materials. The foam is made of little bubbles filled with the blowing agent, which provides good insulation properties. The polyurethane matrix is in charge of holding all the cells together: the higher the amount of polymer that holds together the structure, the higher the density. In fact, in 1 cubic meter of foam, only 4% of the total volume is occupied by the polymer while the remaining 96% is filled by the blowing agent (this applies to a typical 40–45 kg/m^3 foam) The lightness of the foam allows for easy transportation, handling, and installation.

- **Low water absorption and low water permeability**

  Water has a thermal conductivity that is 10 to 20 times higher than commonly used insulating materials, so it is evident how important it is to keep water out of the insulation package. The presence of water, besides causing the loss of insulation efficiency, leads to an increase in weight, the risk of corrosion for metal surfaces, and ice formation whenever temperatures go below the freezing point.

  In this last case the risk of deterioration of the insulation package is possible, thus negatively affecting the insulation properties. The closed cell structure of rigid polyurethane foams guarantees low water absorption; the incorporation of a moisture vapor barrier is nevertheless provided for, with the aim of enabling the insulation to withstand the most stringent requirements.

- **Dimensional stability**

  A dimensionally stable material is a basic requirement to achieve proper insulation performance. A size change in the insulating material can be reversible or irreversible: size changes due to simple thermal contraction/expansion are usually reversible, while size changes due to the combined effects of extreme temperatures, water, moisture, and mechanical loads constitute an irreversible component.

  All materials, in fact, change size when heated up or cooled down: the amount of change depends on the chemical composition of the material. Thus every material has its own coefficient of thermal expansion: this parameter measures how much materials shrink or expand when they are exposed to a temperature change. Size variations due to the coefficient of thermal expansion are reversible.

  Because of their chemical composition, good mechanical properties, reduced moisture pickup, closed cell structure, and chemical resistance, rigid polyurethane foams demonstrate significant size stability.

- **Chemical resistance**

  The chemical composition of rigid polyurethane foam provides excellent resistance to a wide range of chemicals, solvents, and oils.

- **Compatibility**

   Rigid polyurethane foam is compatible with a large number of auxiliary materials, including paper, foil, glass fiber, aluminum, and bitumen. The combination of rigid polyurethane foam with these materials enhances the overall properties, enabling it to be used as semi-structural panels and cladding. Furthermore, proper choice of plaster or foil improves the insulating performance of the foam by forming protective moisture barriers, useful when conditions of high humidity are present.

- **Range of service temperatures**

   Rigid polyurethane foam can be used in applications that experience exceptional extremes of temperature, from −200°C to +130°C. Nevertheless, every polyurethane foam has its own temperature range of application so it is important to double-check the indications on the technical data sheets before selecting the most convenient solution.

- **Fire properties**

   Polyurethane rigid foams are organic compounds. Organics are all combustible materials, although the ignitability and rate of burning of polyurethane rigid foams can be improved to suit a variety of insulating applications and the foams can be formulated to meet the most stringent fire protection standards.

4.1.2. PU/PIR Typical Applications

PIR and PU foams are widely used in the gas liquefaction industry thanks to their excellent behavior at critical temperatures such as those in cryogenic applications.

As they are strongly resistant to thermal shocks, as well as chemically inert, the materials can withstand a wide range of operating temperatures from strong cryogenic (liquid nitrogen, −196°C) up to about 130°C, granting the best compromise in terms of cost/efficiency. For these reasons they are frequently used in the most critical cryogenic applications where the saving of frigories is vital for the efficiency of the running plant. The main categories of industrial plants where PIR/PU foams can be used are the following:

- **Liquefied natural gas (LNG)**

   PIR and PU are main contributors to the achievement of good results for LNG operations in all aspects of the supply chain, from liquefaction through insulation of the tanks on LNG carriers, to the regasification terminal where ships download it into various kinds of reservoirs.

- **Ethylene**

  This is one of the most important hydrocarbons, as it is the first building block for other plastics. Ethylene is the most widely produced organic compound in the world; it was initially manufactured directly close to the market. Since gas liquefaction technology has developed reliable solutions for the transportation of the liquefied gases (in this case, the storage temperature is around $-104°C$), it has become more convenient to process the natural gas close to the gas fields. PU/PIR, thanks to their versatility, have contributed significantly to the success of the development of local industries in several developing countries.

- **Liquefied petroleum gas (LPG)**

  LPG is synthesized by refining petroleum or "wet" natural gas, and is usually derived from fossil fuel sources; it is manufactured during the refining of crude oil, or extracted from oil or gas streams as they emerge from the ground. Varieties of LPG include mixtures that are mainly propane based, mixtures that are mainly butane based and most commonly mixtures including both propane and butane. The development of the LNG business, whose production process consists of various purification phases, enables LPG gases to be obtained as secondary product with low production costs. LPG plants are therefore often located near large LNG plants. LPG is normally considered "soft cryogenic," as the storage temperatures are about $-10/-45°C$ depending on the ratio of the various components.

- **Ammonia/fertilizer**

  Ammonia is a compound of nitrogen and hydrogen with the formula NH_3. It is a gas with a characteristic pungent odor. Ammonia contributes significantly to the nutritional needs of terrestrial organisms by serving as a precursor to food, being a fertilizer. Ammonia, either directly or indirectly, is also a building block for the synthesis of many pharmaceuticals.

- **Underground oil ducts**

  Oil pipelines are made from steel or plastic tubes with inner diameter typically from 10 to 120 cm (about 4 to 48 inches). Most pipelines are buried at a typical depth of about 1 to 2 meters (about 3 to 6 feet). The oil is kept warm enough (about $80°C$) to be fluid, and this allows pumps to move it at about 1 to 6 m/s. Thanks to the great resistance to organic attacks, PU/PIR is often applied for the insulation of these underground ducts.

4.2. MULTILAYER INSULATION: SUPER INSULATION

Vessels that require a high level of thermal isolation are typically enclosed in an outer vessel with a separating space that is vacuum evacuated. With an ambient vacuum 24-hour settle pressure in the 10^{-4} torr range, convective heat transfer across this space is virtually eliminated.

Creating a very small heat path from the outer to the inner vessel typically controls conductive heat transfer. Selecting a material for this heat path that has very low thermal conductivity properties normally works well. Such materials include fiberglass and/or low-density ceramics. Radiated heat transfer is typically controlled by the barrier placed around the inner vessel. Its mission is to prevent heat from radiating into the inner vessel.

One common radiation barrier used in cryogenic applications is known as multilayer insulation (MLI), or super insulation. The space program encouraged the development of MLI around 1960. The MLI generally contains multiple layers of reflective material separated by spacers having low conductivity.

MLI consists of many radiation shields stacked in parallel as close as possible without touching one another. MLI will typically contain about 60 layers per inch. MLI is anisotropic by nature, making it difficult to apply to complex geometries. MLI is generally very sensitive to mechanical compression and edge effects, requiring careful attention to detail during all phases of its installation. Accordingly, performance in practice is not typically as good as theoretically possible.

Each layer is isolated from the other by spacer material such as polyester, nylon, or Mylar. The aluminum foil is carefully wrapped around the container such that it covers the entire surface of the inner vessel. Spacer material, as described, is placed between the layers to completely prevent the separate coverings of foil from contacting. Should they touch, a thermal short circuit will occur and increase the heat transfer. The layers can be applied manually as blankets.

These are hand cut to fit and wrapped over the vessel and vessel ends. Tape that has low out-gassing properties is then used to hold the blanket layers in place. Another method of applying the layers is by "orbital wrapping." This method is used where high-volume vessels are being manufactured. Special equipment is required that wraps the alternating layers much like the wrapping of a spool of string.

As the number of layers increase, the insulation capability is also increased. Typically layers adding up to about one inch in total thickness is applied in the liquid nitrogen temperature range described.

MLI is designed to work under high-order vacuum, i.e., pressure below about 1×10^{-4} torr. Obtaining this vacuum generally requires lengthy pumping along with heating and purging cycles. Chemical materials are required to absorb the out-gassed molecules to maintain the vacuum over extended periods.

Super insulation advancements are being made at the Kennedy Space Center. A new-layered cryogenic insulation system is being developed. This insulation is different from others due to its superior thermal performance in "soft" vacuum conditions. This system overcomes some of the typical shortcomings of super insulation discussed above. Those shortcomings include exhibition of different insulation properties when measured in different directions and sensitivity to mechanical compression.

4.3. FOAM INSULATION

Foam insulation requires no vacuum. Foams generally provide a barrier to heat conduction due to their low density. Furthermore, foams inhibit convective heat transfer by limiting convection to the individual cells, fissures, or other spaces within the foam structure. Foam insulation generally includes some form of moisture barrier. When moisture is allowed to accumulate within the spaces of the foam structure, the thermal conductivity rapidly increases. Typical foam insulation includes polyurethane foam, polyamide foam, and foam glass.

Foam insulation is generally not favored in cryogenic applications. Such insulation is likely to crack due to thermal cycling and environmental exposure. Cracks permit incursions of moisture and humid air that will form ice and greatly increase the surface area for heat transfer.

4.4. PERLITE

The unique properties of perlite have found wide acceptance in various insulation applications including cryogenic low-temperature storage (particularly for the storage of LNG), shipping containers, cold boxes (air separation units), test chambers, and food processing.

Storage chambers that are $-100°C$ and below are considered cryogenic, and super cold or extremely cold cryogenic fluids such as hydrogen and helium are normally stored in spherical, double-walled vessels with evacuated annular spaces using excavated perlite.

Advantages include:
- Perlite insulation exhibits low thermal conductivity and has excellent thermal properties.
- Low cost, and easy to handle and install.
- Noncombustible, meets fire regulations and therefore lowers insurance rates.
- Does not shrink, swell, warp, or slump.
- Inorganic material that is vermin and rot resistant.
- As a result of its closed cell structure, the material does not retain moisture.

4.5. CETRAFOAM

Cetrafoam is a molded rigid insulating foam specifically designed for LNG insulation applications. It is quick and easier to install than traditional PIR rigid foam insulation. Unlike traditional rigid foam insulation, Cetrafoam remains elastic at $-165°C$, which allows it to be applied in a single layer without contraction joints. Its compressive strength and the compressive modulus of the foam also allow it to absorb the linear thermal contraction of the pipe with a safety factor in excess of 10.

Due to the simplicity of Cetrafoam's installation process, labor requirements are reduced by up to 60%. This means fewer installers are required on site at the peak of the process, allowing for installation to be completed months earlier than when using traditional PIR rigid foam insulation. There is also less need for highly skilled workers, which is a critical factor in today's tight labor market.

Waste and waste disposal is another key issue for LNG companies and others operating within the energy sector. There are environmental issues to consider, as well as the cost of the waste material and the cost of its disposal. Currently, cryogenic rigid low density PIR foam insulation is cut from rectangular blocks, which have an average material recovery rate of 50%. The remaining material goes to waste as landfill.

For a standard 4MMTPA LNG liquefaction plant, the volume of waste generated by cutting cylindrical pipe sections from rectangular blocks is approximately 5000 cubic meters. The Cetrafoam molded range of pipe insulation products are molded to the required size and shape in precision molds. Zero cutting waste is generated, saving up to 5000 cubic meters of imported high-grade polyurethane foam pipe insulation from going to landfill as waste. Cetrafoam's unique structural makeup also generates minimal on-site dust, allowing multitasking in common work areas.

4.6. OTHER INSULATION SYSTEMS

Other types of cryogenic insulation systems include those where the evacuated annular spaces (spaces between an inner and outer vessel) contain bulk filled materials, e.g., glass fiber, silica aerogel, or composites. As with MLI, these systems require vacuum levels of around 1×10^{-4} torr to be effective.

4.7. CRYOGENIC INSULATION SYSTEM PERFORMANCE

Cryogenic insulation system performance is often reported for large temperature differences in terms of an apparent thermal conductivity, or "k-value." Boundary temperatures of 77 K (liquid nitrogen) and 295 K (room temperature) are common.

The following k-values apply generally to these boundary conditions.
- MLI systems can produce k-values of below 0.1 mW/(mK) (R-value of approximately 1440) when properly operating at cold vacuum pressure below about 1×10^{-4} torr.
- For bulk filled insulation systems operating at a cold vacuum pressure below about 1×10^{-3} torr, k-values of about 2 mW/(mK) (R-value of approximately 72) is typical.
- Foam and other similar materials at ambient pressures typically produce k-values of about 30 (R-value of approximately 4.8).

All of the values given are for 1-inch insulation thickness. The R-value is an industry standard unit of thermal resistance for comparing insulation values of different materials. The R-value is a measure of resistance to heat flow in units of degree F-hour-square foot/BTU-in.

The relationship between k- and R-values is described as follows:

$$k = d/R$$

where d is the heat flow distance.

4.7.1. Comparative Performance of Insulation

At a vacuum pressure of 0.02 torr many of the basic insulation designs give about the same thermal performance. Tests run to develop a cheap replacement for perlite that does not pack down and break the inner vessel

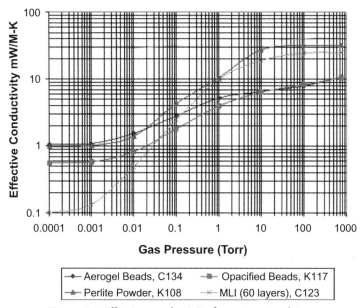

Figure 4.1 Effective conductivity for various insulators.

supports has shown that a new silica aerogel seems to work quite well. Test results compare various insulating materials as a function of vacuum pressure. This is shown in the plot of effective conductivity in Figure 4.1. All tests were performed with 1 inch of insulating material (60 layers of MLI) in a test dewar.

4.7.2. Insulation Performance: Heat Transfer

MLI insulation obeys the radiation equation for a grey body where the heat leak depends on the emittance divided by the number of surfaces times temperature to the fourth power. A comparison of several designs shows a wide range of performance values (see Figures 4.2 and Figures 4.3).

The actual use of MLI can be evaluated by looking at the conductivity per unit thickness as a function of vacuum pressure. MLI will not work at pressures above $1E^{-3}$ torr, whereas perlite and aerogel will work quite well up to a pressures of $2E^{-2}$ torr.

A summary of this information is shown in Table 4.1.

The table shows that MLI is only useable if the vacuum pressure is below 1 milliTorr (1 micron) whereas Perlite and Aerogel work at pressures up to 10 milliTorr.

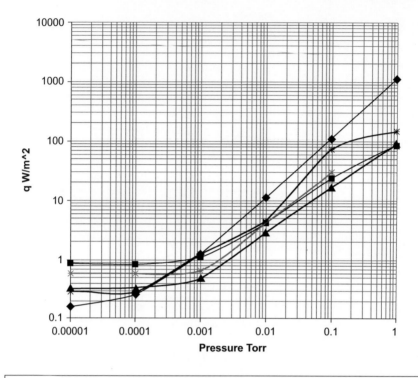

Figure 4.2 Heat leak for a wide range of performance values.

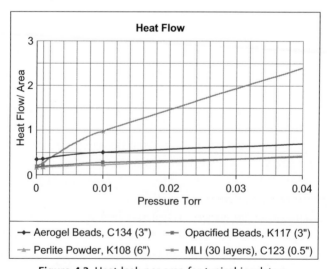

Figure 4.3 Heat leak per area for typical insulators.

Table 4.1 Summary of Information for Typical Insulation Materials

Material	Conductivity mW/(mK)	Vacuum MilliTorr	Relative Heat Leak
Perlite (6")	1	1	1 (Design)
	1.35	10	1.35
Aerogel + Carbon (3")	0.55	1	1.1
	0.85	10	1.7
MLI (30 layers = 1/2")	0.1	0.1	1.2
	0.13	1	1.56
	0.5	10	6

4.8. ARMAFLEX CRYOGENIC SYSTEMS

The storage and transportation of cryogenic media such as LNG, methanol, or ethylene demands high-performance insulation materials. With Armaflex Cryogenic Systems, the multilayered systems ensure exceptional thermal insulation, reduce the risk of corrosion under insulation (CUI), and allow considerable cost savings during the installation process.

4.8.1. Flexible at Cryogenic Temperatures

Armaflex Cryogenic Systems provide in a single material the key performance qualities required for insulation materials in low-temperature environments. These could previously only be achieved through a combination of several different materials (for example rigid foams with separate vapor retarders and contraction joints). The new cryogenic foams have low thermal conductivity and retain their flexibility even at low temperatures. Due to this flexibility, vibrations and impact are absorbed and the risk of cracking as a result of extreme temperature cycles or external mechanical strain is reduced dramatically. An important advantage of Armaflex Cryogenic Systems is the fact that they need neither additional expansion joints nor vapor barriers.

4.8.2. Built-in Expansion Joint

Because of the low-temperature flexibility of the material, the system itself acts as a contraction and expansion joint. Thus the pipe is protected against thermal stress and crack formation. Traditional systems made of rigid foam utilize glass wool or mineral fiber as a buffer against contraction and expansion. This not only increases the installation costs, but also, due to the open-cell material structure of the expansion joints, the risk of moisture infiltrating the system and compromising the insulation effect rises.

4.8.3. Built-in Vapor Barrier

Moreover, Armaflex Cryogenic Systems do not need additional vapor barriers. Armaflex is a closed-cell material with a high resistance to water vapor transmission and as such has a "built-in" vapor barrier. Unlike traditional systems where each insulation layer must be protected against moisture infiltration by a separate vapor barrier, in the case of Armaflex the resistance to water vapor transmission is built up throughout the entire thickness of the insulation. This not only increases the reliability of the insulation system, but also reduces the installation costs significantly as there is no need for the time-consuming process of installing vapor barriers with sealants and special adhesive tapes.

4.8.4. Field of Application

Armaflex Cryogenic Systems can be used in the temperature range of −200°C to +125°C. The inner system layer is a specially developed dieneterpolymer, which prevents thermal stress. With its closed-microcell structure, low thermal conductivity, and high resistance to water vapor transmission, this material provides installations with long-term protection against moisture infiltration and energy losses. The metal cladding ensures high resistance to mechanical impact.

At application temperatures below −110°C, the system is completed by the Armaflex anti-abrasive foil, which provides the inner insulation layer with added surface strength. If noise control is also required on the refrigeration pipework, the combined use of Armacell's acoustic insulation materials, ArmaSound Industrial Systems, is recommended. The systems are coordinated with each other and meet the thermal–acoustic requirements with significantly reduced insulation thickness and weight compared to traditional insulation systems.

Due to their high flexibility, Armaflex Cryogenic Systems are much easier and quicker to install than rigid materials and a good fit can be achieved even on awkwardly shaped installations.

Armaflex Cryogenic Systems are specially designed for use on pipelines, tanks, and production equipment in the oil, gas, and chemical industries. They are also particularly suitable for insulating LNG installations.

4.9. AEROGELS

Aerogels have been in existence for more than 80 years. They consist of lightweight silica solids derived from a gel in which the liquid component has been replaced with gas. The silica solids, which are poor conductors, consist of very small, three-dimensional, intertwined clusters that comprise only 3% of the volume. Conduction through the solid is therefore very low. The remaining 97% of the volume is composed of air in extremely small nanopores. The air has little room to move, inhibiting both convection and gas-phase conduction.

These characteristics make aerogel the world's lowest-density solid and most effective thermal insulator. The outstanding thermal properties of aerogels have been studied for decades, but Aspen Aerogels® has developed a technically and economically viable form of aerogel for industrial insulation uses. Aspen's unique process integrates aerogel into a carrier to create flexible, resilient, durable aerogel blankets with superior insulating properties.

4.9.1. Environmentally Friendly

Strict environmental regulations and increased awareness have led to the requirement for environmentally friendly insulation materials for use in industry. Aerogels pose no chemical threat to the environment. They are silica based, which is essentially sand. Two sample aerogels (Cryogel® Z and Pyrogel® XT) contain no respirable fibers and do not require blowing agents, so they are free of CFCs and HCFCs. These products can be safely disposed and, since the installed volume is considerably less than competing materials, there is less waste going to landfills.

4.9.2. Fire Resistant

Aerogels offer excellent resistance to flame spread and smoke emission. In actual hydrocarbon fires, they protect piping and equipment longer, which provides additional time to respond to a catastrophic event.

4.9.3. Light Weight

Aerogels are lighter than other insulation materials. This enables them to be easily and safely handled on the job site. They can be installed in longer lengths than traditional insulations, which improves installation rates. Their light weight also reduces overall loading of the pipe and equipment support structure.

4.9.4. Durable

Aerogels are flexible materials that deform under compression. They have excellent bounce-back properties, even when exposed to compression forces of hundreds of psi, and they can resist high impact loads with no damage and no compromise in performance. This is unlike rigid insulation, which, although stiff, is friable and susceptible to cracking. This creates thermal short circuits and paths for moisture intrusion. Rigid insulations also are at risk of breakage during shipping and installation.

4.9.5. Hydrophobic

Aerogels are extremely hydrophobic and therefore have outstanding resistance to moisture. One type of aerogel has the lowest k-value of any cryogenic insulation material in the world, reducing thicknesses by 50%–75%. This aerogel's flexible blanket form, with a factory-applied, integral vapor barrier, is both faster to install and more durable once in service, resulting in lower-cost, higher-performing designs.

4.9.6. Applications

Applications of aerogels include subambient piping and equipment, cryogenic storage and transport, industrial gases, LNG import/export pipelines and process areas, and chilled water systems. The service temperature range is $-460°F$ ($-270°C$) to $195°F$ ($90°C$).

4.9.7. Thermal Performance

Aerogels have the lowest thermal conductivity of any material used for cryogenic service. They are therefore much thinner compared to other cold insulation materials. The minimal thickness of aerogels results in a smaller surface area and reduced heat gain compared to other insulation materials. This heat gain "safety factor" improves process control, reduces boiloff, and saves energy. Also, aerogels do not have blowing agents that diffuse out over time, so their thermal performance remains constant.

Permeability to water and water vapor are critical to any insulation system operating at cryogenic temperatures. Aerogels use a factory-applied Mylar vapor barrier to achieve a zero-perm system.

4.9.8. Structural Integrity

Aerogels are well suited for subambient and cryogenic applications. Under these severe conditions, they remain totally flexible and resist thermal shock.

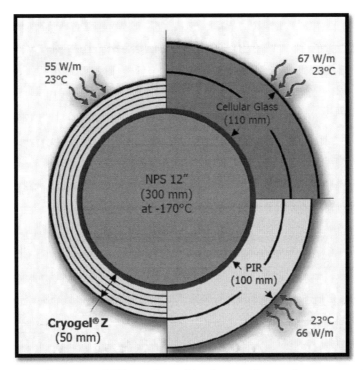

Figure 4.4 Performance of different thermal insulators with regard to energy savings at cryogenic fluid temperature. *(Source: Aspen Aerogels, Inc).*

This is unlike rigid, cellular insulation materials, which experience contraction, thermal shock, damaged structure, freeze-thaw crack propagation, and degraded insulation performance under the same conditions.

4.9.9. Dimensional Stability

Cryogel® Z (Aerogel) insulation has a coefficient of thermal expansion similar to that of stainless steel, so there is minimal differential movement of the insulation system. Its low contraction rate and flexible wrap application eliminate the need for costly and labor-intensive expansion/contraction joints required by traditional rigid insulation systems. Figure 4.4 shows the performance of different thermal insulators including Cryogel® Z with regard to energy savings at cryogenic fluid temperature.

4.9.10. Aerogel Insulation System Advantages

Thinness creates more space in and around pipe racks and equipment. Thinness can decrease the overall size of a production facility, resulting in

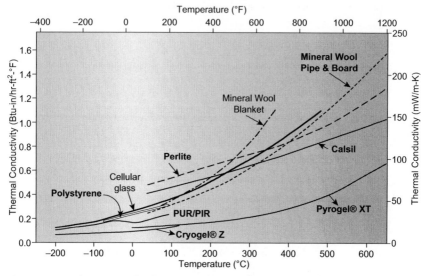

Figure 4.5 Performance of different thermal insulators. *(Source: Aspen Aerogels, Inc).*

major material reductions and cost savings. Thinness results in volume and freight savings, decreased accessory materials, minimal site storage, and simplified logistics.

A unique flexible form and wrap application makes installation faster, easier, and less costly. Rigid insulation systems require numerous segments that must be effectively sealed.

Aerogel insulation systems will not break during shipment. They are competitive with other insulation systems on an installed basis due to decreased material requirements, logistics improvements, reduced installation time, and shorter construction schedules. Figure 4.5 illustrates the performance of different thermal insulators.

4.10. CRYOGENIC VACUUM PIPE

As stated at the beginning of the chapter, since LNG is normally carried by ship at $-163°C$, the functional requirements of LNG shipping are cryogenic reliability due to thermal cyclic stresses high thermal insulation performance for safe and efficient transportation of LNG. Recently, the LNG fuel system has been employed for efficient transportation. The LNG fuel system is operated in cryogenic condition and various aspects should be taken into account in accordance with the cryogenic characteristics. It can be

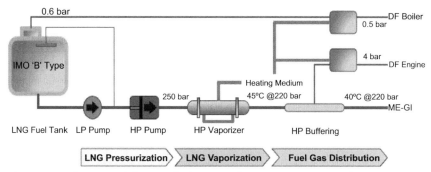

Figure 4.6 Schematic of the LNG fuel system. *(Source: Kwang-Jun Park, Myung-Ji Sim, Jeong-Don Kim, Yoo-Hong Kwon, and Jae-Myung Lee, 9th International Conference on Fracture & Strength of Solids, June 9–13, 2013, Jeju, Korea).*

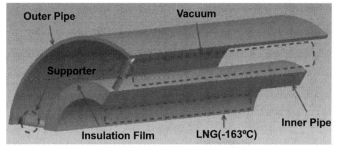

Figure 4.7 Schematic of the cryogenic vacuum pipe with multilayer insulation. *(Source: Kwang-Jun Park, Myung-Ji Sim, Jeong-Don Kim, Yoo-Hong Kwon, and Jae-Myung Lee, 9th International Conference on Fracture & Strength of Solids, June 9–13, 2013, Jeju, Korea).*

classified into three categories, which are storage, vaporization, and supply technology. Figure 4.6 shows the schematic of the LNG fuel system.

In general, cryogenic vacuum pipes in LNG fuel systems consist of inner pipe, outer pipe, insulation film, and supporters such as in Figure 4.7. In order to increase the insulation property as well as thermal–structural safety, cryogenic vacuum pipes with multilayer insulation have been developed recently.

CHAPTER FIVE

Thermal Insulation for Offshore Installations in Deep Water

Contents

Ultra-deep water (up to 3000 m) is one of the next frontiers for offshore oil exploitation. It requires the use of conduits that must resist the severe mechanical and environmental requirements for a long run (about 25 years). One of the key factors is the thermal insulation of the structure to avoid the formation of hydrates and paraffin plugs inside of the steel pipe.

The conservation of fluid flow and the ability to restart production is a prime concern for the operation of a deep-water hydrocarbon production system. The last 5–10 years have seen thermal management of the well

Alireza Bahadori, Thermal Insulation Handbook for the Oil, Gas, and Petrochemical Industries
© 2014 Elsevier Inc.
http://dx.doi.org/10.1016/B978-0-12-800010-6.00005-8

stream emerge as the single most used engineering tool for preventing blockages from obstructing the production lines.

It is generally recognized that keeping the well fluids from cooling down will prevent paraffin from depositing on the pipe walls during fluid flow, as well as forestall the formation of gas hydrates, which are stable forms of ice occurring when water and small gas molecules (methane, ethane, hydrogen disulfide, etc.) combine at temperatures around 20°C and pressures of 10–20 MPa and above.

This overall flow assurance strategy is generally associated with a more specialized treatment of localized elements, such as spools, jumpers, and manifolds, which are more difficult to insulate efficiently, by injecting chemicals that will prevent the water from combining with the gas molecules.

Apart from these very local considerations, it is of interest to treat separately the risers and the flowlines, both because of the difference in flow characteristics, and because the construction issues differ significantly.

Thermal insulation is a critical element in the design and operation of flowlines in deep waters due to a combination of low temperatures and high pressure; this combination creates stringent requirements for optimal insulation.

The thermal performance of a subsea production system is controlled by the hydraulic behavior of fluid in the flowline; conversely, it also impacts the hydraulic design indirectly through the influence of temperature on fluid properties such as gas–oil ratio (GOR), density, and viscosity.

Thermal design, which predicts the temperature profile along the flowline, is one of the most important parts in the flowline design; the information based on the thermal design is required for numerous pipeline analyses including expansion analysis, upheaval or lateral buckling analysis, corrosion protection, hydrate prediction, and wax deposition analysis. In most cases, the solids management (hydrates, wax, asphaltenes, and scales) determines the requirements of hydraulic and thermal designs. In order to maintain a minimum temperature of fluid to prevent hydrate and wax deposition in the flowline, insulation layers may be added to the flowline.

Thermal design includes both steady-state and transient heat transfer analyses. In steady-state operation, the production fluid temperature decreases as it flows along the flowline due to the heat transfer through the pipe wall to the surrounding environment. The temperature profile in the whole pipeline system should be higher than the requirements for the prevention of hydrate and wax formation during normal operation and is determined from steady-state flow and heat transfer calculations.

If the steady flow conditions are interrupted due to a shut-in or restarted again during operation, the transient heat transfer analysis for the system is required to make sure the temperature of fluid is out of the solid formation range within the required time. It is necessary to consider both steady–state and transient analyses in order to ensure that the performance of the insulation coatings will be adequate in all operational scenarios.

The most severe operational hazards of offshore pipelines are the risks associated with the transportation of multiphase fluids. When water, oil, and gas are flowing simultaneously inside the pipeline, there are quite a few potential problems that can occur: water and hydrocarbon fluids can form hydrates and block the pipeline; wax and asphaltene can deposit on the wall and may eventually block the pipeline; with high enough water cut, corrosion may occur; with pressure and temperature changes along the pipeline and/or with incompatible water mixing, scales may form and deposit inside the pipeline and restrict the flow; and severe slugging may form inside the pipeline and cause operational problems to downstream processing facilities.

The challenge that engineers will face is, thus, how to design the pipeline and subsea system to assure that multiphase fluids will be safely and economically transported from the bottom of the wells all the way to the downstream processing plant. The practice of identifying, quantifying, and mitigating all of the flow risks associated with offshore pipelines and subsea systems is called flow assurance. Flow assurance is critical for deep-water pipeline and system operations. In deep water, the seawater temperature is usually much colder than the surface air temperature. When pipeline is submersed in deep water, if there is no thermal insulation layer surrounding the pipe wall, the fluid heat can be quickly lost to the water.

This is especially true if the water current around the pipeline is strong. With an un-insulated pipeline, the heat transfer coefficient at the outer pipe wall can be significant due to the forced convection by the seawater movement. If the fluid temperature inside the pipeline becomes too low due to the heat loss, water and hydrocarbon (oil and gas) may form hydrates and block the flow. Furthermore, if the fluid temperature is low enough, wax may start to precipitate and deposit on the pipe wall. Thus, effective preservation of fluid heat is one of the most important design parameters for offshore pipelines.

Coated insulation systems have been used on existing pipelines in water depths of up to 1000 m. For wells at greater depths, new insulation products are required that are capable of withstanding the high water pressure together with the thermal gradient due to the difference in temperature between the production fluid and the external environment. In addition, the long-term behavior of the system becomes increasingly important as creep rates increase significantly at higher stress levels. This can result in levels of densification of the insulation that reduce the thermal performance of the system.

In deep water, the pipeline is normally followed by a production riser, which goes from the sea bottom to the surface processing facilities (topside). The deeper the water is, the longer the production riser is. With a long riser, the pipeline operating pressure will be higher due to the hydrostatic head in the riser.

For the same fluid temperature, with higher operating pressure, it is easier for the fluids to form hydrates. With a pipeline and riser production system, if the flow conditions are such that severe slugging occurs, the slugs will be proportional to the riser length (the longer the riser, the longer the severe slugs). Effective management of the system thermal properties is crucial to the success of a deep-water field development.

To ensure fit-for-purpose design, all available technologies are considered and, in general, for less stringent requirements, wet insulation on rigid pipeline, or insulated flexible flowlines can be used. However, for more stringent specifications a dry environment will be necessary to provide the required insulation performance.

As new developments are moving progressively into deeper water, where the ambient temperature at the seabed becomes even lower, successful operation becomes more heavily dependent on the thermal management strategy employed.

Thermal management strategy and insulation generally include the following: overall heat transfer coefficient, steady state heat transfer, and transient heat transfer.

Oil field flowlines are insulated mainly to conserve heat. The need to keep the product in the flowline at temperatures higher than the ambient temperature could exist for several reasons including the following:
- Preventing formation of gas hydrates;
- Preventing formation of gas hydrates, wax, or asphaltenes;
- Enhancing product flow properties;
- Increasing cooldown time after shutting down.

In liquefied-gas pipelines, such as those for liquefied natural gas, insulation is required to maintain the cold temperature of the gas to keep it in liquid state.

Polypropylene, polyethylene, and polyurethane are three base materials widely used in the petroleum industry for pipeline insulation. Depending on the application, these base materials are used in different forms, resulting in different overall conductivities.

A three-layer polypropylene applied to pipe surface has conductivity of 0.225 W/(m°C), while a four-layer polypropylene has conductivity of 0.173 W/(m°C). Solid polypropylene has higher conductivity than polypropylene foam. Polymer syntactic polyurethane has conductivity of 0.121 W/(m°C) while glass syntactic polyurethane has conductivity of 0.156 W/(m°C).

These materials have lower conductivities in dry conditions. Because of their low thermal conductivities, more and more polyurethane foam is used in deep-water flowline applications. The physical properties of polyurethane foam include density, compressive strength, thermal conductivity, closed-cell content, leachable halides, flammability, tensile strength, tensile modulus, and water absorption.

The requirements for flowline insulation vary from field to field. Flow assurance analysis needs to be performed to determine the minimum requirements for a given field.

These analyses include the following:
- Flash analysis of the production fluid to determine the hydrate formation temperature in the range of operating pressure;
- Global thermal hydraulic analysis to determine the required overall heat transfer coefficient at each location in the flowline;
- Local heat transfer analysis to determine the type and thickness of insulation to be used at the location;
- Local transient heat transfer analysis at special locations along the flowline to develop cooldown curves and time to the critical minimum allowable temperature at each location.

5.1. STEADY AND TRANSIENT SOLUTIONS FOR FLOWLINE TEMPERATURE

The following assumptions are made in model formulation:
1. Friction-induced heat is negligible.
2. Heat transfer in the radial direction is fully controlled by the insulation fluid.
3. The specific heat of fluid is constant.

5.1.1. Governing Equations

Consider the heat flow during a time period of Δ. The heat balance is given by

$$q_{in} - q_{out} - q_R = q_{acc} \tag{5.1}$$

where

q_{in} = heat energy brought into the pipe element by fluid due to convection

q_{out} = heat energy carried away from the pipe element by fluid due to convection

q_R = heat energy transferred through the insulation layer due to conduction

q_{acc} = heat energy accumulation in the pipe element

These terms can be further formulated as

$$q_{in} = \rho_f C_p v A_f T_{f,L} \Delta t_f \tag{5.2}$$

$$q_{out} = \rho_f C_p v A_f T_{f,L+\Delta L} \Delta t_f \tag{5.3}$$

$$q_R = 2\pi R_n k_n \Delta L \frac{\partial T_f}{\partial r} \Delta t_f \tag{5.4}$$

$$q_{acc} = \rho_f C_p A_f \Delta L \Delta \overline{T}_f \tag{5.5}$$

where

ρ_f = fluid density, kg/m^3

C_p = specific heat at constant pressure, J/kg°C

v = the average flow velocity of fluid in the pipe, m/s

A_f = cross-sectional area of pipe open for fluid flow, m^2

$T_{f,L}$ = temperature of the flowing–in fluid, °C

Δt_f = flow time, s

$T_{f,L+\Delta L}$ = temperature of the flowing–out fluid, °C

S = insulation layer thickness

R_n = inner radius of insulation layer, m

k_n = thermal conductivity of the insulation layer, W/m°C

ΔL = length of the pipe segment, m

$\frac{\partial T_f}{\partial r}$ = radial–temperature gradient in the insulation layer, °C/m

$\Delta \overline{T}_f$ the average temperature increase of fluid in the pipe segment, °C

Substituting Eqs. (5.2) through (5.5) into Equation (5.1), we obtain

$$\rho_f C_p v A_f T_{f,L} \Delta t_f - \rho_f C_p v A_f T_{f,L+\Delta L} \Delta t_f - 2\pi R_n k_n \Delta L \frac{\partial T_f}{\partial r} \Delta t_f$$
$$= \rho_f C_p A_f \Delta L \Delta \overline{T}_f \qquad (5.6)$$

$$\rho_f C_p v A_f \Delta t_f \left(T_{f,L} - T_{f,L+\Delta L} \right) - 2\pi R_n k_n \Delta L \frac{\partial T_f}{\partial r} \Delta t_f = \rho_f C_p A_f \Delta L \Delta \overline{T}_f$$
$$(5.7)$$

Dividing all the terms of this equation by $\Delta L \Delta t_f$ yields

$$\rho_f C_p v A_f \frac{\left(T_{f,L} - T_{f,L+\Delta L} \right)}{\Delta L} - 2\pi R_n k_n \frac{\partial T_f}{\partial r} = \rho_f C_p A_f \frac{\Delta \overline{T}_f}{\Delta t_f} \qquad (5.8)$$

For infinitesimal values of ΔL and Δt_f this equation becomes

$$v \frac{\partial T_f}{\partial L} + \frac{\partial \overline{T}_f}{\partial t_f} = -\frac{2\pi R_n k_n}{\rho_f C_p A_f} \frac{\partial T_f}{\partial r} \qquad (5.9)$$

The radial-temperature gradient in the insulation layer can be formulated as

$$\frac{\partial T_f}{\partial r} = \frac{T_{f,L} - \left(T_{f,0} - G\cos(\theta)L \right)}{s} \qquad (5.10)$$

where

$T_{f,0} =$ temperature of the medium outside the insulation layer at $L = 0$, °C
$G =$ geothermal gradient, °C/m
$\theta =$ inclination time, degree
$s =$ thickness of the insulation layer, m

Substituting Eq. (5.10) into Eq. (5.9) yields

$$v \frac{\partial T_f}{\partial L} + \frac{\partial \overline{T}_f}{\partial t_f} = a T_f + bL + c \qquad (5.11)$$

where

$$a = -\frac{2\pi R_n k_n}{\rho_f C_p s A_f}$$

$$b = aG\cos(\theta)$$

$$c = -a T_{f,0}$$

The resultant equations are summarized in this section. The internal temperature profile under steady fluid flow conditions is expressed as

$$T_f = \frac{1}{\alpha^2}\left[\beta - \alpha\beta L - \alpha\gamma - e^{-\alpha(L+C)}\right] \tag{5.12}$$

where the constant groups are defined as

$$\alpha = \frac{2\pi Rk}{v\rho C_p sA}$$

$$\beta = \alpha G\cos\theta$$

$$\gamma = -\alpha T_{f,0}$$

$$C = -\frac{1}{\alpha}\ln\left(\beta - \alpha^2 T_{f,s} - \alpha\gamma\right)$$

where T_f is the temperature inside the pipe, L is the longitudinal distance from the fluid entry point, R is the inner radius of the insulation layer, k is the thermal conductivity of the insulation material, v is the average flow velocity of fluid in the pipe, ρ is the fluid density, C_p is the heat capacity of fluid at constant pressure, s is the thickness of insulation layer, A is the inner cross-sectional area of pipe, G is the principal thermal gradient outside the insulation, θ is the angle between the principal thermal gradient and the pipe orientation, $T_{f,0}$ is the temperature of outer medium at the fluid entry location, and $T_{f,s}$ is the temperature of the fluid at fluid entry point.

The rate of heat transfer across the insulation layer over the whole length of the flowline is expressed as

$$q = \frac{2\pi Rk}{s}\left(T_{f,0}L - \frac{G\cos\theta}{2}L^2 - \frac{1}{\alpha^2}\left\{(\beta - \alpha\gamma)L - \frac{\alpha\beta}{2}L^2\right.\right.$$

$$\left.\left. + \frac{1}{\alpha}\left[e^{-\alpha(L+C)} - e^{-\alpha C}\right]\right\}\right) \tag{5.13}$$

where q is the rate of heat transfer (heat loss). The internal temperature profile after starting up a fluid flow is expressed as follows:

$$T_f = \frac{1}{\alpha^2}\left\{\beta - \alpha\beta L - \alpha\gamma - e^{-\alpha[L+f(L-vt)]}\right\} \tag{5.14}$$

where the f is given by

$$f(L - vt) = -(L - vt) - \frac{1}{\alpha} \ln\{\beta - \alpha\beta(L - vt) - \alpha\gamma$$
$$- \alpha^2 [T_{f,s} - G\cos\theta(L - vt)]\}$$

(5.15)

where

$$\alpha = \frac{2\pi R k}{v' \rho C_p s A}$$

$$\beta' = \alpha' G\cos\theta$$

$$\gamma' = \alpha' T_{f,0}$$

and t is time. Suppose after increasing or decreasing the flow rate, the fluid has a new velocity v' in the flowline. The internal temperature profile is expressed as follows:

$$T_f \frac{1}{\alpha'^2}\left\{\beta' - \alpha'\beta'L - \alpha'\gamma' - e^{-\alpha'[L + f(L - v't)]}\right\}$$

(5.16)

Here, the function f is given by

$$f(L - v't) = -(L - v't) - \frac{1}{\alpha'} \ln\left\{\beta' - \alpha'\beta'^{(L - v't)} - \alpha'\gamma'\right.$$
$$\left. - \left(\frac{\alpha'}{\alpha}\right)^2 \left[\beta - \alpha\beta(L - v't) - \alpha\gamma - e^{-\alpha\{(L - v't) + C\}}\right]\right\}$$

(5.17)

For single-pipe flowlines and risers, the mechanical loads as well as the thermal insulation requirements normally increase with deeper waters. Hence, the traditional thermal insulation foam technology used in shallow waters and the associated design and test methodology may not be applicable to deep-water projects.

Polymer foams change mechanical and thermal properties as a function of foam density. Higher density normally means better mechanical properties and reduced density improves insulation capacity. For deep-water thermal designs, this could lead to build up of excessively thick coatings that may cause manufacturing concerns as well as reduce installation vessel capacity. In addition, excessive coating thickness may reduce seabed stability for the flowline and increase drag on a steel catenary riser (SCR).

5.2. MATERIALS DEVELOPMENT

The foaming process for polymers generally leads to a trade–off between mechanical properties and thermal insulation properties. The increased hydrostatic head associated with deeper waters calls for higher compressive strength for polypropylene foam. Higher compressive strength also improves creep characteristics and can generally be attributed to higher polymer stiffness and the final foam structure.

5.2.1. Polypropylene Block Copolymers

The new generation polypropylene (PP) materials are based on a unique balance between stiffness, toughness, and good long-term creep resistance. This heterophasic PP material is a highly crystalline material with a finely distributed and dispersed ethylene-polypropylene rubber phase. Good mechanical properties are shown over a wide temperature range, in addition to high abrasion and good chemical resistance.

5.2.2. High Melt Strength Polypropylene (HMS-PP)

High strength combined with improved melt elongation are the main characteristics for HMS-PP. A long–chain branched polymer is introduced into the PP, thus improving foaming conditions. Because of the polymer modifications, controlled bubble growth can be observed, leading to stable foam with a uniform closed cell foam structure. Because of the loads introduced by extrusion, improving melt strength and melt elongation considerably improves foam makeup.

By combining a stiff linear copolymer polypropylene with HMS-PP the benefits of high melt strength and high melt elongation result in excellent foam quality, characterized by evenly distributed bubbles with a closed cell bubble structure in the pipe foam layer. This leads to both higher compression strength and improved creep resistance.

5.2.3. Foam Structure and Scanning Electron Microscope (SEM) Characterization

A foam based on the material combination of stiff PP and HMS-PP contains bubbles that are smaller and more evenly distributed in the foam layer. High melt elongation affects both foam stability and foam structure, manifested as even bubble size and distribution throughout the pipe insulation layer.

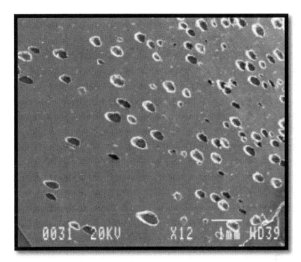

Figure 5.1 SEM photograph: Reference PP foam. *(Source: A. Boye Hansen and C. Rydin, SPE 2002 Offshore Technology Conference, Houston, Texas, USA, May 6–9, 2002, OTC 14121).*

Figure 5.2 SEM photograph: Novel PP foam. *(Source: A. Boye Hansen and C. Rydin, SPE 2002 Offshore Technology Conference, Houston, Texas, USA, May 6–9, 2002, OTC 14121).*

The difference in foam structure is illustrated in Figures 5.1 and 5.2, where foam from a reference PP copolymer is compared to the novel PP using scanning electron microscope (SEM) photography of a foamed cross section. High melt strength allows improved stability and foaming closer to the outer surface of the pipe. This is seen in the SEM photograph, where the lower left-hand corner corresponds to the outer surface of the pipe.

The foam structure is directly correlated to the mechanical strength of the foam and is therefore a valuable tool when evaluating the mechanical properties.

5.2.4. Test and Design Methodology

The most critical parameter when designing with the viscoelastic behavior of plastics is associated with creep. Creep in foamed structures relates both to water depth (hydrostatic pressure) and the associated temperature gradient. The temperature gradient is dependent on the layer thickness, thermal conductivity of each layer, and internal and external fluid temperatures.

Creep is the most important long-term design consideration, as creep will result in changes of the insulation properties over time. Creep in foamed structures will produce an increase in density, which in turn will increase the thermal conductivity. Hence, it is important to understand the creep mechanisms in the foam, use these mechanisms in the design stages, and compensate for this creep in the design.

Secondly, it is important to characterize the actual compressive load that the foam will experience during in-service conditions. On the pipe, the foam benefits from support axially and tangentially. Therefore, the creep will show as radial displacement of the PP foam.

The current standard in the industry calls for determining the compressive strength of at 5% strain, e.g., ASTM D695M. Use of this standard encompasses testing of rods machined from the PP-foamed structure. These rods are then exposed to a compressive load unsupported. Hence, the rods will displace the load axially, radially, and tangentially. This will produce extremely conservative results and does not produce valid data regarding actual behavior on a pipe.

Also, test methods exist that call for hydrostatic compression of small-scale samples. In most cases, this introduces full hydrostatic loads in all three directions on the sample, but again does not provide actual means for introducing stresses and strains equivalent to actual applications.

By using a finite element program, the stresses in a submerged flowline can be studied. Figure 5.3 shows the principal stress direction and the calculated average stress values in the foam for a submerged pipe at a water depth of 2000 meters. Also, the figure compares the calculated stress levels for three typical small-scale test methods.

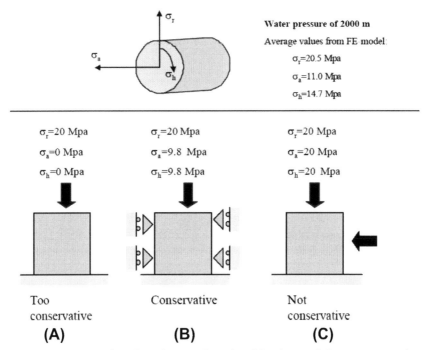

Figure 5.3 Stress analysis for submerged insulated flowline and three corresponding test methods (A through C). (*Source: A. Boye Hansen and C. Rydin, SPE 2002 Offshore Technology Conference, Houston, Texas, USA, May 6–9, 2002, OTC 14121).*

In the figure, the stress levels are shown for (A), which is equivalent to the uniaxial compressive strength test method as described in ASTM D695; (B), which represents the stress for the triaxial creep test; and (C), which shows the stress levels for a hydrostatic test case.

The triaxial creep test provides conservative results compared to a simulated service test as the test is carried out on cylindrical specimen, machined to tight tolerances to fit into the autoclaves. Differences in test results between the triaxial creep test and a full-scale simulated service test can be attributed to the coating hoop stiffness and the wedge effect (decrease in coating diameter towards steel pipe).

The triaxial cylinder creep test seems to be the small-scale test that provides the most accurate state of stress and at the same time a conservative compression estimate.

5.2.5. Test Method Description

The test method is triaxial creep test. A lubricating mixture of silicon grease and water entrapped along the specimen surface ensures a low friction between the cylinder wall and the PP foam test specimen. Friction is easily controlled and held at a minimum by rotating the cylinder wall relative to the test specimen.

5.2.6. Limit State Design

Commonly used design practice in the industry would allow the use of safety factors, as has been applied to steel pipe design in the past. As an example, the use of a uniaxial compressive strength test according to ASTM D695 and a safety factor of 2 would normally mean that PP foam would not be deployed deeper than approximately a water depth of 600 meters. However, this design approach does not reflect the viscoelastic behavior of PP foam, nor does it reflect on-pipe behavior, or show performance changes with time, pressure, and temperature.

As creep is the major criterion when designing with PP systems, the use of limit state design is most applicable when performing a design with the PP system. Although this approach requires a detailed understanding of the foam behavior as a function of foam density, hydrostatic pressure, temperature, and time, this design philosophy reflects the structural response to an actual load scenario and will not produce overly conservative designs.

5.2.7. Qualification Program and Test Results

In order to qualify the PP foam for water depths down to 2000 meters, a rigorous qualification program was developed and performed to meet the defined performance criteria. The main objectives for the qualification program have been:
- To qualify the insulation system for subsea use;
- To develop data for service life prediction (small scale and full scale);
- To use the generated data for design and engineering of thermal insulation systems for deep–water service;
- To establish operator acceptance and deploy the first commercial installation.

The overall philosophy has been to execute the program so those loads reflect the actual conditions on the insulation coating and to establish acceptance criteria. Performed tests reflect loads during:
- Manufacturing;
- Storage (stacking of pipes);
- Installation (e.g., reeling);
- Operation.

The framework for the qualification was existing international standards for polymers and foams as well as established procedures, requirements, and specifications defined by operators and the manufacturer. Because of the novelty in the product, the manufacturer also developed special test equipment and test procedures to perform the creep tests.

5.2.8. Aging Tests

All polymeric products experience degradation reactions in the presence of oxygen. The main factors influencing polymer degradation are concentration of oxygen and exposure to heat and light. To be able to use polymers in an application over years of service, it is very important to understand the degradation mechanism and its prevention. To prevent degradation, the polymer must be stabilized so it is effectively protected during processing, transport, storage, installation, and operation. An advanced stabilization package has therefore been developed to meet both short-term and long-term thermal stability requirements for offshore pipeline applications.

Long-term thermal stability of PP coating materials has been evaluated through a series of oven aging tests. Degradation of PP because of thermal stability is a key issue when elevated temperatures are combined with the presence of oxygen. Thermal insulation coatings on flowlines and risers have limited exposure to oxygen; therefore, degradation slows down in such an environment. To simulate the environment for these subsea flowlines, oven tests have been carried out in an inert atmosphere (circulated nitrogen).

5.2.9. Mechanical Tests

The increased stiffness of the new PP materials clearly reflects the mechanical properties of the foam. By increasing the stiffness (tensile modulus) by 40%, the compression strength increases from 13 to 19 MPa, measured by uniaxial compression of 5% according to ASTM D695.

This improvement in compression strength is valid, although the density for the novel PP foam is lower than the reference foam. This means that both compression strength and thermal insulation capacity is improved for the novel foam. With equivalent densities, the novel foam demonstrates an 80% improvement in compression strength as compared to the reference foam, as shown in Figure 5.4.

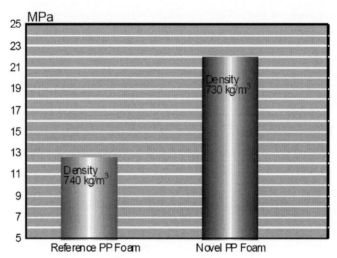

Figure 5.4 Compressive strength at 5% strain comparison. *(Source: A. Boye Hansen and C. Rydin, SPE 2002 Offshore Technology Conference, Houston, Texas, USA, May 6–9, 2002, OTC 14121).*

The improved creep resistance can be attributed to the increase in both the tensile and compression modulus. Long-term creep tests with the triaxial method are currently ongoing. An extrapolation of the results in Figure 5.5 indicates that the total compression and creep at 12 MPa (water depth of 1200 meters) at 20°C will be 7.7% after 20 years.

The homogeneity of the foam structure is of critical importance to reach an even insulation layer and thereby optimal mechanical properties of the pipe coating. The most important factors to reach a uniform foam structure are the material selection in combination with optimal process parameters during pipe coating manufacturing.

5.2.10. Thermal Conductivity

Thermal conductivity represents a key property for insulation foams. Lower k-values mean improved insulation and reduced layer thickness and costs. Although the base polymer for the novel PP insulation system has slightly higher thermal conductivity than the original PP (0.23 W/mK versus 0.22 W/mK), the improved mechanical properties of the novel foam also improve the insulation properties in deep water. Figure 5.6 is a plot of thermal conductivity (k-value) versus uniaxial compressive strength for the novel PP foam and the reference PP foam.

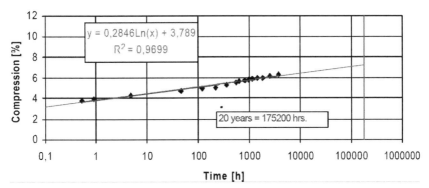

Figure 5.5 Creep curve, novel PP foam (at 12 MPa and 20°C). *(Source: A. Boye Hansen and C. Rydin, SPE 2002 Offshore Technology Conference, Houston, Texas, USA, May 6–9, 2002, OTC 14121).*

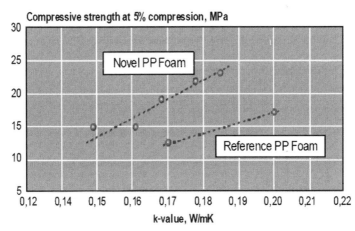

Figure 5.6 Thermal conductivity vs. compressive strength. *(Source: A. Boye Hansen and C. Rydin, SPE 2002 Offshore Technology Conference, Houston, Texas, USA, May 6–9, 2002, OTC 14121).*

5.2.11. Bending Tests

Bending tests were performed to determine suitability for two different conditions:

• Suitability for reeling (static bending test);
• Suitability for use on steel catenary risers (dynamic bending test).

Tests include field joints and results show that the PP insulation system is well suited for reeling and for use on SCRs.

5.2.12. Simulated Service Tests

This test comprises inserting a coated section of pipe in an autoclave and pressurizing the system to the working hydraulic pressure. The pipe then has hot oil passed through the bore and the external pressurized water is kept at a steady low temperature in line with seabed temperatures. Sensors are attached to the coating to monitor the heat flux and the compression of the coating. Through a number of these tests it has been established that the coating stabilizes between 2 to 3 days.

5.2.13. Design Example

Based on the results from small-scale testing and the simulated service tests, logarithmic creep curves can be plotted. These creep curves are important when designing a PP insulation system for various water depths and temperatures.

Below is a design example of a typical deep-water case study. The functional requirements for this case are a water depth of 2000 meters (20 MPa), a wellhead temperature of 60°C, and an overall heat transfer coefficient requirement of 5.3 W/m²K.

When using creep-resistant foam for deep-water fields it is important to reduce the exposure temperature on the foam. A load case of 20 MPa is close

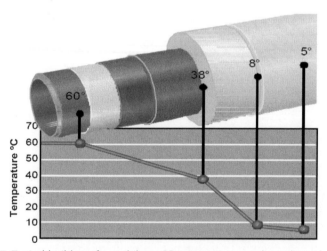

Figure 5.7 Typical buildup of a multilayer PP temperature gradient through the layer thickness. (Source: A. Boye Hansen and C. Rydin, SPE 2002 Offshore Technology Conference, Houston, Texas, USA, May 6–9, 2002, OTC 14121).

to the yield strength of the material and it is necessary to derive a design that would reduce the creep rate as a result of temperature.

These effects are known for all plastics that show viscoelastic behavior. From the creep tests and the full-scale simulated service test, it is important to reduce the temperature to up to 40°C. Figure 5.7 shows a typical buildup of a multilayer PP temperature gradient through the layer thickness.

5.2.14. Manufacturing and Installation

Manufacturing of the novel PP insulation systems is executed by cross-head extrusion. This method provides the opportunity to build multilayer structures where foam properties can be tailor-made by varying process parameters. Such a method can also be made mobile and transported to various sites to improve logistics and reduce costs.

5.3. FLOW ASSURANCE: THERMAL REACH

There are generally three temperatures of importance when designing a hydrocarbon production system:
- Wellhead inlet temperature;
- Process inlet temperature;
- Minimum preservation temperature for shutdown.

The sea-bottom temperature is also of importance as it defines the temperature gradient across the insulating materials, but this is usually a constant 4°C for all deep-water projects, except for some very special cases. The two first temperatures are related to the system in production and define a "temperature budget," i.e., how many degrees (of temperature) can be lost during the transportation of the fluid from the wellhead to the process facility (e.g., a floating production, storage, and offloading (FPSO) unit).

This will give a first value for the minimum insulation to be applied to the flowlines. The latter temperature is related to the appearance of hydrates: the operator will choose to work with fluids at temperatures above the hydrate appearance temperature and allow himself a margin to take into account contingent cooling in case of a production stop.

Operationally, this margin is defined as a number of hours ("cooldown time") that the operator requires before having to start up a preservation scheme to replace the hydrate-prone fluids inside the flowlines with inert ("dead") oil.

This duration results from a risk–based analysis and is usually in the 12 to 24 hour range, depending on flowline volume and topside pumping capacity, but it may go as high as 52 hours (author's experience).

This cooldown time defines another insulation value that may very well be the driving criterion for insulation, especially if the cooldown is to be calculated with a gas-filled flowline, which will not pack nearly as much energy as a liquid-filled pipeline at typical pressures of 100 bars and less.

A last point that also should be considered is the internal cooling of the fluid. This is related to the expansion of the associated gas as it travels down the flowline (Joule-Thomson effect) and the vaporization of hydrocarbon liquids as the pressure decreases in the flowline. These phenomena are very often lumped together under the term "Joule-Thompson cooling." They obviously are most present in the parts of the flowline experiencing the largest pressure drops, i.e., the risers. Thermal insulation has no influence on this effect.

In short, the thermal design of a pipeline has to consider:
1. Thermal losses to the environment in flowing conditions;
2. Preservation time in case of shutdown;
3. Thermal losses due to internal processes.

The different elements the thermal budget can be evaluated as follows:
1. Thermal losses in flowing conditions along the pipeline:

$$T_{out} = T_{sea} + (T_{in} - T_{sea}) \exp^{\frac{U\pi D}{Q_m C_{pi}} x} \qquad (5.18)$$

2. Upon a shutdown, the pipe will cool with time according to the following equation:

$$T_{final} = T_{sea} + (T_{initial} - T_{sea}) \exp^{\frac{U\pi D}{H} t} \qquad (5.19)$$

3. In the riser section, internal processes will dominate the temperature evolution: typically 6 to 10°C can be lost due to pressure decrease. For a cylindrically homogeneous construction, the overall heat transfer coefficient value, (U-value), is evaluated according to the following equation:

$$U = \frac{\lambda}{R_{ref} \ln \frac{R_{ins} + th_{ins}}{R_{ins}}} \qquad (5.20)$$

For pipes where different materials are layered on top of each other (typically wet-insulated pipes), the cumulative U-value is obtained by adding the individual U-values of each layer according to the following equation:

$$U^{-1} = \sum_i \frac{1}{U_i} \qquad (5.21)$$

where

T_{in} = inlet temperature of well stream typically wellhead temperature, °C

T_{out} = outlet temperature from pipe, for example at foot of riser, °C

T_{sea} = sea temperature, 4°C

$T_{initial}$ = temperature at start of cooldown, °C

T_{final} = temperature at the end of cooldown

U = overall heat transfer coefficient for the pipe, W/(m²K)

D = pipe diameter, m

Q_m = mass flow rate, kg/s

C_p = thermal capacity of well stream, J/(kgK)

x = distance along pipeline, m

H = combined thermal capacity of 1 m section of pipeline and well stream, J/(mK)

t = time from onset of cooldown, s

R_{ref} = reference radius for calculation of U-value (usually flowline inner or outer surface), m

R_{ins} = insulation inner radius, m

th_{ins} = thickness of insulation, m

λ = thermal conductivity of insulation, W/(mK)

The equations assume homogeneous material properties for the produced fluids and a homogeneous insulation for pipes, which is generally sufficient for setting up a first model.

Equation (5.18) is usually specified for the system operating at reduced capacity, i.e., at a "degraded" flow rate, such as 50% of the nominal flow rate. Heat losses per kilometer of 0.5°C or less can then be obtained with the use of high-performance insulation materials that provide U-values of less than 1 W/(m²K).

The corollary is that if a wellhead temperature of 60°C is expected, and 20°C of reserve heat are required to face a nominal shutdown period of 24 hours, then a thermal reach of the system of up to 20 to 30 km can be used with U-values down to 0.5 W/(m²K).

5.4. INSULATION MATERIAL SELECTION

The design should be a particularly efficient implementation of a high-performance insulation package specifically developed for pipeline insulation. Some insulation provides a thermal conductivity of less than 7 mW/(mK) against 14 mW/(mK) for the best competing materials such as nanoporous products. This translates into a design requiring smaller outer pipes because less insulation material is needed. Table 5.1 shows the thermal conductivity for different insulation materials.

For single-pipe flowlines and risers, the mechanical loads as well as the thermal insulation requirements normally increase with deeper waters. Hence, the traditional thermal insulation foam technology used in shallow waters and the associated design and test methodology may not be applicable to deep-water projects.

The mechanical and thermal properties of polymer foams vary as a function of foam density. Higher density normally means better mechanical properties and reduced density improves insulation capacity. This is also true in the case of foams for subsea applications, where the increased hydrostatic

Table 5.1 Insulation Material Comparison Table

Material	Thermal Conductivity
Wet Insulation	
Syntactic PP	150−185 mW/(mK)
Pipe-in-Pipe	
Fiberglass	30−40 mW/(mK)
PU foam	25 mW/(mK)
Nanoporous material	14 mW/(mK)
Izoflex™	7 mW/(mK)

Notes:
1. Izoflex material is resistant to compression: no spacers are needed as it will keep the inner and outer pipes separated by virtue of its own compression strength. This provides the following advantages:
• The construction is simplified.
• There are no spacers with local point loads.
• There are no additional heat losses due to thermal bridges at spacers.
• There is no risk of spacers slipping and creating local cold points.
2. Syntactic material thermal insulation is tailored to project-specific temperature and water-depth requirements by the inclusion of hollow glass microspheres the polypropylene matrix. The requirement for hardier spheres at large water depths (thicker wall) will affect the con-ductivity negatively.

head associated with deeper waters calls for higher compressive strength and better creep properties of the PP foam. Higher compressive strength also improves creep characteristics and can be attributed to higher polymer stiffness and the final foam structure.

For deep-water thermal designs, this could lead to build up of excessively thick coatings that may cause manufacturing concerns as well as reduce installation vessel capacity. In addition, excessive coating thickness may reduce seabed stability for the flowline and increase drag forces on a steel catenary riser (SCR).

Combining a stiff linear copolymer polypropylene (PP) with a branched homopolymer PP, the benefits of high melt strength and high melt elongation result in excellent foam quality, characterized by evenly distributed bubbles with a closed cell bubble structure in the pipe foam layer. This leads to both higher compressive strength and improved creep resistance. At the same time, this novel combination of polypropylene technology retains tensile properties and impact resistance.

By combining the unique PP foam properties stemming from a combination of material characteristics and processing techniques with a strain-based design methodology, this PP foam can be deployed in deep waters as will be described in this section.

5.4.1. Material Development and Characteristics

• **Characteristics of high melt strength PP**

   High melt strength combined with improved melt elongation are the main characteristics for the branched homopolymer called HMS-PP. A long-chain branched polymer is introduced into the PP, thus improving foaming conditions. Because of the polymer modifications, controlled bubble growth can be observed, leading to stable foam with a uniform closed cell foam structure. Combining HMS with other PP grades and proper extrusion and mixing can lead to the production of considerably improved foam structures.

• **Characteristics of stiff polypropylene**

   A branched homopolymer will by chemical nature show more brittle behavior than regular PP grades. For the overall mechanical properties to match all requirements during manufacturing, installation, and operation, a high stiffness copolymer PP is mixed with the HMS polymer. In general, the stiff PP shows the following properties compared to a regular grade (see Table 5.2).

Table 5.2 Typical Stiff PP Property Differences

Properties	Typical PP Grade	New PP Grade
Thermal conductivity (W/mK)	0.22	0.23
Tensile stress at yield (50 mm/min), MPa	28	31
Tensile strain at yield (50 mm/min), MPa	6	8
Flexural modulus (2 mm/min), MPa	1300	1750

Therefore, combining the high stiffness PP and the branched PP, the mixture exhibits the following properties compared to traditional structural PP foam:

- High melt strength;
- High melt elongation;
- Finer foam cell structure (see Figure 5.8);
- Higher stiffness and better creep resistance.

These characteristics are also strongly linked to the extrusion process and subsequent die design and morphology. By discharging the polymer melt into an annulus, the bubble growth and formation of foam cells can be better controlled than if the melt is applied onto the pipe by means of wrapping.

The SEM photomicrographs in Figure 5.8 represent the same foam density; however, mechanical properties are improved. Use of the improved

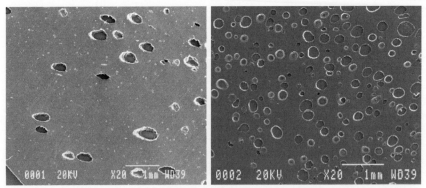

Figure 5.8 Standard PP foam on left and improved foam on right. (Source: *Jackson, A. and Boye Hansen, A., et al. Design parameters for single pipe thermal insulation systems for offshore flow assurance, Rio Pipeline 2005, October 17–19*).

Table 5.3 Mechanical Properties Comparison

Parameter	Reference Foam	Novel PP Foam
Density (kg/m^3)	820	650
Tensile stress @ yield (MPa)	16	13
Tensile strain @ break (%)	65	26
Young's modulus (MPa)	800	830
Compression modulus (MPa)	480	470
Thermal conductivity (W/mK)	0.2	0.13

mechanical properties and strain-based design makes it possible to deploy closed cell PP foam in water depths beyond 1500 meters (see Table 5.3).

Durability is also an important feature. Creep is the dominant mechanism that governs the long-term properties. PP shows typical viscoelastic behavior, where short-term deformation relates to the elastic behavior of the foam (recoverable deformation) and long-term deformation relates to viscous behavior (nonrecoverable deformation). With improved elasticity in the novel foam, long-term creep resistance of the novel foam is better than the reference foam.

5.4.2. Syntactic Polypropylene (SPP)

SPP is a filled compound where the filler is hollow glass microspheres. Syntactic polypropylene has been used in the offshore industry since 1995, when Shell selected this product for the Mars SCR insulation. Up until approximately two years ago, all of the executed projects relied on applying ready-made SPP.

However, the need to reduce thermal history (reduce material degradation) imposed on SPP compounds and drivers to reduce costs led to development of a mobile turnkey compounding plant for this process. Figure 5.9 shows a schematic of direct extrusion versus pellet extrusion.

The main characteristics of SPP are compressive and creep behavior similar to solid PP, a lower k-value than solid PP, which reduces thickness (typically 22% less than solid PP), and that it can be used as a thermal barrier.

Hollow glass microspheres are fragile and susceptible to point loads. Although the hydrostatic crush strength in many cases exceeds the needs related to serviced water depths, broken spheres as a result of loads in the process may deteriorate mechanical and thermal properties of the final compound. These loads include capillary pressure effects, shear, excessive glass loading rate, sinkers from microsphere production, and on-line change of melt behavior.

Extrusion from Pellets:

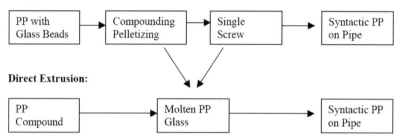

Figure 5.9 Schematic of direct extrusion versus pellet extrusion.

5.4.3. Manufacturing Processes

Schematically, Figure 5.10 shows a typical deep-water insulation system buildup.

Such a system relies on use of solid PP as a thermal barrier, syntactic PP as a thermal barrier and insulation, and closed cell foamed PP as an insulation system. Layer thickness will be determined by the thermal gradient through the coating thickness and subsequent exposure temperatures. As an example, this would typically expose the SPP to temperatures higher than 100°C at 1500 to 2500 meter water depth and the closed cell foam to a maximum

Figure 5.10 Multilayer PP insulation system. (*Source: Jackson, A. and Boye Hansen, A., et al. Design parameters for single pipe thermal insulation systems for offshore flow assurance, Rio Pipeline 2005, October 17–19*).

Table 5.4 Process Characteristics

Item	SPP	Closed Cell Foam
Melt strength	Poor	Good
Die swell	Low	High
Die pressure	Medium	High
Melt viscosity	High	Medium
Melt temperature	High	Medium
Shear sensitivity	High	Medium

40 to 60°C. This will be discussed in more detail in the design part of this section.

Such exposure temperatures in combination with external pressures in the actual service means that proper care has to be taken when selecting resin candidates and material combinations for both the SPP and the closed cell foam. Table 5.4 shows the different parameters that distinguish and compare processing of the SPP and the closed cell PP foam.

5.4.4. Qualification Schemes

In order to qualify the PP foam for water depths down to 2000 meters, a rigorous qualification program was developed and performed to meet the defined performance criteria. The main objectives for the qualification program were:

- To qualify the insulation system for subsea use;
- To develop data for service life prediction (small scale and full scale);
- To use the generated data for design and engineering of thermal insulation systems for deep water service.

The overall philosophy has been to execute the program so the loads reflect the actual conditions on the insulation coating and to establish acceptance criteria. Performed tests reflect loads during:

- Manufacturing;
- Storage (stacking of pipes);
- Installation (e.g., reeling);
- Operation.

The framework for the qualification was existing international standards for polymers and foams as well as established procedures, requirements, and specifications defined by operators and the manufacturer. In many instances, the InSpec Specifications and Recommended Practice form the basis for qualifying new products and systems.

For the two distinct PP qualities described in this section, it is extremely important to recognize the intrinsic properties that may reduce performance over time, and that need to be compensated for in the design as described below.

- **Matrix candidates for syntactic PP**

    Although the SPP grades in many cases behave as solid materials (limited creep), there will be inherent capacity in different matrix PP materials that may limit the possibility of using one single grade for all combinations of service temperature and water depth. As inclusion of hollow glass microspheres is not straightforward, resin modifications may be necessary. Such modifications may limit depth and/or temperature rating as modifying resins changes mechanical and thermal properties. Applying the triaxial creep test method and determining possible extrusion can screen such differences in properties.

- **Closed cell PP foam**

    The most critical parameter when designing with the viscoelastic behavior of plastics is associated with creep. Creep in the foamed structures relates both to water depth (hydrostatic pressure) and the associated temperature gradient. The temperature gradient is dependent on the layer thickness, thermal conductivity of each layer, and internal and external fluid temperatures.

    Creep is the most important long-term design consideration, as creep will result in changes of the insulation properties over time. Creep in foamed structures will produce an increase in density, which in turn will increase the thermal conductivity. Hence, it is important to understand the creep mechanisms in the foam, use these mechanisms in the design stages, and compensate for this creep in the design.

    Secondly, it is important to characterize the actual compressive load that the foam will experience during in-service conditions. On the pipe, the foam benefits from support axially and tangentially. Therefore, the creep will show as radial displacement of the PP foam. Figure 5.11 shows results from a 10,000 hour creep test. Such a long-term test will reveal any signs of secondary or tertiary creep and further document the long-term durability.

As can be seen from the tests plotted in the figure, the general recommendations for foam densities for deep waters ($>$ 1000 meters) would typically be above 700 kg/m^3.

The system design of subsea thermal insulation systems with foamed materials requires detailed knowledge of the relationship between

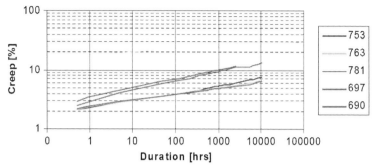

Figure 5.11 Long-term durability tests (creep at 12 MPa and 45°C).

compressive strength and temperature, and the temperature dependency of creep. Although the initial compressive response of the system can conveniently be measured for a given set of operating conditions in a simulated service test, such testing is relatively short term, and does not provide reliable information concerning creep.

For this reason it has become now common practice within an industry leader to perform long-term (> 5000 hours) triaxial compression and creep testing in addition to standard uniaxial tests when qualifying new products and as a benchmark for durability. Using this approach, the compressive creep rate is established under geometrically realistic conditions. Armed with such knowledge, and in combination with correctly calibrated finite-element analysis design tools it is possible to derive limit state system designs to meet both transient and steady-state performance requirements at the end of the design field life.

A good knowledge of the response of the system to hydrostatic loading at temperature also allows for the determination of layer thicknesses for individual materials in the system such as PP foam, such that reliability and system cost are optimized.

The thermal and hydrostatic aspects of insulation coating are designed using appropriate physical laws representing the conditions at the seabed, and using material geometries and thermophysical properties describing the start and modeled end of life conditions. Effects of possible degradation mechanisms on the thermophysical aspects of the system such as water ingress are factored into the design using appropriate functions, fitted to long-term experimental data. The mechanical response to temperature and pressure are inextricable from a design standpoint, and are in essence calculated simultaneously in an integrated mechanical/thermal model.

The methodology used for simulating the mechanical response is to construct and discretize an initial multilayer model, and to apply temperature dependent stress-strain and creep laws into each node. For each time step the combined effects of the hydrostatic and thermal loading are calculated and these effects translated into densification of the coating, changes in thermophysical properties, and changes in geometry.

A new through thickness temperature profile is calculated, and this is then used in the next time step. This process is repeated until the end of life situation is achieved. The mechanical calculation is subsequently verified against the results of triaxial compression and creep testing in order to ensure that a rational solution has been derived. It is important to use the temperature-dependent moduli of materials, as their response is dependent not only on the absolute values of the internal and external temperature, but also the through thickness temperature profile.

The heat transfer equation is solved numerically using an implicit finite difference scheme. The following equation details the solution of the finite difference equation for a transient heat conduction problem in multidimensional directions:

$$C_i \frac{T_i^{k+1} - T_i^k}{\Delta t} = \sum_n \frac{T_n^{k+1} - T_i^{k+1}}{R_{ni}} = \dot{Q}_i''' \Delta V_i \qquad (5.22)$$

where

T = temperature

C_i = energy capacitance

R = resistance

i = volume element

n = number of dimensions

k = time step

The pipe geometry is discretized into a network of small finite volume elements. The above equation is applied to each of the elements. Matrix inversion is used to solve the system of implicit numerical equations that constitutes the pipe geometry.

A reciprocal U-value is found by adding all the subsequent layer resistances, for which the following formula applies:

$$U = \frac{1}{\pi \cdot OD \cdot R_{total}} \qquad (5.23)$$

where

R_{total} = thermal serial resistance from the inside of the coating to the outside of the coating

OD = reference diameter for the U-value

U_i = heat transfer coefficient of the layer relative to the referenced surface

It has formerly been practice within the subsea thermal insulation industry to use single figure thermal conductivities and heat capacities; however, there has been a movement away from this practice in certain cases, with the inclusion now of the temperature dependence of these important factors. This allows for a more rigorous description of the thermal performance of the system. Transient state performance is calculated based on the end of life system, using the appropriately modified geometries and thermophysical properties.

5.4.5. Syntactic Foam

Syntactic foam materials based on hollow glass microspheres embedded in an organic matrix traditionally find application in the marine environment for buoyancy. They are now being used for the thermal insulation of offshore pipelines and bundles that convey oil and gas resources. As a matter of fact, syntactic foams can provide substantial heat loss resistance while withstanding external deep-water pressure (30 MPa at 3000 m deep) thanks to their closed cell microstructure.

The humid aging of syntactic foams under hydrostatic pressure, up to 100 MPa, has been widely studied. But the combination of high-pressure/high-temperature conditions (up to 30 MPa/130°C) is a recent concern: a pioneering research program including aging in hot wet conditions has shown that traditional industrial syntactic foams were undergoing rapid and severe degradation in seawater at temperatures as low as 60°C. There is experimental evidence that a key factor may be the chemical degradation of glass microspheres due to the "entrance" of water.

The syntactic foams are made with a model diepoxydiamine matrix. The polymerization kinetics are controlled thanks to the low reactivity of the diamine. Good dispersion of the microspheres is ensured by checking the viscosity when casting the foam.

Multilayered coating systems incorporating polymer foams have been found to provide a cost-effective solution to the insulation of offshore pipelines. These systems have typically five or seven layers, depending on the water depth. A cross-section through a seven-layer system incorporating polypropylene (PP) foam and solid layers is shown in Figure 5.12. The foam

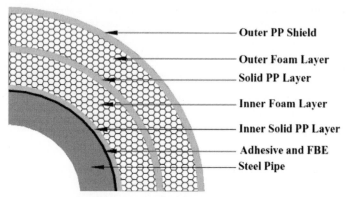

Figure 5.12 Typical seven-layer insulation system. *(Source: A.M. Harte et al.* Journal of Materials Processing Technology *155–156 (2004) 1242–1246).*

layers provide the main thermal resistance of the system. These foam layers may have different densities and may incorporate reinforcement. The other layers have varied functions. The outer solid PP shield layer provides impact resistance and prevents water ingress to the foam layers. The internal solid PP layers are used as a thermal barrier to the inner foam layer and as a transition between the different foam layers. The inner fusion bonded epoxy layer, which forms a corrosion barrier for the steel pipe, and the adhesive layer are assumed to provide negligible contribution to the thermal and structural capacity of the system. Five-layer systems are used in shallower waters. These have a similar configuration to the seven-layer system but with one of the foam and solid PP layers removed.

Definitions and Terminology

Abrasion Resistance The ability of a material to withstand abrasion without appreciative erosion.

Absorption The process of drawing fluid or gas into a porous material, such as a sponge soaking up water.

Abuse Coverings and Finishes Jackets, mastics, or films used to protect insulation from mechanical and personnel abuse.

Acoustics The study of sound, and its transfer, reflection, and absorption.

Acoustic Insulation Refers to the measures taken to reduce the transmission of noise from its source to the places that must be protected or insulated.

Added R-value Thermal resistance added to a construction element by insulation.

Adhesive A substance used to bond materials by surface attachment.

Adsorption Refers to the surface retention or adhesion of a very thin layer of water molecules to the surfaces of a material (such as insulation fibers) with which they are in contact.

Aerogel A homogeneous, low-density solid-state material derived from a gel, in which the liquid component of the gel has been replaced with a gas. The resulting material has a porous structure with an average pore size below the mean free path of air molecules at standard atmospheric pressure and temperature.

Air Barrier A layer of material resistant to air flow, usually in the form of polyolefin (i.e., Typar, Tyvek, and other housewraps). A material that is applied in conjunction with a building component (such as a wall, ceiling, or sill plate) to prevent the movement of air through that component.

Air Barrier System The assembly of components used in building construction to create a plane of air tightness throughout the building envelope and to control air leakage.

Air Changes per Hour (ACH) An expression of ventilation rates – the number of times in an hour that a home's entire air volume is exchanged with outside air.

Air Duct Ducts, usually made of sheet metal, that carry cooled or heated air to all rooms.

Air Filters Adhesive filters made of metal or various fibers that are coated with an adhesive liquid to which particles of lint and dust adhere. These filters will remove as much as 90% of the dirt if they do not become clogged. The more common filters are of the throwaway or disposable type.

Air Hawks Rectangular-shaped vents located on attic rooftops.

Air Infiltration The amount of air leaking in and out of a building through cracks in walls, windows, and doors.

Air Intake An opening in a building's envelope whose purpose is to allow outside air to be drawn in to replace inside air.

Air Space The area between insulation facing and the interior of exterior wall coverings. Normally a 1-inch air gap.

Air Tightness Describes the leakiness of the building fabric.

Airway A space between roof insulation and roof boards provided for movement of air.

Aluminum Foil Thin sheet of rolled aluminum (0.15 mm thick and under).

Anti-Abrasive Coating Cushioning material applied where insulation contacts the pipe, duct, vessel, or adjacent insulation to prevent eroding of either or both.

Asbestos The generic name for those silicate minerals that cleave naturally into fibers, the three important forms being chrysotile (white asbestos), crocidolite (blue asbestos), and amosite.

ASTM C-739 Standard Specifications for Cellulosic Fiber (wood base) Loose-Fill Thermal Insulation.

ASTM C-1149 Standard Specification for Self-Supported Spray-Applied Cellulosic Thermal/Acoustical Insulation. Covers 10 material attributes (density, thermal resistance, surface burning characteristics, adhesive/cohesive strength, smoldering combustion, fungi resistance, corrosion, moisture vapor absorption, odor, and flame resistance permanency).

Attic An attic is a space found directly below the pitched roof of a house.

Attic Fan A fan that blows heated air out of the attic of a building.

Attic Insulation The installation of approved insulation products (rock wool, fiberglass, and cellulose) evenly across the unconditioned attic area to achieve desired levels of thermal resistance.

Attic Insulation Removal The removal of visible soot and char off the surface of insulating materials sufficient enough to achieve an acceptable appearance. One of the most difficult jobs is working in attics and attempting to remove settled loose soot and char particles off of insulation.

Attic Vent A passive or mechanical device used to ventilate an attic space, primarily to reduce heat buildup and moisture condensation.

Attic Ventilation The process of ventilating the attic in order to remove moisture and heat from the attic

Attic Ventilators In houses, screened openings provided to ventilate an attic space. They are located in the soffit area as inlet ventilators and in the gable end or along the ridge as outlet ventilators. They can also consist of power-driven fans used as an exhaust system.

Backer Board A rigid, non-vapor-barrier-forming material such as drywall, treated cardboard, plywood, etc., that is used to cover the open side of an existing wall and forms a cavity that may be filled with loosefill insulation. Must have sufficient strength to withstand the pressure developed when filling the cavity.

Baffles Device used to achieve a 1-inch ventilation space between insulation and roof sheathing. Helps assure airflow from eave vents in attics and cathedral ceilings.

Bands Strapping used to fasten insulation and/or jacketing in place.

Basement That portion of a building that is partly or completely below grade.

Basement Wall A wall of a building that is mostly below grade.

Batt A piece of flexible to semi-rigid insulation of specified width and length with or without vapor retarder facing.

Beveled Lags Lags similar to plain lags, but with one or more edges beveled.

Binder Substance contained in insulation material that stabilizes the fibers (sometimes called a thermal setting resin).

Blanket Insulation of the flexible type, formed into sheets or rolls, usually with a vapor barrier on one side and with or without a container sheet on the other side.

Blanket Insulation Insulation of flexible type, formed into sheets or rolls, usually with a vapor barrier on side and with or without a container sheet on the other side.

Blanket Insulation, Metal Mesh Blanket insulation covered by flexible metal-mesh facings attached on one or both sides.

Block Insulation Semi-rigid insulation formed into sections, rectangular both in plan and cross section, usually 90–120 cm long, 15–60 cm wide, and 2.5–15 cm thick.

Block Insulation (Slab) Semi-rigid insulation formed into sections, rectangular both in plan and cross section, usually 90–120 cm long, 15–60 cm wide, and 2.5–15 cm thick.

Block (Slab) Rigid or semi-rigid insulation formed into sections, rectangular both in plan and cross section, usually 90–120 cm long, 15–60 cm wide, and 2.5–15 cm thick.

Blocking A material used to retain the insulation in place in open areas.

Blower Door Diagnostic equipment consisting of a fan, removable panel, and gauges, used to measure and locate air leaks.

Blower Door Test A test used to determine the "tightness" (energy leakage) of your home.

Blowing Agent A gas or a substance capable of producing a gas used in making foamed materials.

Blown Insulation Fiber insulation in loose form used to insulate attics and existing walls where framing members are not exposed.

Board Rigid or semi-rigid insulation formed into sections, rectangular both in plan and cross section, usually more than 120 cm long, 60–75 cm wide, and up to 10 cm thick.

Board Insulation Semi-rigid insulation formed into sections, rectangular both in plan and cross section, usually more than 120 cm long, 60–75 cm wide, and up to 10 cm thick.

Boric Acid A nontoxic chemical additive that is used as a fire retardant in cellulose insulation.

Breather Coating A weather barrier coating designed to prevent water (rain, snow, sleet, spillage, wash water, etc.) from entering the insulation system, while still allowing the escape of small quantities of water vapor resulting from heat applied to the moisture entrapped in the insulation.

BTU, British Thermal Unit The amount of energy that is required to raise 1 lb. of water up 1°F. Fuel values, heat loss, and heat gain are measured in BTUs.

BTUH A rate of energy transfer – can be expressed as BTUs/hour.

Building Envelope The external elements (walls, floor, ceiling, roof, windows, and doors) of a building that enclose the conditioned space; the building shell.

Bulk Insulation Insulation depending for its performance upon thickness and thermal conductivity to achieve material R-value.

Butt Joints The end joints of pipe insulation.

Butt Strip Strips of similar jacket material applied around pipe insulation butt joints.

C-Value (Thermal Conductance) A measure of a material's ability to allow heat to pass through it. The same as U-value, but without air film resistances.

Calcium Silicate Insulation Insulation composed principally of hydrous calcium silicate, and which usually contains reinforcing fibers.

Caulk To seal and make water- and/or airtight.

Canvas A closely woven fabric of cotton, flax, hemp, or jute characterized by strength and firmness.

Cavity Empty space between studs/joists in which insulation is placed.

Cellular Elastomeric Insulation composed principally of natural or synthetic elastomers, or both, processed to form a flexible, semi-rigid, or rigid foam that has a closed-cell structure.

Cellular Glass Insulation composed of glass processed to form a rigid foam having a predominately closed cell structure.

Cellular Glass (Foamed Glass) A lightweight expanded glass with small cells, preferably nonintercommunicating, produced by a foaming process.

Cellular Insulation Insulation composed of small, individual cells separated from each other. The cellular material may be glass or plastic such as polystyrene, polyurethane, polyisocyanurate, or elastomeric.

Cellulose Insulation Insulation made from recycled newspaper, making it a natural, environmentally friendly alternative to other insulation products.

Cellulosic Fiber Insulation composed principally of cellulose fibers usually derived from paper, paperboard stock, or wood, with or without binders.

Cement, Insulating A mixture of dry granular, fibrous, or powdery (or both) materials that when mixed with water develops a plastic consistency, and when dried in place forms a coherent covering that affords substantial resistance to heat transmission.

Ceramic Fiber Fibrous material, loose or fabricated into convenient forms, mainly intended for use at appropriate elevated temperatures. The fibers may consist of silica (SiO_2) or of an appropriate metal silicate, e.g., aluminosilicate. Alternatively, they may be formed synthetically from appropriate refractory metal oxides, e.g., alumina, zirconia.

CIMA The Cellulose Insulation Manufacturers Association is the trade association for the cellulose segment of the thermal/acoustical insulation industry.

CIMAC The Cellulose Insulation Manufacturers Association of Canada is the trade association for the cellulose segment of the thermal/acoustical insulation industry.

Closed Cell Foam A material comprised predominantly of individual noninterconnecting cellular voids.

Commercial Insulation The insulation of commercial spaces like schools, colleges, metal buildings, etc.

Compressive Strength The property of an insulation material that resists any change in dimensions when acted upon by a compaction force.

Concealed Spaces Spaces not generally visible after the project is completed such as furred spaces, pipe spaces, pipe and duct shafts, spaces above ceilings, unfinished spaces, crawl spaces, attics, and tunnels.

Condensation Changing a substance from a vapor to a liquid state by removing the heat. The condensate shows up on surfaces as a film or drops of water.

Conditioned Space Building area supplied with conditioned air that is heated or cooled to a certain temperature and may be mechanically controlled to provide a certain humidity level.

Conduction Transmission of energy (heat/sound) through a material or from one material to another by direct contact. Materials with low rates of conductive heat transfer make good insulation.

Conductivity, Thermal (Lambda Value) The rate at which heat is transmitted through a material, measured in watts per square meter of surface area for a temperature gradient of one Kelvin per meter thickness, simplified to W/mK.

Convection Transmission of energy (heat/sound) from one place to another by movement of a fluid such as air or water.

Cork The elastic, tough outer tissue of the cork oak that is used specially for stoppers and insulation.

Corkboard Preformed material composed of granulated cork bonded by heating under pressure, with or without added adhesive.

Corrosion The process of wearing away gradually, usually by chemical action.

Cover To place insulation and/or finish materials on, over, or around a surface so as to insulate, protect, or seal.

Crawl Space A shallow open area between the floor of a building and the ground, normally enclosed by the foundation wall.

Critical Radiant Heat Flux A test used to simulate conditions in a hot attic, designed to test the surface burning characteristics of insulation.

Cryogenic Insulation Insulation for extremely low-temperature processe surfaces from $-100°F$ to $-459°F$ (absolute zero).

Damming The use of a substance to support fire-stopping materials until cured.

Delamination The separation of the layers of material in a laminate.

Density Determined by the weight expressed in kg of a cubic meter of a material.

Dew Point The temperature at which a vapor begins to condense out of the air.

Diatomaceous Silica Insulation composed principally of diatomaceous earth with or without binders, and which usually contains reinforcing fibers.

Diffusion The movement of water vapor from regions of high relative humidity (RH) toward regions of lower RH driven by a higher to lower temperature differential.

Double Sided Reflective foil on both faces of reflective insulation.

Duct A passageway made of sheet metal or other suitable material used for conveying air or other gas.

Eave Vents Vent openings located in the soffit under the eaves of a house to allow the passage of air through the attic and out the roof vents.

Elastomeric A closed-cell foam insulation containing elastomers that provide the property of high elasticity.

Emission The manner by which substances are discharged through the air.

EPS EPS (Expanded PolyStyrene) is a lightweight, rigid, plastic foam insulation material produced from solid beads of polystyrene.

Exhaust Duct A duct carrying air from a conditioned space to an outlet outside the building.

Expanded Metal Metal network made by suitably stamping or cutting sheet metal and stretching it to form open diamond-shaped meshes.

Expansion Joint An arrangement in an insulation system to minimize the risk of cracking due to thermal movement.

Exposed Spaces Those spaces not referred to as concealed or as defined by the specifier.

Faced Insulation Insulation with a facing already attached. Kraft paper and foil-backed paper are common facings.

Facing A protective or decorative (or both) surface applied as the outermost layers of an insulation system.

Fiberglass A fibrous material made by spinning molten glass. Used as an insulator and heat loss retardant.

Fiberization The manufacturing process of turning molten raw material (e.g., glass and stone for insulation) into fibers.

Fibrous Glass A synthetic vitreous fiber insulation made by melting predominantly silica sand and other inorganic materials, and then physically forming the melt into fibers. To form an insulation product, there are often other materials applied to the mineral wool such as binders, oils, etc. Commonly referred to as fiberglass.

Fibrous Insulation Insulation composed of small diameter fibers that finely divide the air space. The fibers used are silica, rock wool, slag wool, or alumina silica. Insulation constructed from fiber, naturally occurring or manufactured, that incorporates single or composite filaments generally circular in cross section and of length considerably greater than the diameter.

Fireblocking Building materials installed to resist the free passage of flame and gases to other areas of the building through concealed spaces.

Fish–Mouth A gap between layers of sheet materials caused by warping or bunching of one or both layers. Typically seen when stapling Kraft face insulation or the jacketing on pipe insulation.

Flange A projecting collar attached to a pipe for the purpose of connecting to another pipe, valve, or fitting.

Flange Cover The insulation for a pipe flange composed of the specified thickness of insulation material, which may be preformed. Also, a preformed jacketing. Foam board plastic foam insulation manufactured most commonly in 4" × 8" sheets in thickness of 1/4" to 3".

Flexible Insulation A material that tends to conform to the shape of the surface against which it is laid, or is so designed as to alter its manufactured shape to accommodate bends and angles.

Foamed In–Situ Plastics Cellular plastics produced in situ and foamed by physical or chemical means.

Furnace An appliance for heating a medium to distribute heat throughout the dwelling unit.

Gable Vents A louver vent mounted in the top of the gable to allow the passage of air through the attic (the hole near the pointy part of the wall).

Gas Heating System A heating system that uses natural gas or bottled liquid propane gas as fuel.

Girt A girt is a horizontal structural member usually located in a framed wall. Used as a term in metal building construction.

Glass Cloth Fabric woven from continuous filament or staple glass fiber.

Glass Fiber (Glass Wool) A material consisting of glass fibers used in making various products, including yarns, fabrics, insulation, and structural objects or parts. Glass fiber is resistant to heat and fire. Mineral fiber produced from molten glass.

Granular Insulation Insulation composed of small nodules that contain voids or hollow spaces. The material may be calcium silicate, diatomaceous earth, expanded vermiculite, perlite, cellulose, or microporous insulations.

Greenest of the Green The Cellulose Insulation Manufacturers Association (CIMA) is actively promoting the environmental benefits of cellulose insulation through their Greenest of the Green campaign.

Greenhouse Gas Emissions Gases that trap heat in the atmosphere are often called greenhouse gases. Some greenhouse gases such as carbon dioxide occur naturally and are emitted to the atmosphere through natural processes and human activities.

Green Insulation Insulation of spaces using recyclable materials.

Heat Flow Rate The rate at which heat moves from an area of higher temperature to an area of lower temperature, in Btu/hr (W/hr). Heat flow is generally used to quantify the rate of total heat gain or heat loss of a system.

Heated Space Building area supplied directly with heat.

Heat Loss Heat that is lost from a building through air leakage, conduction, and radiation. To maintain a steady interior temperature, heat losses must be offset by a combination of heat gains and heat contributed by a heating system.

Heat Recovery Ventilation System A mechanical ventilation system that recovers energy from exhausted indoor air and transfers it to incoming air. This system usually incorporates an air-to-air heat exchanger that transfers the heat from exhaust air to the incoming air or vice versa.

Heat Transfer Heat flow from a hot to a cold body.

High-Performance Insulation Fiberglass insulation with densely packed fibers, resulting in higher R-values for a given thickness. Most commonly used in confined spaces such as walls or cathedral ceilings.

Home Audit or Energy Audit An assessment performed by an energy specialist in order to identify how a structure's energy efficiency can be improved. Many incentives or rebates require an audit be conducted before and after the improvements in order to verify savings. Often conducted free of charge by utility companies.

Humidity A measure of the amount of water vapor in the atmosphere.

Impact Resistance Capability of an insulation material and/or finish to withstand mechanical or physical abuse.

Infiltration Uncontrolled leakage of air into a building through cracks around doors, windows, and electrical outlets and at structural joints.

Insulate To cover with a material of low conductivity in order to reduce the passage or leakage of heat, reduce the surface temperature, or reduce the noise emanating from the object.

Insulated Ceiling (I.C.) Marking on recessed lighting fixture indicating it is designed for direct insulation contact.

Insulating Cement A mixture of various insulating fibers and binders with water to form a moldable paste insulation for application to fittings, irregular surfaces, or voids.

Insulation Materials with low thermal conductivity characteristics that are used to slow the transfer of heat.

Insulation (Thermal) Density Those materials or combination of materials that slow the flow of heat. One factor determining R-value; higher density equates to better insulating properties.

Insulation Removal The process of removing installed insulation due to the formation of mildew and mold.

Jacket A form of facing applied over insulation. It may be integral with the insulation, or field applied using sheet materials. Jacket is a covering placed over insulation for various functions.

Kilowatt-Hour (kWh) Standard unit for measuring electrical energy consumption: kilowatts × hours.

k-Value (Conductivity) The measure of heat in BTUs that pass through one square foot of a homogeneous substance, 1-inch thick, in an hour, for each degree F temperature difference. The lower the k-value, the higher the insulating value.

Knee Walls Walls of varying length. Used to provide additional support to roof rafters with a wide span or to finish off an attic.

Lacing A method of joining or securing insulation materials, reinforcements, or finishes for insulation materials using eyelets, hooks, wire, cord, etc.

Lacing Wire (Tie Wire) Light gauge wire, single or multistrand, used for lacing together adjacent edges of mattresses or of metal covering, or for securing insulating material on substantially flat surfaces.

Lag Preformed rigid insulation for longitudinal application to cylinders larger than those for which pipe sections are available. There are three type as follows:

- Plain lags
    Lags having rectangular cross section, for use on cylinders of such diameter that this shape conforms sufficiently closely to the surface.
- Beveled lags
    Lags similar to plain lags, but with one or more edges beveled.
- Radiused and beveled lags
    Beveled lags with faces curved to fit the surface of the cylinders (sometimes known as curved and beveled lags).

Lagging Insulation A block material for insulating tanks and boilers, usually curved or tapered; it can be made from any of several insulation materials.

Lambda Value Lambda value (thermal conductivity) is a physical coefficient that measures the heat transmission behavior of a material. The lower the lambda value, the better the thermal efficiency of the material. Unit: W/mK.

Life Cycle Assessment A life cycle assessment (LCA) is a technique to assess environmental impacts associated with all the stages of a product's life from cradle to grave (i.e., from raw material extraction through materials processing, manufacture, distribution, use, repair and maintenance, and disposal or recycling).

Loosefill Insulation material (usually mineral wool, vermiculite, or cellulose) used for pouring or blowing into the space to be insulated.

Loosefill Insulation Material in the form of powder, granules, foamed, expanded or exfoliated aggregate, or loose or pelleted fibers used in the dry state as a filling for cavities, casings, or jackets. Insulation in granular, nodular, fibrous, powdery, or similar form designed to be installed by pouring, blowing, or hand placement.

Mastic A relatively thick consistency protective finish capable of application to thermal insulation or other surfaces, usually by spray or trowel, in thick coats, greater than 0.75 mm.

Material Safety Data Sheet (MSDS) A standard-format information sheet, prepared by a material manufacturer, describing the potential hazards, physical properties, and procedures for safe use of a material.

Material R-Value Thermal resistance determined by dividing thickness by thermal conductivity, excluding surface air film resistances.

Mattress A flexible construction comprising an insulating material faced on one side or both sides, or totally enclosed with fabric, film, paper, wire netting, expanded metal, or similar covering attached mechanically to the insulating material.

Metal Cleading/Jacketing Sheet metal fitted as a protective finish over insulation.

Metal Flue Metal chamber through which hot air, gas, steam, or smoke may pass.

Microporous Insulation Material in the form of compacted powder or fibers with an average interconnecting pore size comparable to or below the mean free path of air molecules at standard atmospheric pressure. Microporous insulation may contain opacifiers to reduce the amount of radiant heat transmitted.

Mineral Fiber A generic term for all nonmetallic inorganic fibers.

Mineral Wool A generic term for mineral fibers of a woolly consistency, normally made from molten glass, rock, or slag. Insulation composed principally of fibers manufactured from rock, slag, or glass with or without binders.

Mitered Joint A joint made by cutting (mitering) preformed pipe sections to fit around bends in a pipeline.

Moisture Moisture refers to the presence of water, often in trace amounts. Excessive moisture is usually undesirable in buildings as it can cause decay in timber or other organic material, corrosion in metals, and electrical short circuits.

Mold and Mildew Resistance The property of a material that enables it to resist the formation of fungus growth.

Natural Ventilation An air space bounded by one or more permeable surfaces allowing a degree of air movement (e.g, an attic space below an unsarked tiled roof).

Noncombustible Construction Buildings in which walls, partitions, structural elements, floors, ceilings, roofs, and exits are made of noncombustible materials and which require higher fire resistance ratings than combustible construction.

Noncombustibility The property of a material that enables it to withstand high temperatures without ignition.

Nonventilated Air space enclosed by nonpermeable building materials.

One-Coat Cement A mixture of various insulation fibers, fillers, and binders with hydraulic-setting cement. The material can be applied directly to fittings to match adjacent insulation thickness in one application and smoothed to provide a hard finish.

Open Cell Foam A material comprised predominantly of interconnecting cellular voids.

Organic Compounds containing carbon.

Overspray Airborne spray loss of polyurethane foam that leads to undesirable depositions of spray foam insulation on nearby surfaces.

Panel Insulation A prefabricated unit of insulation and lagging.

Perimeter Insulation Insulation installed on the sidewalls of a crawl space to reduce heat loss. Also protects plumbing in the space from freezing temperatures. Perimeter insulation should only be used at the express approval of your utility.

Perlite A glossy volcanic rock that expands when heated. Processed perlite is used as loosefill insulation material or bound into slabs.

Perm A unit of water vapor transmission defined as 1 grain of water vapor per square foot per hour per inch of mercury pressure difference (1 inch mercury = 0.49 psi). Metric unit of measure is ng/m^2sPa. 1 perm = 55 ng/m^2sPa.

Permeability The time rate of water vapor transmission through unit area of a material of unit thickness induced by unit vapor pressure difference between two specific surfaces, under specified temperature and humidity conditions.

Permeance A measure of the transmission of water vapor through a material or combination of materials, measured in perms.

pH A measure of acidity/alkalinity of aqueous mixtures. A measure of pH 7 is neutral, lower is more acidic, higher is more alkaline.

Phenolic Foam A foamed insulation made from resins of phenols condensed with aldehydes.

Pipe Insulation Insulation in a form suitable for application to cylindrical surfaces.

Pipe Sections Sections of insulating material in cylindrical form suitable for application to pipes.

Plain Lags Lags having rectangular cross sections, for use on cylinders of such diameter that this shape conforms sufficiently closely to the surface.

Plastic Composition Insulating material in loose, dry form, prepared for application as a paste or dough by mixing with water, usually on site, and normally setting under the influence of heat applied to the internal surface.

Polyethylene A closed-cell, thermoplastic material used for insulation.

Polyurethane Polyurethane, commonly abbreviated PU, is any polymer consisting of a chain of organic units joined by urethane links. Polyurethane polymers are formed by reacting a monomer containing at least two isocyanate functional groups with another monomer containing at least two alcohol groups in the presence of a catalyst.

Power Vent A vent that includes a fan to speed up airflow. Often installed on roofs.

Preformed Insulation Thermal insulating material fabricated in such a manner that at least one surface conforms to the shape of the surface to be covered and which, when handled, will maintain its shape without cracking breaking, crumbling, or permanent deformation.

ProCell Blue ProCell Blue is an all-borate Type 2 thermal and acoustical cellulose fiber insulation manufactured from elected recycled paper and paperboard stock.

ProCell Gold ProCell Gold is a revolutionary Type 2 thermal and acoustical spray-applied insulation that is ideal for both commercial and residential structures. ProCell Gold's excellent coverage, perfect fit, and greater density applies quickly and easily and creates no seams or voids, minimizes air leakage and infiltration, and provides uniform coverage with a density over twice that of fiberglass batts.

Radiant Barrier Radiant barriers or reflective barriers inhibit heat transfer by thermal radiation.

Radiant Barrier Foil Radiant barrier foil is a thin aluminum film that works to reflect heat gain. It is typically fastened to the underside of roofs or in attic spaces, where it prevents the heat absorbed by hot roofs from transferring to the home's interior.

Radiant Barrier Spray Barrier spray is a coating applied to structures for insulation and related purposes. The most widely used barrier spray is spray polyurethane foam (SPF), which provides a sealed thermal barrier for residential and commercial applications. SPF products are engineered for a variety of uses including roofing and fire control. Radiant barrier spray is a product used to block radiant heat from the sun.

Radiation Transfer of energy (heat/sound) from one object to another through an intermediate space. Only the object receiving the radiation, not the space, is heated. The heat is in the form of low frequency, infrared, invisible, light energy, transferring from a "warm" object to a "cold" object. It is known as the "black body effect." Heat flow transfer by electromagnetic radiation (infrared waves).

Radiused and Beveled Lags Beveled lags with faces curved to fit the surface of the cylinder (sometimes know as curved and beveled lags).

Reflective Insulation Insulation depending for its performance upon reduction of radiant heat transfer across air spaces by use of one or more surfaces of high reflectance and low emittance.

Relative Humidity The ratio of the mol fraction of water vapor present in the air to the mol fraction of water vapor present in saturated air at the same temperature and barometric pressure. Approximately it equals the ratio of the partial pressure or density of the water vapor in the air to the saturation pressure or density, respectively, at the same temperature. The ratio expressed as a percentage of the amount of moisture air actually contains to the maximum amount it could contain at that temperature.

Resilient Channels Metal channels used to inhibit sound transmission from wood studs through drywall.

Resin Component B in spray foam chemistry. This component is mixed with the A component on-site to make spray foam insulation.

R (R-Value) R-value is a measure of resistance to heat flow. Insulation materials have tiny pockets of trapped air. These pockets resist the transfer of heat through material. The ability of insulation to slow the transfer of heat is measured in R-values. The higher the R-value, the better the insulation material's ability to resist the flow of heat through it. Unit: $ft^{2o}F/BTU.in.$ (imperial).

Retrofit The application of additional insulation over existing insulation, new insulation if old insulation has been removed, or new insulation over existing, previously uninsulated surfaces.

Ridge Vents A vent mounted along the entire ridgeline of the roof to allow the passage of air through the attic or cathedral ceiling.

RSI A unit of measurement of resistance to heat flow in $m^{2o}C/W$ per 25 mm (metric). R = 0.176 RSI.

Rock Wool Mineral wool produced from naturally occurring igneous rock. Thermal insulation material composed of threads or filaments of slag, produced by reprocessing the residual materials from metals smelting. A synthetic vitreous fiber insulation made by melting predominantly igneous rock and other inorganic materials, and then physically forming the melt into fibers.

Roof Vents A louver or small dome mounted near the ridge of the roof to allow the passage of air beneath the roof sheathing or through the attic.

Securing Bands Bands of metal (suitably treated as may be necessary to minimize corrosion), or of plastic material, used for securing insulation to pipes or other structures.

Self-Setting Cement Finishing material, based on Portland cement, that is supplied as a dry powder and, when mixed with water in suitable proportions, will set without the application of heat.

Slag Wool Mineral fiber produced from steel mill slag.

Slag Wool Insulation Man-made material made primarily from iron ore blast furnace slag that is spun into a fibrous form.

Smoldering Combustion A test to assess the fire resistance within the insulation layer.

Soaking Heat A test condition in which the specimen is completely immersed in an atmosphere maintained at a controlled temperature.

Soffits Soffit (from French *soffite*, formed as a ceiling; to fix underneath), in architecture, describes the underside of any construction element. Examples of soffits include the underside of a flight of stairs.

Spray-Applied Cellulose An installation method in which water is added to cellulose insulation to make it stick when blown into wall cavities. Also known as Type 2 cellulose insulation.

Sprayed-On Insulation Insulation of the fibrous or foam type that is applied to a surface by means of power spray devices.

Spray Foam Spray foam is a man-made chemical that is composed of both organic and nonorganic materials. When these materials are combined at a specific temperature, form a polyurethane material that expands several times its original size. This is where the compound is named.

Spray Foam Insulation The use of spray foam insulation on cracks and crevices creates an airtight and watertight seal. One of the advantages of spray foam is its ability to expand and reach elusive spaces.

Sprayed Insulation An adherent coating of insulating material.

Stabilized Stabilized cellulose is used most often in attic/roof insulation. It is applied with a very small amount of water to activate an adhesive.

Standard Testing Laboratory test methodology for determining relative properties of materials at specific conditions.

STC (Sound Transmission Class) STC is an integer rating of how well a building partition attenuates airborne sound. It is widely used to rate interior partitions, ceilings/ floors, doors, windows, and exterior wall configurations.

Stone Wool Stone wool is an effective insulation, manufactured from intertwined fibers of molten volcanic rock.

Stud Upright post in the framework of a wall to support an approved interior material such as gypsum wallboard.

Thermal Break or Barrier A nonmetallic material positioned between metallic components of windows to prevent a direct path of heat loss through thermal conduction. Also, a material applied over spray foam insulation designed to slow the temperature rise of the foam during a fire situation and delay its involvement in the fire.

Thermal Bridge A thermally conductive material that penetrates or bypasses an insulation system, such as a metal fastener or stud.

Thermal Conductivity The thermal transmission through unit area of a slab of a uniform material of unit thickness when unit difference of temperature is established between its faces [W/(mK)].

Thermal Insulation A material or system that has the property of resisting the transfer of heat. Insulation applicable within the general temperature range of 300°F to 1800°F.

Thermal Insulation System Applied or installed thermal insulation including any accessories, vapor retarder, and facing required.

Thermal Resistance (R) An index of a material's resistance to heat flow. See R and RSI.

Thermal Shock A building material's reaction to rapid changes in temperature.

Thermography A building energy diagnostic technique using an infrared camera for locating areas of temperature differential in a building.

Thermostat Temperature-sensitive control device that signals a heating or cooling system to operate if the temperature in the building reaches a preset limit.

Transmission, Heat The quantity of heat flowing through unit area due to all modes of heat transfer induced by the prevailing conditions.

Type 1 Insulation Dry cellulose insulation intended for pneumatic application into open areas with slopes up to 4.5:12, or injection application into closed cavities, such as walls, floors, and cathedral ceilings. Type 1 insulations may also be manually applied.

Type 2 Insulation Intended for spray application with water or liquid adhesive into open areas regardless of slope (e.g., attics), exposed surfaces (e.g, walls or ceilings) and/or into any open cavity (wall, floor, or ceiling) that may be closed later. This type of product may also contain internal binders to increase the adhesive/cohesive capabilities of the sprayed fibers in order to reduce settlement and/or ensure it remains in place.

U-Value Overall thermal conductance. The U-value is equal to the inverse of the sum of the R-values in a system (U = 1/R total).

U-Value (Transmittance) The combined thermal value of all the materials in a building section, air spaces, and surface air films. It is the time rate of heat flow per unit (sq. ft.), per degree F temperature difference with units in (BTU/h·ft^2·°F).; The lower the overall U-value, the more energy efficient the assembly.

Underground or Buried Insulation Insulation applied on piping and equipment located below grade and usually in direct contact with the surrounding soil.

Unfaced Insulation Insulation with no attached vapor barrier.

Vapor Barrier A vapor check with water vapor permeance not exceeding 0.067 g/(sMN), when tested in accordance with BS 2972.

Vapor Permeable Underlay A vapor permeable underlay repels water that penetrates a roofing finish but is permeable to water vapor escaping from the structure. It is usually defined as a material with a vapor resistance of not more than 0.25 MNs/g.

Vapor Retarder/Barrier A layer of moisture resistant material that usually controls moisture diffusion (defined as less than 1 perm) to prevent moisture buildup in the walls.

Ventilation Ventilating (the V in HVAC) is the process of "changing" or replacing air in any space to provide high indoor air quality (i.e., to control temperature, replenish oxygen, or remove moisture, odors, smoke, heat, dust, airborne bacteria, and carbon dioxide).

Vermiculite An expanded mineral insulation consisting of a mica-like substance that expands when heated. The resulting granules are generally used as loosefill insulation

Viscosity The thickness or resistance to flow of a liquid. Viscosity generally decreases as temperature increases; application temperatures of spray foam components are specified in part to control viscosity at the spray gun.

Volatile Organic Compounds (VOC) Any compound containing carbon and hydrogen or containing carbon and hydrogen in combination with other elements.

Waterproof Impervious to prolonged exposure to water or water entry.

Weather Barrier (Weather Coat) A breather jacket or coating that allows passage of water vapor and protects from atmospheric conditions. A material or materials which when installed on the outer surface of thermal insulation, protects the insulation from the ravages of weather, such as rain, snow, select, wind, solar radiation, atmospheric contamination, and mechanical damage.

Weather Stripping Material such as vinyl, foam, or metal strips installed to seal small cracks around the moving parts of doors and windows. Weather stripping is designed to block uncontrolled infiltration of cold air through these spaces and sometimes to repel wind-driven rain and moisture. The term weather stripping often applies to caulking, a compound used to fill in joints and cracks in the house exterior.

Water Vapor Permeability (Permeance) The time rate of water-vapor transmission through unit area of flat material of unit thickness induced by unit vapor pressure difference between two specific surfaces, under specified temperature and humidity conditions.

Water Vapor Retarder (Barrier) A material or system that adequately impedes the transmission of water vapor under specified conditions.

Water Vapor Transmission Rate The steady water vapor flow in unit time through unit area of a body, normal to specific parallel surfaces, under specific conditions of temperature and humidity at each surface.

Whole House Plenum An enclosed (nonventilated) and insulated crawl space used as a return or supply duct for a forced air heating/cooling system.

Windwashing The phenomenon of air movement driven by wind passing through or behind the thermal insulation within enclosures, causing significant loss of heat flow control and potentially causing condensation. Typically occurs at exposed building edges, such as at the outside corners and roof eaves because of the large pressure gradients at these locations.

Wood Fiber Insulation composed of wood/cellulosic fibers, with or without binders.

Bibliography, References, and Further Reading

Abusoglu A, Kanoglu M. Exergetic and thermoeconomic analyses of diesel engine powered cogeneration: Part 1 – formulations. Appl Therm Eng 2009;29(2–3):234–41.

Ahmed M, Meade O, Medina MA. Reducing heat transfer across the insulated walls of refrigerated truck trailers by the application of phase change materials. Energy Convers Manage 2010;51:383–92.

Alawadhi EM. Thermal analysis of a pipe insulation with a phase change material: Material selection and sizing. Heat Transfer Eng 2008;29(7):624–31.

Al-Ajlan SA. Measurements of thermal properties of insulation materials by using transient plane source technique. Appl Therm Eng 2006;26:2184–91.

Al-Homoud Mohammad S. Performance characteristics and practical applications of common building thermal insulation materials. Build Environ 2005;40:353–66.

Al-Turki AM, Zaki GM. Cooling load response for building walls comprising heat storage and thermal insulation layers. Energy Convers Manage 1991;32(3):235–47.

Anand S, Bansal NK, Park SR, Tyagi SK. Comparative study of different insulating materials for reducing the heat losses in steam pipes: A technical study. Int J Sustainable Energy 2012;31(2):133–41.

Angelescu N, P unescu L. Researches in view of heat losses reduction through the thermal insulation of the casting ladles. Metalurgia Int 2009;14(8):33–8.

Anon. Shipboard thermal-oil lines protected with cellular glass insulation. Oil Gas J 1999;97(17):63–4.

Arslan O, Kose R. Thermoeconomic optimization of insulation thickness considering condensed vapor in buildings. Energy Build 2006;38(12):1400–8.

Arulanantham M, Kaushika ND. Coupled radiative and conductive thermal transfers across transparent honeycomb insulation materials. Appl Therm Eng 1996;16(3): 209–17.

ASTM (AMERICAN SOCIETY FOR TESTING AND MATERIALS) C168-97 "Standards Terminology Relating to Thermal Insulation Material." C450-94 "Recommended Practice for Prefabrication and Field Fabrication of Thermal Insulating Fitting Covers for NPS Piping, Vessel Lagging, and Dished Head Segments." C585 "Practice for Inner and Outer Diameters of Rigid Thermal Insulation for Nominal Sizes of Pipe and Tubing (NPS System)."

ASTM C390 "Criteria for Sampling and Acceptance of Preformed Thermal Insulation Lots." C356 "Test Method for Linear Shrinkage of Preformed High Temperature Thermal Insulation Subjected to Soaking Heat." C335 "Test Method for Steady State Heat Transfer Properties of Horizontal Pipe Insulation."

ASTM C871 "Test Methods for Chemical Analysis of Thermal Insulation Materials for Leachable Chloride, Fluoride, Silicate, and Sodium Ions." C692 "Test Methods for Evaluating the Influence of Thermal Insulation Wicking-Type on the External Stress Corrosion Cracking Tendency of Austenitic Stainless Steel."

Augustinovicz, S.D. ,Fesmire J.E.,Cryogenic insulation system for soft vacuum, Dynacs Engineering Company, NASA, Kennedy Space Center, NASA SP5027 Cryogenic Systems by Randall Barron.

Aydemir NU, Magapu VK, Sousa ACM, Venart JES. Thermal response analysis of LPG tanks exposed to fire. J Hazard Mater 1988;20:239–62.

Babac G, Sisman A, Cimen T. Two-dimensional thermal analysis of liquid hydrogen tank insulation. Int J Hydrogen Energy 2009;34(15):6357–63.

Bahadori A, Mokhatab S. Estimating thermal conductivity of hydrocarbon compounds. Chem Eng 2008;115(13):52–4.

Bahadori A, Vuthaluru HB. A simple method for the estimation of thermal insulation thickness. Appl Energy 2010a;87:613–9.

Bahadori A, Vuthaluru HB. A simple correlation for estimation of economic thickness of thermal insulation for process piping and equipment. Appl Therm Eng 2010b;30:254–9.

Bahadori A, Vuthaluru HB. Estimation of heat losses from process piping and equipment. Petroleum Technol Q 2010;15(3):121–3.

Barnhart JM. Economic thickness of thermal insulation. Chem Eng Prog 1974;70(8):50–4.

Beckmann MM, Volkert BC, McMullen ND. Testing, analysis solve insulation problem on gulf flowline. Oil Gas J 2001;99(38):52–8.

Bejan A, Tsatsaronis G, Moran M. Thermal Design andOoptimization. New York: John Wiley and Sons Inc; 1996.

Bejan A. Fundamentals of exergy analysis, entropy generation minimization, and the generation of flow architecture. Int J Energy Res 2002;26(7):545–65.

Bejan A. Heat Transfer. New York: John Wiley and Sons; 1993.

Bejan A. Shape and Structure, from Engineering to Nature. Cambridge, UK: Cambridge University Press; 2000.

Bektas Ekici B, Aytac Gulten A, Aksoy UT. A study on the optimum insulation thicknesses of various types of external walls with respect to different materials, fuels and climate zones in Turkey. Appl Energy 2012;92:211–7.

Berti E. Syntactic polypropylene coating solution provides thermal insulation for Bonga risers. Offshore Mag 2004;64:2.

Boehm RF. Design Analysis of Thermal Systems. New York: John Wiley and Sons; 1987.

Bojic M, Yik F, Leung W. Thermal insulation of cooled spaces in high rise residential buildings in Hong Kong. Energy Convers Manage 2002;43:165–83.

Bolatturk A. Determination of optimum insulation thickness for building walls with respect to various fuels and climate zones in Turkey. Appl Therm Eng 2006;26(11–12):1301–9.

Bolatturk A. Optimum insulation thicknesses for building walls with respect to cooling and heating degree-hours in the warmest zone of Turkey. Build Environ 2008;43(6): 1055–64.

Bolattürk A. Optimum insulation thicknesses for building walls with respect to cooling and heating degree-hours in the warmest zone of Turkey. Energy Build 2008;43:1055–64.

Bouchonneau NV, Sauvant-Moynot D, Choqueuse F, Grosjean E, Poncet D Perreux. Experimental testing and modelling of an industrial insulated pipeline for deep, sea application. J Petroleum Sci Eng 2010;73:1–12.

Bouchonneau N, Sauvant-Moynot V, Grosjean F, Choqueuse D, Poncet E, Perreux D. Thermal insulation material for subsea pipelines: benefits of instrumented full scale testing to predict the long term thermo-mechanical behaviour. Proceedings of the Offshore Technology Conference — OTC 18679. Houston: Texas, USA; 2007.

Boye Hansen A, Duncan J, Simonsen E. Polypropylene Deepwater Insulation Experience to Date and Future Trends in Combination with Direct Heating Flow Control Systems", 2nd Workshop on Subsea Pipelines. Rio de Janeiro; 26–27 April 1999.

Boye Hansen A, Friberg R. Thermal Insulation of Non-Jacketed Deep-Water Flowlines and Risers Based on Mobile Manufacturing Units. Rio Oil Gas Expo October 2000:16–9.

Boye Hansen A, Lechner F, et al. Design of a containerized mobile turnkey compounding plant for the direct extrusion of PP with hollow glass micropsheres. ANTEC; 2004.

Boye Hansen A, Rydin C. Development and Qualification of Novel Thermal Insulation Systems for Deepwater Flowlines and Risers Based on Polypropylene; May 2002. Offshore Technology Conference, OTC 14121, 6–9.

BS 2654. Manufacture of Vertical Steel Welded Non-Refrigerated Storage Tanks with Butt Weld Shell for Petroleum Industry; 1989.

BS 2972. Methods of Test for Inorganic Thermal Insulating Materials; 1989.

BS 3974. Specification for Pipe Support; 1986.

BS 4275. Recommendation for the Selection and Maintenance of Respiratory Protective Equipment; 1974.

BS 4508. Thermally Insulated Underground Piping Systems; 1989.

BS 5422. Specification for the Use of Thermal Insulating Material; 1990.

BS CP 3009. Thermally Insulated Underground Piping Systems; 1970.

BSI (BRITISH STANDARD INSTITUTION) BS 5970. Code of Practice for Thermal Insulation of Pipework and Equipment (in the Temperature Range -100°C to +870°C); 1981.

Byrne A, Byrne G, Davies A, Robinson AJ. Transient and quasi-steady thermal behaviour of a building envelope due to retrofitted cavity wall and ceiling insulation. Energy Build 2013;61:356–65.

Campo A. Quick algebraic estimate of the thickness of insulation for the design of process pipelines with allowable heat losses to ambient air. Heat Transfer Eng 2002;23(3):25–34.

Cengel YA, Boles MA. Thermodynamics: An Engineering Approach. New York: McGraw Hill Inc; 1994.

Chen W. Thermal analysis on the cooling performance of a wet porous evaporative plate for building. Energy Convers Manage 2011;52:2217–26.

Cheng X, Fan J. Simulation of heat and moisture transfer with phase change and mobile condensates in fibrous insulation. Int J Therm Sci 2004;43(7):665–76.

Choi SW, Roh JU, Kim MS, Lee WI. Analysis of two main LNG CCS (cargo containment system) insulation boxes for leakage safety using experimentally defined thermal properties. Appl Ocean Res 2012;37:72–89.

Choqueuse D, Chauchot P, Sauvant-Moynot V, Lefebvre X. Recent progress min analysis and testing of insulation and buoyancy materials. Proceedings of the Fourth International Conference on Composite Materials for Offshore Operations — CMOO-4. Houston: Texas USA; 2005.

Choqueuse D, Chomard A, Bucherie C. Insulation materials for ultra deep seaflow assurance: evaluation of the material properties. In: Proceedings of the Offshore Technology Conference — OTC 14115. Houston: Texas, USA; 2002.

Choqueuse D, Chomard A, Chauchot P. How to provide relevant data for the prediction of long term behavior of insulation materials under hot/wet conditions. In: Proceedings of the Offshore Technology Conference — OTC 16503. Houston: Texas, USA; 2004.

Civan F. Use exponential functions to correlate temperature dependence. Chem Eng Progress 2008;104(7):46–52.

Curtis JP. Optimisation of homogeneous thermal insulation layers. Int J Solids Structures 1983;19(9):813–23.

de Sousa FVV, da Mota RO, Quintela JP, Vieira MM, Margarit ICP, Mattos OR. Characterization of corrosive agents in polyurethane foams for thermal insulation of pipelines. Electrochim Acta 2007;52(27 SPEC. ISS):7780–5.

Dincer I. On thermal energy storage systems and applications in buildings. Energy Build 2002;34(4):377–88.

Ding S-F, Tang W-Y, Zhang S-K. Research on temperature field and thermal stress of liquefied natural gas carrier with incomplete insulation. J Shanghai Jiaotong Univ (Sci) 2010;15(3):346–53.

Doi Takuya, Tanaka Tadayoshi. Transient heat loss and temperature response in thermal insulation pipe - numerical analysis and simplified analysis. Heat Transfer –Jpn Res 1993;22(3):305–24.

Dombayci OA. The environmental impact of optimum insulation thickness for external walls of buildings. Build Environ 2007;42(11):3855–9.

Dombayci Ö A, Gölcü M, Pancar Y. Optimization of insulation thickness for external walls using different energy-sources. Appl Energy 2006;83:921–8.

Dong B-S. Corrosion investigation of underground heat insulating pipelines. Corrosion Prot 2006;27(12):640–1.

Droste B, Schoen W. Full scale fire tests with unprotected and thermal insulated LPG storage tanks. J Hazard Mater 1988;20:41–53.

Ebert H-P, Laudensack B, Hemberger F, Nilsson O, Fricke J. Thermal insulations based on carbon black. High Temperatures High Pressures 1998;30(3):261–7.

Ek C-G. The new generation polypropylene block copolymer for non-pressure pipe applications. Munich: paper at Plastics Pipes XI conference; September 3-6, 2001.

Elliott SR, Blaner M, Kepko W, Doty JR. Energy savings at V&M star with microporous insulation. Iron Steel Technol 2007;4(2):27–33.

Eriksen Gunnar, Christiansen ER. Elastomer for line insulation coating passes North Sea test. Oil Gas J 1990;88(42):48–50.

Evans LB, Stephany NE. An experimental study of transient heat transfer to liquids in cylindrical enclosures. Los Angeles, USA: Proceedings of the Heat Transfer Conference; 1965.

Eyglunent B. Manuel de thermique — Théorie et pratique. Paris: HERMES Science Publications; 1997.

Incropera FP, Dewitt DP. Fundamentals of Heat and Mass Transfer. 4th ed. Wiley; 1996.

Filler DM, Carlson RF. Thermal insulation systems for bioremediation in cold regions. J Cold Regions Eng 2000;14(3):119–29.

Findley WN. Combined stress creep of non-linear viscoelastic material. In: Smith AI, Nicolson AM, editors. Advances in Creep Design, The A.E. Johnson Memorial Volume. London: Applied Science Publishers Ltd; 1971.

Fu Z-G, Yu B, Zhu J, Xie J, Li W. Thermal effect of buried oil pipeline in permafrost regions. J Eng Thermophysics 2012;33(12):2163–6.

Gao Y. The pressure drop calculation of sea water pipe with no heat oil and gas transferring. Adv Mater Res 2013;634-638(1):3613–7.

Giakoumis EG. Cylinder wall insulation effects on the first- and second-law balances of a turbocharged diesel engine operating under transient load conditions. Energy Convers Manage 2007;48:2925–33.

Gimenez N, Sauvant-Moynot V, Sautereau H. Wet ageing of syntactic foams under high pressure/high temperature in deionized and artificial sea water. Halkidiki, Greece: Proceedings of the Twenty-Fourth International Conference on Offshore Mechanics and Artic Engineering; 2005.

GPSA. Gas Processors and Suppliers Association Engineering Data Book. 12th ed. USA: Tulsa, OK; 2004.

Haldane D, Graff FVD, Lankhorst AM. A direct measurement system to obtain the thermal conductivity of pipeline insulation coating systems under simulated service conditions. Proceedings of the Offshore Technology Conference — OTC 11040. USA: Houston, Texas; 1999. pp. 1–16.

Haldane D, Scrimshaw KH. Development of an alternative approach to the testing of thermal insulation materials for subsea applications. Barcelona, Spain: Proceedings of the Fourteenth International Conference on Pipeline Protection; 2001. 29–39.

Hansen AB, Friberg R. Thermal insulation of non-jacketed deepwater flowlines and risers based on mobile manufacturing units. In: Proceedings of the Rio Oil and Gas Expo and Conference. Brazil: IBP41900; October 2000.

Haralambopoulos DA, Paparsenos GF. Assessing the thermal insulation of old buildings – the need for in situ spot measurements of thermal resistance and planar infrared thermography. Energy Convers Manage 1998;39:65–79.

Harris FS. Safety features on LNG ships. Cryogenics 1993;8(33):772–7.

Harte A, Williams D, Grealish F. Short and long-term performance of polymer foam insulation systems in deepwater offshore applications. In: Drechsler K, editor Advanced Composites: The Balance between Performance and Cost. Paris: Proceedings of the 24th International SAMPE Europe Conference of the Society for the Advancement of Materials and Process Engineering; 2003. pp. 515–23.

Harte AM, et al. A coupled temperature–displacement model for predicting the long-term, performance of offshore pipeline insulation systems. J Mater Processing Technol 2004:155–6. 1242–1246.

Hartviksen G, Melve B, Rydin C. Full scale testing of Åsgard flowline PP insulation at 140°C–ageing and long-term performance. London: Oilfield Engineering with Polymers Conference; 28–29 November 2001.

Hasan A. Optimum insulation-thickness for buildings using life-cycle cost. Appl Energy 1999;63(2):115–24.

Helmisyah AJ, Abdullah S, Ghazali MJ. Effect of thermal barrier coating on piston crown for compressed natural gas with direct injection engine. Appl Mech Mater 2011:52–4. pp. 1830–1835.

Holman J. Heat Transfer. McGraw-Hill Series in Mechanical Engineering; 1983.

Huang J. Theoretical analysis of three methods for calculating thermal insulation of clothing from thermal manikin. Ann Occup Hyg 2012;56(6):728–35.

Ito K, Akagi S, Nishikawa M. A Multiobjective Optimization Approach to Design Problem of Heat Insulation for Thermal Distribution Piping Network Systems; 1982. ASME Paper 82-DET-57.

Jackson A, Boye Hansen A, et al. Design parameters for single pipe thermal insulation systems for offshore flow assurance, 17–19 October. Rio Pipeline; 2005.

Jeon S-J, Jin B-M, Kim Y-J, Chung C-H. Consistent thermal analysis procedure of LNG storage tank. Struct Eng Mech 2007;25(4):445–66.

Jerman M, Černý R. Effect of moisture content on heat and moisture transport and storage properties of thermal insulation materials. Energy Build 2012;53:39–46.

Junshiro I, Kiyokazu K, Hidetoshi M, Hidefumi I, Yoshihiro S. Building of advanced large sized membrane type LNG carrier. Mitsubishi Heavy Industires, Ltd. Tech review 2004;41(6):1–7.

Kalyon M, Sahin AZ. Application of optimal control theory in pipe insulation. Numerical Heat Transfer Part A Appl 2002;41(4):391–402.

Karabay H. The thermo-economic optimization of hot-water piping systems: a parametric study of the effect of the system conditions. Strojniski Vestnik J Mech Eng 2007;53(9): 548–55.

Kaynakli O. A study on residential heating energy requirement and optimum insulation thickness. Renew Energy 2008;33:1164–72.

Kecebas A, Kayveci M. Effect on optimum insulation thickness, cost and saving of storage design temperature in cold storage in Turkey. Energy Educ Sci Technol Part A 2010;25(1–2):117–27.

Kecebas A, Alkan MA, Bayhan M. Thermo-economic analysis of pipe insulation for district heating piping systems. Appl Therm Eng 2011;31(17–18):3929–37.

Khalifa AJN, Hamood AM. Effect of insulation thickness on the productivity of basin type solar stills: an experimental verification under local climate. Energy Convers Manage 2009;50:2457–61.

Khalifa AJN, Hamood AM. Effect of insulation thickness on the productivity of basin type solar stills: An experimental verification under local climate. Energy Convers Manag 2009;50(9):2457–61.

Kim JG, Yoon SH, Park SW, Lee DG. Optimum silane treatment for the adhesively bonded aluminum adherend at the cryogenic temperature. J Adhesion Sci Technol 2010;24:775–87.

Kim D, Silva C, Marotte E, Fletcher L. Characterization/modeling of wire screen insulation for deep-water pipes. J of Thermophysics Heat Transfer 2007;21(1):145–52.

Kokorev LS, Kharitonov VV, Bol'shakov VI, Sysoev YM, Plakseev AA. Natural convection in fibrous thermal insulation; 1980. HIGH TEMP. 18 (2, Mar.-Apr. 1980, p. 281–285.

Korotkov YuF, Diarov RK, Makarov NA, Korotkova EYu. Calculation of thermal insulation of equipment for oil, gas and water preparation. Khimicheskoe I Neftegazovoe Mashinostroenie 2004;(1):18–9.

Kovylyanskij YaA, Umerkin GKh, Rojshtejn LI. On problem of development of thermal insulation designs for heat pipelines. Teploenergetika 1992;(11):65–9.

Krarti M. Steady-state heat transfer from partially insulated basements. Energy Build 1993;20(1):1–9.

Krarti Moncef, Choi Sangho. Simplified method for foundation heat loss calculation. ASHRAE Trans 1996;102(1):140–52.

Kulkarni M, Nelson K. Introduction of a new crossover radius for the guaranteed heat transfer reduction in radial thermal insulation systems. Heat Transfer Eng 2006;27(7):23–8.

Kutuev SB, Chaplitskaya VL. Effective thermal insulation and power efficiency of external protective structures. Geliotekhnika 1991;(3):42–6.

Labude J. Study on the insulation of drying ranges in the textile industry. Textil Praxis Int 1993;48(6):515–20.

Lefèbvre K, Sauvant-Moynot V, Choqueuse D, Chauchot P. Durabilité des matériaux syntactiques d'isolation thermique et de flottabilité: des mécanismes de dégradation à la modélisation des propriétés long terme. France: Proceedings of the Conference: Matériaux 2006, Dijon; 2006.

Li P, Cheng H. Thermal analysis and performance study for multilayer perforated insulation material used in space. Appl Therm Eng 2006;26:2020–6.

Li YF, Chow WK. Optimum insulation-thickness for thermal and freezing protection. Appl Energy 2005;80(1):23–33.

Li Z, Xu L, Zhang J, Sun H. Design of thermal insulation for LNG tanker. Nat Gas Industry 2004;24(2):10–1. +85.

Lim TK, Axcell BP, Cotton MA. Single-phase heat transfer in the high temperature multiple porous insulation. Appl Therm Eng 2007;27:1352–62.

Lin EI, Stultz JW, Reeve RT. Effective emittance for Cassini multilayer insulation blankets and heat loss near seams. J Thermophysics Heat Transfer 1996;10(2):357–63.

Liu G, Zhang G, Zhang Y. Study on the hot oil pipeline's cooling process. Adv Mater Res 2012:433–40. pp. 4396–4400.

Liu J, Pei G, Ji Y. Analysis of heat loss of ground pipeline of thermal production oilfield. Appl Mech Mater 2011:90–3. pp. 3057–3060.

Liu L, Yu P, Xu Y. The method of computing the thermal conductivity of heat-insulation oil pipe coupling. Appl Mech Mater 2013:278–80. pp. 251–255.

Lutskanov S. Saving fuel with efficient crown insulation. Glass Int 2003;26(6):26.

Maillet D, André S, Batsale JC, Degiovanni A, Moyne C. Thermal Quadrupoles: Solving the Heat equation through Integral Transforms. John Wiley & Sons, Inc; 2000.

Mal'kovskii VI, Dashevskii YuM, Zhuze VB. Optimising the thermal insulation of long pipelines. Therm Eng (Eng Transl Teploenergetika) 1989;36(6):328.

Maref W, Swinton MC, Kumaran MK, Bomberg MT. Three-Dimensional Anal Thermal Resist Exterior Basement Insulation Syst (EIBS)Build Environ 2001;36(4):407–19.

Marks JB, Holton KD. Protection of thermal insulation. Chem Eng Prog 1974;70(8):46–9.

Medved S, Černe B. A simplified method for calculating heat losses to the ground according to the EN ISO 13370 standard. Energy Build 2002;34(5):523–8.

Moiseev VA, Andrienko VG, Klokotov YuN, Petrov YuV, Yalov YuN. New generation equipment for thermal-steam treatment of formations to enhance their oil recovery. Neftyanoe Khozyaistvo - Oil Industry 2011;(3):118–21.

Mollison Michael I. Foam insulation gets first reeled installation off Australia. Oil Gas J 1992;90(20):80–4.

Norsworthy R, Dunn PJ. Corrosion under thermal insulation. Mater Perform 2002;41(9):38–43.

Ochs F, Heidemann W, Müller-Steinhagen H. Effective thermal conductivity of moistened insulation materials as a function of temperature. Int J Heat Mass Transfer 2008;51(3-4):539–52.

Omori T, Tanabe SI, Itagaki M. Influence of building insulation performance and heating systems on thermal environment and energy performance. J Environ Eng 2011;76(661):231–8.

Ong KS. Temperature reduction in attic and ceiling via insulation of several passive roof designs. Energy Convers Manage 2011;52:2405–11.

Ovando-Castelar R, Martínez-Estrella JI, Garciá-Gutiérrez A, Canchola-Félix I, Miranda-Herrera CA, Jacobo-Galván VP. Estimation of steam pipeline network heat losses at the Cerro Prieto geothermal field based on pipeline thermal insulation conditions. Trans Geothermal Resources Counc 2012;36(2):1111–8.

Ozel M, Pihtili K. Optimum location and distribution of insulation layers on building walls with various orientations. Build Environ 2007;42(8):3051–9.

Özel M. Dynamic approach and cost analysis for optimum insulation thicknesses of the building external walls. J Fac Eng Archit Gazi Univ 2008;23(4):879–84.

Ozel M. Cost analysis for optimum thicknesses and environmental impacts of different insulation materials. Energy Build 2012;49:552–9.

Ozkahraman HT, Bolatturk A. The use of tuff stone cladding in buildings for energy conservation. Constr Build Mater 2006;20(7):435–40.

Özkan DB, Onan C. Optimization of insulation thickness for different glazing areas in buildings for various climatic regions in Turkey. Appl Energy 2011;88(4):1331–42.

Ozturk IT, Karabay H, Bilgen E. Thermo-economic optimization of hot water piping systems: a comparison study. Energy 2006;31(12):2094–107.

Panzer U, Rätzsch M. Novel High Melt Strength Polypropylene, SPO; 1998.

Pattee FM, Kopp F. Impact of Electrically-Heated Systems on The Operation of Deep Water Subsea Oil Flowlines. OTC 11894. Houston, TX: Offshore Technology Conference; May 2000. 1–4.

Pavlík Z, Černý R. Hygrothermal performance study of an innovative interior thermal insulation system. Appl Therm Eng 2009;29:1941–6.

Pedersen Carsten R, Courville George E. Computer analysis of the annual thermal performance of a roof system with slightly wet fibrous glass insulation under transient conditions. J Therm Insulation 1991;15:110–36.

Peng LC, Peng TL. Thermal insulation and pipe stress. Hydrocarbon Process 1998;77(5):111–3.

Petrov-Denisov VG, Kovylyanskij YaA, Pichkov AM, Gordeeva VN, Rojtshtejn LI. Estimation for durability of thermal insulation designs of heat pipelines in laying them by the underground non-contact method. Teploenergetika 1992;(11):56–9.

Polivoda FA. Calculation of heat conduction of anisotropic thermal insulation of heat pipe-lines and solar collectors. Teploenergetika 1993;(6):67–9.

Reiss H. A coupled numerical analysis of shield temperatures, heat losses and residual gas pressures in an evacuated super-insulation using thermal and fluid networks: Part I: Stationary conditions. Cryogenics 2004;44(4):259–71.

Robertson S, Macfarlan G, Smith M. Deep water expenditures to reach $20 billion/year by 2010. Offshore Mag 2005;65:12.

Rydin C, Sjöberg L. Polypropylene thermal insulation systems for offshore pipeline applications. Barcelona: 14th Pipeline Protection Conference; October 2001. 29–31.

Timoshenko S. Strength of Materials, Part II. 3rd ed. New Jersey: Van Nostrand Reinhold; 1956.

Sahin AZ, Kalyon M. The critical radius of insulation in thermal radiation environment. Heat Mass Transfer 2004;40(5):377–82.

Sambou V, Lartigue B, Monchoux F, Adj M. Theoretical and experimental study of heat transfer through a vertical partitioned enclosure: application to the optimization of the thermal resistance. Appl Therm Eng 2008;28:488–98.

Serednyts'kyi YaA. Polyurethane materials as anticorrosive coatings of pipelines. Mater Sci 2000;36(3):415–21.

Skrinska Alfonsas, Vegyte Nijole. Exploration of heat in operating industrial buildings. Adv Eng Heat Transfer 1995:595–600.

Skujans J, Vulans A, Iljins U, Aboltins A. Measurements of heat transfer of multi-layered wall construction with foam gypsum. Appl Therm Eng 2007;27:1219–24.

Song S-Y, Yi J-S, Koo B-K. Insulation plan of aluminum curtain wall-fastening unit for high-rise residential complex. Build Environ 2008;43(7):1310–7.

Soylemez MS, Unsal M. Optimum insulation thickness for refrigeration applications. Energy Convers Manage 1999;40(1):13–21.

Srinivas N. Energy management through upgraded thermal insulation systems. Chem Age India 1984;35(12):923–7.

Stehfest H. Algorithm 368: Numerical inversion of Laplace transform; 1970. Commun. ACM 13, 1.

Suman JC, Karpathy Sandor A. Design method addresses subsea pipeline thermal stresses. Oil Gas J 1993;91(35):85–9.

Tanaka Tadayoshi, Kamoshida Junji, Okada Masashi, Katayama Kohzo. Transient heat loss from thermal insulation pipe. Trans Jpn Soc Mech Eng Part B 1991;57(536):1442–50.

Ucar A. Thermoeconomic analysis method for optimization of insulation thickness for the four different climatic regions of Turkey. Energy 2010;35(4):1854–64.

Ucar A, Balo F. Effect of fuel type on the optimum thickness of selected insulation materials for the four different climatic regions of Turkey. Appl Energy 2009;86:730–6.

Uygunoğlu T, Kecebas A. LCC analysis for energy-saving in residential buildings with different types of construction masonry blocks. Energy Build 2011;43(9):2077–85.

Vafai K. Characterization of thermal performance of fibrous insulations subject to a humid environment. J Thermophysics Heat Transfer 1993;7(1):187–92.

Wahlgren P. Convection in loose-fill attic insulation. Doktorsavhandlingar vid Chalmers Tekniska Hogskola 2001;(1762).

Wahlgren P. Measurements and simulations of natural and forced convection in loose-fill attic insulation. J Therm Envelope Build Sci 2002;26(1):93–109.

Wang X, Javora P, Qu Q, Pearcy R. A new thermal-insulating fluid and its application in deepwater-riser insulation in the Gulf of Mexico. SPE Prod Facil 2005;20(1):35–40.

Wang X, Qu Q, Javora P, Pearcy R. New trend in oilfield flow-assurance management: A review of thermal insulating fluids. SPE Prod Operations 2009;24(1):35–42.

Wei W, Li X, Wang R, Li Y. Effects of structure and shape on thermal performance of perforated multilayer insulation blankets. Appl Therm Eng 2009;29:1264–6.

Wei W, Wang R-S. Performance test on multilayer thermal insulation blanket and industrial application to LNG container. Nat Gas Industry 2007;27(6):109–11. A15.

Weizhong LI. Development and application of K331RT-150 high-temperature thermal production packer. Drilling Prod Technol 2004;27(3). pp. 74+5+82.

Westgate ZJ, White DJ, Randolph MF. Field observations of as-laid pipeline embedment in carbonate sediments. Geotechnique 2012;62(9):787–98.

Whalen J. Performance of molded expanded polystyrene (EPS) thermal insulation in below-grade applications. ASTM Special Tech Publ 2002;(1426):366–73.

Williams D. An analytical tool for the design and analysis of thermal insulation systems for risers and subsea flowlines. M.Eng.Sc. Thesis. Galway: National University of Ireland; 2002.

Williams D, Harte A, Grealish F. Development of an analytical tool for the design of deep water riser/flow line thermal insulation systems. J Offshore Mech Arctic Eng 2005;127(2):96–103.

Wong KL, Salazar JLL, Prasad L, Chen WL. The inaccuracy of heat transfer characteristics for non-insulated and insulated spherical containers neglecting the influence of heat radiation. Energy Convers Manage 2011;52:1612–21.

Wu Q, Dong X, Xu Z, Liu X, Liu L, Li X, Xu Y. Diagnosis for thermal defects of pipeline insulation in refining process. Appl Mech Mater 2011:71–8. pp. 2231–2234.

Xamán J, Lira L, Arce J. Analysis of the temperature distribution in a guarded hot plate apparatus. Appl Therm Eng 2009;29:617–23.

Yöldöz A, Gurlek G, Erkek M, Ozbalta N. Economical and environmental analyses of thermal insulation thickness in buildings. Therm Sci Technol 2008;28:25–34.

Yoon J, Lee EJ, Krarti M. Optimization of Korean crop storage insulation systems. Energy Convers Manage 2003;44:1145–62.

Yu J, Yang L, Tian L, Liao D. A study on optimum insulation thicknesses of external walls in hot summer and cold winter zone of China. Appl Energy 2009. http://dx.doi.org/10.1016/j.apenergy.2009.03.010.

Yu YH, Kim BG, Lee DG. Cryogenic reliability of composite insulation panels for liquefied natural gas (LNG) ships. Composite Structures 2012;94(2):462–8.

Yu YH, Kim BG, Lee DG. Cryogenic reliability of the sandwich insulation board for LNG ship. Composite Structures 2013;95:547–56.

Yun SK, Park DH. Development and analysis of the highly efficient support system in a liquid hydrogen vessel. J Korean Soc Marine Eng 2007;31(4):363–9.

Zhang J, Qu G, Jin H. Estimates on thermal effects of the China–Russia crude oil pipeline in permafrost regions. Cold Regions Sci Technol 2010;64(3):243–7.

Zhuang T. Construction technologies of anti-corrosion and thermal insulation of buried steel pipelines in Sichuan and Chongqing gas fields. Nat Gas Industry 2010;30(4):102–6.

Index

Note: Page numbers with "*b*" denote boxes; "*f*" figures, "*t*" tables.

Printed and bound by CPI Group (UK) Ltd, Croydon, CR0 4YY

13/05/2025

01869806-0001